AF352806

Automation and Robotics: Latest Achievements, Challenges and Prospects

Automation and Robotics: Latest Achievements, Challenges and Prospects

Editors

Pavol Božek
Tibor Krenicky
Yury Nikitin

MDPI • Basel • Beijing • Wuhan • Barcelona • Belgrade • Manchester • Tokyo • Cluj • Tianjin

Editors
Pavol Božek
Slovak University of Technology
in Bratislava
Faculty of Materials Science and
Technology in Trnava
Institute of Production
Technologies
Slovakia

Tibor Krenicky
Faculty of Manufacturing
Technologies
Technical University of Kosice
Slovakia

Yury Nikitin
Kalashnikov Izhevsk
State Technical
University Russia

Editorial Office
MDPI
St. Alban-Anlage 66
4052 Basel, Switzerland

This is a reprint of articles from the Special Issue published online in the open access journal *Applied Sciences* (ISSN 2076-3417) (available at: https://www.mdpi.com/journal/applsci/special_issues/automation_robotics_achievements).

For citation purposes, cite each article independently as indicated on the article page online and as indicated below:

LastName, A.A.; LastName, B.B.; LastName, C.C. Article Title. *Journal Name* **Year**, *Volume Number*, Page Range.

ISBN 978-3-0365-3313-1 (Hbk)
ISBN 978-3-0365-3314-8 (PDF)

Cover image courtesy of Pavol Božek.

Contents

About the Editors

Pavol Božek graduated in 1978 from the Faculty of Mechanical Engineering of the Slovak Technical University in Bratislava. He is currently working in the position of full professor of informatics, automation and mechatronics at the Faculty of Materials Technology based in Trnava of STU in Bratislava. The research activity of Professor Božek is focused on the area of production technology, robotics, diagnostics, automation and control of production systems. A total of 11 patents and 16 utility models are registered under his name. He is an expert member of national and international associations (e.g., WAIT—World Association of Innovative Technologies), a scientific board member of the union commission for Manufacturing Technics, a member of the Slovak Institute of Technical Standardization—Industrial Process Management, a member and chairman of the scientific journal Management of Companies' scientific board, and a member of the scientific board ECOLETRA.

Tibor Krenicky graduated in Physics at the Faculty of Mathematics and Physics of the Comenius University in Bratislava (Slovakia) in 1999. He completed his PhD in 2005 at the Institute of Experimental Physics of the Slovak Academy of Sciences and the Faculty of Natural Sciences of the UPJS in Kosice. He started teaching in 2005 at the Technical University of Kosice, currently holds an Associate Professor position at the Faculty of Manufacturing Technologies in Presov, Department of Technical Device Design and Monitoring. His work focuses mainly on the simulation, determination and evaluation of operating parameters and the operational status of technical systems using multiparametric monitoring and virtual instrumentation as well as the design and implementation of experimental devices for diagnostics of technical systems. He is an experienced Editorial Board member of several international scientific journals and the Guest Editor of several CCs and indexed Special Issues of scientific papers.

Yury Nikitin was born in Izhevsk, Russia. He graduated with a Computer Engineering degree from the Faculty of Instrumentation Engineering of Izhevsk Mechanical Institute in 1988. He received a Candidate of Technical Science degree from the Izhevsk State Technical University (ISTU) in 1994. Yury Nikitin holds the position of Associate Professor of the Department of Mechatronic Systems at ISTU. He has published over 150 journal and conference papers in his field. He has been the principal investigator of several state projects on the research, modelling and diagnosing of mechatronic systems. Yury Nikitin has been awarded the grade of Honorary Worker of Higher Professional Education of the Russian Federation. He is mainly focused on the identification of mechatronic systems, electric motors and equipment, diagnosing and modelling, robotization and automation.

applied sciences

Editorial

Editorial for Special Issue "Automation and Robotics: Latest Achievements, Challenges and Prospects"

Pavol Božek [1], Tibor Krenicky [2,*] and Yury Nikitin [3]

[1] Institute of Production Technologies, Faculty of Materials Science and Technology, Slovak University of Technology, J. Bottu 25, 91724 Trnava, Slovakia; pavol.bozek@stuba.sk
[2] Department of Technical Systems Design and Monitoring, Faculty of Manufacturing Technologies, Technical University of Kosice, Sturova 31, 08001 Presov, Slovakia
[3] Department of Mechatronic Systems, Institute of Modern Technologies in Mechanical and Automotive Engineering and Metallurgy, Kalashnikov Izhevsk State Technical University, Studencheskaya 7, 426069 Izhevsk, Russia; doc_nikitin@mail.ru
* Correspondence: tibor.krenicky@tuke.sk

Keywords: automation; robotics; drives; actuators; sensors; diagnostics; holography; simulation; virtual instrumentation; control systems; robot navigation; reverse validation; trajectory control

Citation: Božek, P.; Krenicky, T.; Nikitin, Y. Editorial for Special Issue "Automation and Robotics: Latest Achievements, Challenges and Prospects". *Appl. Sci.* **2021**, *11*, 12039. https://doi.org/10.3390/app112412039

Received: 6 December 2021
Accepted: 13 December 2021
Published: 17 December 2021

Publisher's Note: MDPI stays neutral with regard to jurisdictional claims in published maps and institutional affiliations.

The determination of this Special Issue topic in the field of automation and robotics was well received by the community of scientists and researchers. Their successful response arose from the universality of the challenges and prospects in the field of robotics that are solved on the basis of well-known concepts at various research workplaces. The positive response to accept the challenge to compare and disseminate the latest developments within the area resonated and has been confirmed, to some extent surprisingly, in a broad range of research areas. On the other side, the aims of all papers are closely related, mainly in terms of reducing the production preparation time, and accurate identification in space, especially relating to machine vision and imaging.

The most numerous group of published articles is related to the field of camera systems, image recognition, 3D virtual models and robot trajectory planning [1–4]. Their contributions are interconnected within the field of the latest machine vision prospects.

The second strongest group of published papers enhances the field of mechanization, automation and robotics focused on production technologies and mechanical engineering. The authors of this group of papers [5–9] combine results oriented to the design of the mechanical parts with the support of 3D adjustable simulations and other objects designed for kinematic data adaptation and positioning of industrial robots. The authors of these papers deal with advanced theoretical topics related to the statics and dynamics of action elements of robotic systems [10–13].

The thematic range of this Special Issue also includes works focused on the field of robotics and automation methods and applications in a non-engineering environment. The studies focus on automating unusual environments, such as underground mining, machining technologies in engineering or inverse data kinematics on non-standard modular robotic arms. These three specific groups of contributions are particularly enriching due to their originality [7,14–16]. This is demonstrated, e.g., in [7] which also mentions a patent application related to the research activity described in the contribution.

To conclude, the call for researchers and scientists working on the selected topic has been positively accepted. Various research teams contributed and presented their research facilities at a high professional level, offering their results. They presented original results related to the latest production methods and the implementation of the new generation tools. All authors worked with the newest approaches and methods in designing these automated components. Components implemented at automated and robotic workplaces are designed for new applications and future technologies and related generations.

Thus, the main goal of this Special Issue on automation and robotics has been fulfilled. The presented contributions have the potential to contribute to reducing the costs of production processes, and to support tools for the simulation and design of key elements and components to achieve exceptional quality and reliability in the production sphere.

The team of editors thanks all the authors whose contributions were accepted, but also those whose contributions were not included in this SI, which were not accepted based on a lack of scientific rigor, but mainly due to a weaker coupling to the SI focus.

At the end of this Editorial, we would like to thank all the authors who explained and implemented the comments and suggestions of the reviewers and editors with patience and thus contributed to raising the level in their field. We would also like to thank all the reviewers who have taken on this highly responsible role and deserve a place of honor in this Special Issue. Last but not least, our special thanks also go to every member of the MDPI *Applied Sciences* Editorial Team, who were our right hands and with whom this SI was created extremely smoothly, especially the very helpful, patient, communicative and flexible Ms. Maura Pei. Thank you so much for your cooperation.

Author Contributions: Conceptualization, P.B., T.K. and Y.N.; methodology, P.B.; validation, P.B., T.K. and Y.N.; formal analysis, T.K.; resources, P.B.; writing—original draft preparation, P.B.; writing—review and editing, T.K.; visualization, T.K.; supervision, P.B.; project administration, P.B., T.K. and Y.N. All authors have read and agreed to the published version of the manuscript.

Funding: This research received no external funding.

Conflicts of Interest: The authors declare no conflict of interest.

References

1. Huczala, D.; Oščádal, P.; Spurný, T.; Vysocký, A.; Vocetka, M.; Bobovský, Z. Camera-Based Method for Identification of the Layout of a Robotic Workcell. *Appl. Sci.* **2020**, *10*, 7679. [CrossRef]
2. Karrach, L.; Pivarčiová, E.; Bozek, P. Recognition of Perspective Distorted QR Codes with a Partially Damaged Finder Pattern in Real Scene Images. *Appl. Sci.* **2020**, *10*, 7814. [CrossRef]
3. Oščádal, P.; Huczala, D.; Bém, J.; Krys, V.; Bobovský, Z. Smart Building Surveillance System as Shared Sensory System for Localization of AGVs. *Appl. Sci.* **2020**, *10*, 8452. [CrossRef]
4. Židek, K.; Piteľ, J.; Balog, M.; Hošovský, A.; Hladký, V.; Lazorík, P.; Iakovets, A.; Demčák, J. CNN Training Using 3D Virtual Models for Assisted Assembly with Mixed Reality and Collaborative Robots. *Appl. Sci.* **2021**, *11*, 4269. [CrossRef]
5. Blatnický, M.; Dižo, J.; Sága, M.; Gerlici, J.; Kuba, E. Design of a Mechanical Part of an Automated Platform for Oblique Manipulation. *Appl. Sci.* **2020**, *10*, 8467. [CrossRef]
6. Zaidi, L.; Corrales Ramon, J.; Sabourin, L.; Bouzgarrou, B.; Mezouar, Y. Grasp Planning Pipeline for Robust Manipulation of 3D Deformable Objects with Industrial Robotic Hand + Arm Systems. *Appl. Sci.* **2020**, *10*, 8736. [CrossRef]
7. Ondočko, Š.; Svetlík, J.; Šašala, M.; Bobovský, Z.; Stejskal, T.; Dobránsky, J.; Demeč, P.; Hrivniak, L. Inverse Kinematics Data Adaptation to Non-Standard Modular Robotic Arm Consisting of Unique Rotational Modules. *Appl. Sci.* **2021**, *11*, 1203. [CrossRef]
8. Kuric, I.; Tlach, V.; Sága, M.; Císar, M.; Zajačko, I. Industrial Robot Positioning Performance Measured on Inclined and Parallel Planes by Double Ballbar. *Appl. Sci.* **2021**, *11*, 1777. [CrossRef]
9. Xia, H.; Zhang, X.; Zhang, H. A New Foot Trajectory Planning Method for Legged Robots and Its Application in Hexapod Robots. *Appl. Sci.* **2021**, *11*, 9217. [CrossRef]
10. Wu, Y.; Luo, L.; Yin, S.; Yu, M.; Qiao, F.; Huang, H.; Shi, X.; Wei, Q.; Liu, X. An FPGA Based Energy Efficient DS-SLAM Accelerator for Mobile Robots in Dynamic Environment. *Appl. Sci.* **2021**, *11*, 1828. [CrossRef]
11. Trojanová, M.; Čakurda, T.; Hošovský, A.; Krenický, T. Estimation of Grey-Box Dynamic Model of 2-DOF Pneumatic Actuator Robotic Arm Using Gravity Tests. *Appl. Sci.* **2021**, *11*, 4490. [CrossRef]
12. Sinčák, P.; Virgala, I.; Kelemen, M.; Prada, E.; Bobovský, Z.; Kot, T. Chimney Sweeping Robot Based on a Pneumatic Actuator. *Appl. Sci.* **2021**, *11*, 4872. [CrossRef]
13. Matisková, D.; Čakurda, T.; Marasová, D.; Balara, A. Determination of the Function of the Course of the Static Property of PAMs as Actuators in Industrial Robotics. *Appl. Sci.* **2021**, *11*, 7288. [CrossRef]
14. Bołoz, Ł.; Biały, W. Automation and Robotization of Underground Mining in Poland. *Appl. Sci.* **2020**, *10*, 7221. [CrossRef]
15. Saga, M.; Perutka, K.; Kuric, I.; Zajačko, I.; Bulej, V.; Tlach, V.; Bezák, M. Methods of Pre-Identification of TITO Systems. *Appl. Sci.* **2021**, *11*, 6954. [CrossRef]
16. Varga, J.; Tóth, T.; Frankovský, P.; Dulebová, Ľ.; Spišák, E.; Zajačko, I.; Živčák, J. The Influence of Automated Machining Strategy on Geometric Deviations of Machined Surfaces. *Appl. Sci.* **2021**, *11*, 2353. [CrossRef]

 *applied sciences*

Letter

Camera-Based Method for Identification of the Layout of a Robotic Workcell

Daniel Huczala *, Petr Oščádal, Tomáš Spurný, Aleš Vysocký, Michal Vocetka and Zdenko Bobovský

VSB-TU of Ostrava, Faculty of Mechanical Engineering, Department of Robotics, 17. listopadu 2172/15, Poruba, 708 00 Ostrava, Czech Republic; petr.oscadal@vsb.cz (P.O.); tomas.spurny.st@vsb.cz (T.S.); ales.vysocky@vsb.cz (A.V.); michal.vocetka@vsb.cz (M.V.); zdenko.bobovsky@vsb.cz (Z.B.)
* Correspondence: daniel.huczala@vsb.cz

Received: 9 September 2020; Accepted: 20 October 2020; Published: 30 October 2020

Featured Application: A fast and low-cost process for automated identification of positions of workcell components, including robots. Suitable for rapid deployment of robotic applications without a need of previous simulations or CAD modeling.

Abstract: In this paper, a new method for the calibration of robotic cell components is presented and demonstrated by identification of an industrial robotic manipulator's base and end-effector frames in a workplace. It is based on a mathematical approach using a Jacobian matrix. In addition, using the presented method, identification of other kinematic parameters of a robot is possible. The Universal Robot UR3 was later chosen to prove the working principle in both simulations and experiment, with a simple repeatable low-cost solution for such a task—image analysis to detect tag markers. The results showing the accuracy of the system are included and discussed.

Keywords: robotic manipulator; identification; calibration; Jacobian; workcell layout detection

---

1. Introduction

For robotic arms there has always been a trade off between the repeatability and absolute accuracy of the measurement of a robot's positioning in 3D space, as examined by Abderrahim [1] or by Young [2]. Many manufacturers of the industrial robot present only the repeatability parameter in their datasheets, when it is way more precise than the absolute positioning. The general problem of robot accuracy is described with experiments by Salmani [3].

The absolute positioning of a robot examines how accurately the robot can move to a position with respect to a frame. To achieve better results, parameter identification and robot calibration are performed. Identification is the process in which a real robot's kinematic (and possibly dynamic) characteristics are compared with its mathematical model. It includes determination of the error values that are afterwards applied into the control system, which improves the robot's total pose accuracy using a software solution without the need for adjusting the hardware of the robot. A generally suggested method for robot calibration is the use of a laser tracker. The methodology identifies the error parameters of a robot's kinematic structure, as is described by Nubiola [4]. The precision may be even increased, as Wu showed in [5] when trying to filter errors in measurements and finding optimal measurement configurations. In [6] Nguyen added neural network to compensate for non-geometric errors after the parameter identification was performed.

Unfortunately, these solutions are very expensive because of the price of a laser tracker. One may rent a laser tracker if needed, but this is also a time consuming process due to the need to perform precise experiments, measurements and evaluations after every error made during the process that may lead to incorrect final results. Therefore, the wide deployment of laser trackers is ineffective for many manufacturers.

There are other methods of robot calibration that tend to avoid the use of laser tracker. In [7], Joubair proposed a method using planes of a very preciously made granite cube, but acquisition of such a cube is not easy in general. Filion [8] or Moller [9] used additional equipment; in their case it is portable photogrammetry system. In [10] a new methodology is introduced by Lembono, who suggested to use three flat planes in a robot's workplace with a 2D laser range finder that intersects the planes, but the simulation was not verified by an experiment. A very different approach was taken by Marie [11] where the elasto-geometrical calibration method based on finite element theory and fuzzy logic modeling was presented.

On the other hand, the very precise results that the methodologies above wanted to achieve are not always necessary, and some nuances in a robot's kinematic structure that appear during its manufacturing process are acceptable for the users of the robot. The problem they may face is determination of the workplace coordinate system (base frame) in which the robot is deployed and eventually the offset of the tool's center point when a tool is attached to the robot's mounting flange, when they need to position it absolutely in a world frame.

For such applications the typical way to calibrate more robots is to use point markers attached to every robot, as described by Gan [12]. However, one important condition is that the robots need to be close together so they can approach each other with the point markers and perform the calibration. Additionally, there are a few optical methods using a camera to improve a robot's accuracy. Arai [13] used two cameras placed in specific positions to track an LED tag that was mounted on a robot; the method we propose allows us to put the camera in any place, in any orientation that will provide good visibility. In [14] Motta or Watanabe in [15] attached a camera to a robot and performed the identification process, but this cannot be used for other robots or to track positions of other components at the workplace at the same time. Van Albada describes in [16] the process of identification for a single robot. Santolaria presented in [17] the use of on-board measurement sensors mounted on a robot.

To avoid these restrictions, we propose a solution based on the OpenCV libraries [18] for Aruco tag detection by a camera, which adds to the calibration process benefits of simplicity, repeatability and low price. The outcomes may be used in offline robot programming, in reconfigurable robotic workplaces and for tracking of components, with as many tag markers and robots as needed, if the visibility for a camera or multiple cameras is provided.

There are methods for 2D camera calibration already presented, and they can be divided into two main approaches. The eye-on-hand calibration, wherein the camera is mounted on the robot and a calibration plate is static, and the eye-on-base method with the calibration marker mounted on the robot with static cameras around [19]. There are also Robot Operating System (ROS) packages [20,21] providing tools for 2D or 3D camera calibration using these two methods. The ROS is an advanced universal platform that may be difficult for some researchers to be able to utilize. Our approach combines both eye-on-hand and eye-on-base calibration processes, avoids using ROS and can be applied not only to localize the base of a robot, but to also localize other devices or objects in the workplace that are either static or of known kinematic structure (multiple robots) in relation to chosen world frame.

2. Materials and Methods

When an image with a tag is obtained, the OpenCV library's algorithm inserts a coordinate system frame in the tag and can calculate transformation from the camera to the tag. If there are tags placed on all important components of a cell, such as manipulated objects or pallets, the transformation between them may be calculated as well. If there is an industrial robot deployed in a workplace, we can attach an end-effector with Aruco tags to it, perform a trajectory with transformation measurements and using mathematical identification methods calculate the precise position of its base, no matter where it is.

2.1. Geometric Model of a Robot

For such an identification, a geometric model that is as precise as possible of a robot needs to be determined. The Universal Robot UR3 was chosen for demonstrating the function of the proposed solution. Its geometric model used for all calculations is based on modified Denavit–Hartenberg notation (MDH), described in [22] by Craig. Our geometric model consists of 9 coordinate systems. The "b" frame is the reference coordinate system (world frame); later in our measurements it is represented by a tag marker placed on a rod. The "0" frame represents base frame of the robot. The frames from "1" to "6" represent the joints; the 6th frame position corresponds to the mounting flange. The "e" frame stands for the tool offset, in this case a measuring point that was focused by the sensor. The scheme of the model is illustrated in Figure 1. The MDH parameters are noted in Table 1; the o_i stands for offset of the ith joints in this position.

Table 1. MDH parameters of the UR3 robot.

i	α_{i-1} [rad]	x_{i-1} [mm]	z_i [mm]	θ_i [rad]	o_i [rad]
1	0	0	151.90	q_1	0
2	$\pi/2$	0	119.85	q_2	π
3	0	243.65	0	q_3	0
4	0	213.25	-9.45	q_4	0
5	$-\pi/2$	0	83.35	q_5	0
6	$\pi/2$	0	81.90	q_6	π

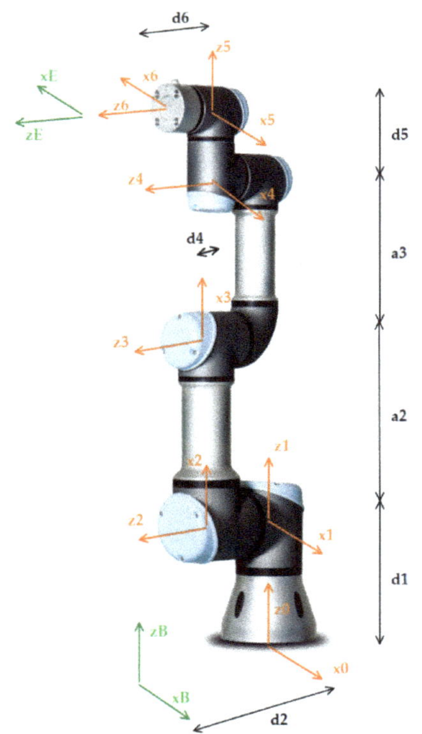

Figure 1. UR3 with coordinate frames according to MDH.

For a vector $q = [q_1, q_2, q_3, q_4, q_5, q_6]^T$ representing the joint variables, a homogeneous transformation matrix $T_{be}(q)$ gives the position and orientation P of the UR3's end-effector tool frame "e" with respect to the base frame "b" of the workplace.

$$P = T_{be}(q) \tag{1}$$

$$T_{be}(q) = A_{b0}A_{01}(q_1)A_{12}(q_2)A_{23}(q_3)A_{34}(q_4)A_{45}(q_5)A_{56}(q_6)A_{6e} \tag{2}$$

According to [22], matrix $A_{i-1,i}$ in MDH notation is obtained by multiplying rotation matrix R_x along x axis, translation matrix T_x along x axis, rotation matrix R_z along z axis and translation matrix T_z along z axis.

$$A_{i-1,i} = R_x(\alpha_{i-1})T_x(x_{i-1})R_z(\theta_i)T_z(d_i) \tag{3}$$

G=The geometric model of the UR3 is mathematically expressed by the transformation matrix $T_{be}(q)$ noted in Equation (2). Matrix A_{b0} is displacement between the reference "b" frame and "0" frame; orientation difference is represented by R_0 rotational matrix. Matrix A_{6e} is displacement between the mounting flange and "e" frame of the end-effector. The objective of this study is to determine the 12 parameters of A_{b0} matrix, so to find the base frame "0" of a robot in a workplace. To be able to achieve this, it is necessary to identify during the calculations also the displacement of the end-effector (x_e, y_e, and z_e); however, the rotational part of the A_{6e} can be freely chosen. The reason is that the transform is static (there is no variable between 6th frame of the robot and end-effector "e" frame). For simplicity, we choose R_e as identity matrix.

$$A_{b0} = \begin{bmatrix} & R_0 & & x_0 \\ & & & y_0 \\ & & & z_0 \\ 0 & 0 & 0 & 1 \end{bmatrix} ; A_{6e} = \begin{bmatrix} 1 & 0 & 0 & x_e \\ 0 & 1 & 0 & y_e \\ 0 & 0 & 1 & z_e \\ 0 & 0 & 0 & 1 \end{bmatrix} \tag{4}$$

$$R_0 = \begin{bmatrix} r_{11} & r_{12} & r_{13} \\ r_{21} & r_{22} & r_{23} \\ r_{31} & r_{32} & r_{33} \end{bmatrix} \tag{5}$$

In general, geometric models are idealized and very difficult to make comply with real conditions due to manufacturing inaccuracy and environmental conditions. Error values can be estimated and included into the mathematical models though. Finding the relations between theoretical and real models is the crucial task of robot calibration. To find such a relation, geometric parameters of a device have to be identified. Robot identification is a process wherein error parameter values are determined using results of a test measurement. In the following simulation the UR3 robot performed a trajectory as described in Section 3. Obtained data of the end-effector position P_c were compared with the robot's position $P(q)$ based on q_i for each joint.

The parameters x_0, y_0, z_0, r_{11}, r_{12}, r_{13}, r_{21}, r_{22}, r_{23}, r_{31}, r_{32}, r_{33}, x_e, y_e and z_e were chosen to be identified. The reason for identification of the end-effector frame is because the Aruco tags may be placed on a low-cost end-effector by 3D printing and the designed CAD model transformation might be different than the real solution. On the other hand, we wanted to make the model as simple as possible, so we avoided MDH parameter identification between particular joints and links of the robot, which would lead to the robot's calibration process. We assume that this simple identification process may compensate for the small errors between the links.

When the transformation matrix $T_{be}(q)$ was defined earlier, the position vector of the end-effector P was represented by 4th column of T_{be} with respect to the reference frame. If everything is ideal, we can consider this coordinate's equal to the coordinate's values calculated using the position sensor, as Equation (6) shows, for a specific q, where T'_{be} means 1st to 3rd elements of the 4th column of the transformation matrix.

$$P_c = T'_{be}(q) = P(q) \tag{6}$$

2.2. Identification with the Jacobian Method

The most common method for parameter identification is the application of a Jacobian, which is also described, for example, in [6,16]. This iterative method utilizes benefits of the Jacobian matrix that is obtained by partial derivative of the position vector (1st to 3rd elements of 4th column) of T_{be} with respect to the parameters in X, the parameters that are going to be identified. Symbolically, the Jacobian is expressed as 3×15 J_i matrix in Equation (7), where p_i stands for a parameter of X.

$$J_i = \begin{bmatrix} \frac{\partial T_{be_x}(q,X)}{\partial p_1} & \cdots & \frac{\partial T_{be_x}(q,X)}{\partial p_n} \\ \frac{\partial T_{be_y}(q,X)}{\partial p_1} & \cdots & \frac{\partial T_{be_y}(q,X)}{\partial p_n} \\ \frac{\partial T_{be_z}(q,X)}{\partial p_1} & \cdots & \frac{\partial T_{be_z}(q,X)}{\partial p_n} \end{bmatrix} \tag{7}$$

For every measured point i the J_i is determined. By applying all measured points a $3n \times 15$ J matrix is obtained, where n is a number of measured points.

$$J = \begin{bmatrix} J_1 \\ J_2 \\ \cdots \\ J_n \end{bmatrix} \tag{8}$$

As a first step of every iteration a position vector Y_m is calculated using values X_j, where j represents the iteration step. For the first iteration, guessed values X_0 are used. The q is the vector of joint variables measured by robot's joint position sensors.

$$Y_m = T_{be}(q, X_i) \tag{9}$$

The next step is to compare and calculate the difference between the position measured by camera Y_c and the previously calculated position Y_m, so ΔY is determined. Y_c is $n \times 3$ matrix, where n stands for number of measured points, and there are three measured coordinates x, y, z.

$$\Delta Y = Y_c - Y_m \tag{10}$$

The key equation of this method is Equation (11). When a position changes, the Jacobian matrix changes too; therefore, Δx can be observed as the change of the parameters in X.

$$\Delta Y = J \Delta x \tag{11}$$

By using matrix operations in Equations (12) and (13), the values of Δx are determined.

$$(J^T J)^{-1} J^T \Delta Y = (J^T J)^{-1} J^T J \Delta x \tag{12}$$

$$\Delta x = (J^T J)^{-1} J^T \Delta Y \tag{13}$$

At the ends of iterations we added the computed values to the X_{j+1}. A convergent check follows to decide whether another iteration step is needed.

$$X_{j+1} = X_j + \Delta x \tag{14}$$

3. Simulations and Experiment

Two types of simulations and one experiment were performed to verify the proposed method of parameter identification. Simulation A was calculated only with absolute positions of the end-effector coordinates determined by CoppeliaSim software with the built-in UR3 model, as shown in Figure 2.

The robot moved along the same path in both simulations and the experiment, consisting of 250 points. The robot stopped at each pose and the measurements were taken. The reason for choosing such a path was to obtain coordinates of the joints that were as different as possible; on the other hand, due to the experiment that was performed with cameras we needed to guarantee the visibility of the Aruco tags, which were used for the simulation B and the experiment.

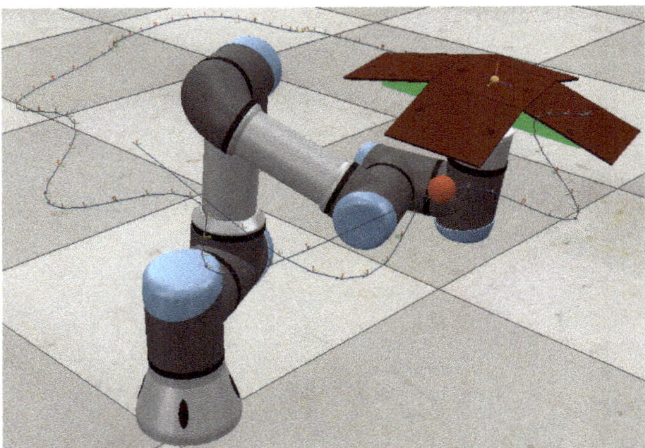

Figure 2. Simulated path of the UR3 in CoppeliaSim, simulation A.

3.1. Simulation A

The robot was moved along a defined path with fixed points to stop at. Once it stopped, the joint coordinates and the position vector of the end-effector related to the world base frame were acquired. With these two sets of input values, the identification was made using the methods described in the previous chapter. Results may be seen in Section 3.5.

3.2. Simulation B

For simulation B and the experiment, there were cameras and OpenCV libraries [18] applied for image processing to detect the Aruco tags. Based on the previous research by Oscadal [23], we used a 3D gridboard with tags, which improves the reliability and accuracy of detection in comparison with basic 2D tags. The gridboard represents a coordinate frame; in our case it was the base frame "b" and the end-effector frame "e" as shown in Figures 3 and 4. The OpenCV library algorithm can calculate transformation from a camera to a tag. In real-time measurements, we used Equation (15) thanks to which the position of any camera was not important. Matrix T_{cb} is the transformation from the camera to the base; matrix T_{ce} is the transformation from the camera to the end-effector. The "c" frame is the camera frame. On the other hand, the position of a camera may be saved for later operations.

$$T_{be} = T_{cb}^{-1} T_{ce} \tag{15}$$

In the simulation B (Figure 3) we deployed the image analysis in CoppeliaSim, with a single camera of resolution 1280×720 px. The virtual camera was self-calibrated and the detection parameters of the OpenCV library for finding the Aruco tags were set similarly as in [23]. The dimensions of the tags were 70×70 mm with a 6×6 bit matrix.

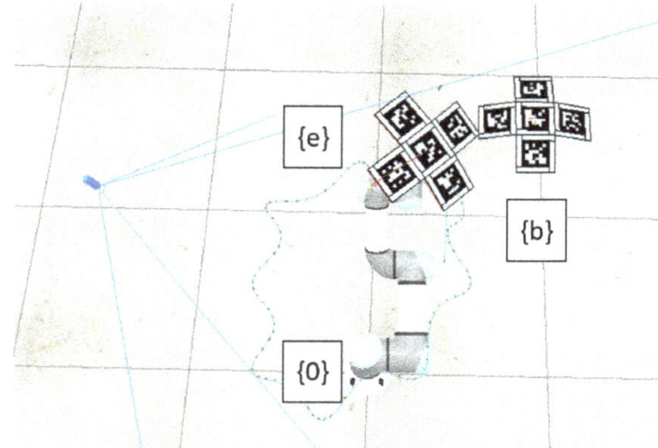

Figure 3. Simulation B with tags and a camera.

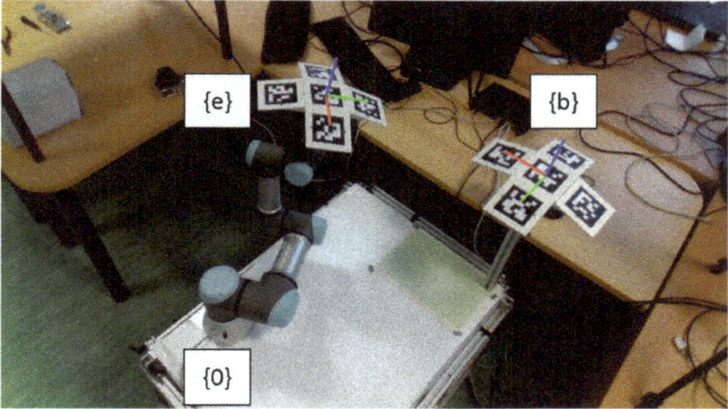

Figure 4. Experiment A setup; point of view of camera 2.

3.3. Experiment A

Three independent cameras were placed around the UR3 robot to observe its trajectory and to calculate transformations in a laboratory during a real experiment. Intel RealSense D435i cameras with 1280×720 px resolution were used, and even though they are depth cameras, only the simple 2D RGB pictures were analyzed. The specifications of the cameras are shown in Table 2. The cameras were self-calibrated following the methodology used in [23].

As already said before, the robot went through 250 positions on the path. To reduce the inaccuracy of the detection in reality, 10 camera frames were taken for every position, which gave us 2500 measured points. In total for the three cameras, 7500 pictures were analyzed during the calibration process.

Table 2. Specifications of Intel RealSense D435i camera [24].

Parameter	Value
Resolution	1920×1080 px
Frame Rate	30 fps
Sensor FOV (H $\times$ V $\times$ D)	$69.4° \times 42.5° \times 77° (\pm 3°)$
Dimensions	$90 \times 25 \times 25$ mm
Connection	USB-C 3.1 Gen 1

3.4. Experiment B

To observe the impact of placement of the world base frame "b," another experiment was performed as shown in Figure 5. Please, notice the difference in position and rotation of the base frame.

Only one camera (the same type, resolution and calibration) was used in this case with robot following a similar trajectory as in the previous experiment A. The robot went through 200 poses; at each pose, five images were taken and analyzed, so 1000 measurements in total were made.

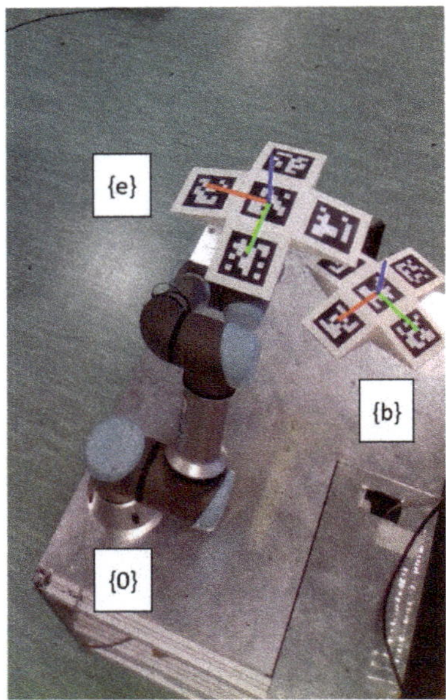

Figure 5. Experiment B setup.

3.5. Results

The calibration results of simulations are presented in Table 3. The data calculated based on simulation A (without tag detection, only end-effector position tracking) show high-precision identification with a very small difference in comparison to the expected results (difference Δ is shown in the brackets). Such an accuracy could be achieved using, for example, a laser tracker in reality. For simulation B, when the tags were detected using simulated camera, the error was higher (maximum 2.59 mm for x_0), which gave us an idea about how accurate the system might be, so the noise from the environment was lowered, but the camera parameters were kept.

In Table 4 are the results of the experiment A. The best values were obtained when all the results of the three cameras were combined and analyzed together. The error for the base frame (x_0, y_0, and z_0) was maximally 7.61 mm in the z_0 direction. We performed other measurements following the same strategy; they all provided similar results, which made it made clear that the position of a camera has an influence on the detection accuracy, which is supported by results of Krajnik's research [25].

In Table 5 are the results of the experiment B. It proves the possibility of placing the base frame freely with respect to the robot.

The trajectories are compared in Figures 6–10. Measured cameras/measured path are plotted points that were obtained by camera/end-effector position sensor; robot path is a self-check plot of the end-effector path after the identified parameters of X were applied in the transformation matrix. Points of origin are points where the world base frame was determined for every point on the path. In Figures 11 and 12 we can observe the error values in box plots.

Table 3. Results of simulations for the X vector after identification.

Parameter	Expected	Simulation A	Simulation B
x_0 [mm]	431.00	430.98 (Δ 0.02)	433.59 (Δ 2.59)
y_0 [mm]	555.00	555.02 (Δ 0.02)	556.01 (Δ 1.01)
z_0 [mm]	-460.00	-460.07 (Δ 0.07)	-458.41 (Δ 1.59)
r_{11} [-]	1.00	1.00 (Δ 0.00)	1.006 (Δ 0.006)
r_{21} [-]	0.00	0.00 (Δ 0.00)	-0.006 (Δ 0.006)
r_{31} [-]	0.00	0.00 (Δ 0.00)	0.001 (Δ 0.001)
r_{12} [-]	0.00	0.00 (Δ 0.00)	0.003 (Δ 0.003)
r_{22} [-]	1.00	1.00 (Δ 0.00)	1.004 (Δ 0.004)
r_{32} [-]	0.00	0.00 (Δ 0.00)	-0.000 (Δ 0.000)
r_{13} [-]	0.00	0.00 (Δ 0.00)	-0.006 (Δ 0.006)
r_{23} [-]	0.00	0.00 (Δ 0.00)	-0.001 (Δ 0.001)
r_{33} [-]	1.00	1.00 (Δ 0.00)	1.000 (Δ 0.000)
x_e [mm]	0.00	-0.05 (Δ 0.05)	0.14 (Δ 0.14)
y_e [mm]	0.00	-0.02 (Δ 0.02)	0.23 (Δ 0.23)
z_e [mm]	66.00	65.98 (Δ 0.02)	64.78 (Δ 1.22)

Table 4. Results of experiment A for the X vector after identification.

Parameter	Expected	Cam 1	Cam 2	Cam 3	Cameras Combined
x_0 [mm]	431.00	432.53	430.81	426.91	430.05 (Δ 0.95)
y_0 [mm]	555.00	551.15	561.06	555.45	555.96 (Δ 0.96)
z_0 [mm]	-460.00	-451.33	-449.27	-456.496	-452.39 (Δ 7.61)
r_{11} [-]	1.00	1.006	1.024	1.013	1.014 (Δ 0.014)
r_{21} [-]	0.00	-0.018	-0.039	-0.015	-0.024 (Δ 0.024)
r_{31} [-]	0.00	-0.004	-0.013	0.008	-0.003 (Δ 0.003)
r_{12} [-]	0.00	0.020	0.014	0.007	0.013 (Δ 0.013)
r_{22} [-]	1.00	1.008	1.019	1.004	1.011 (Δ 0.011)
r_{32} [-]	0.00	0.000	-0.009	0.010	0.000 (Δ 0.000)
r_{13} [-]	0.00	0.010	0.010	0.001	0.007 (Δ 0.007)
r_{23} [-]	0.00	0.012	-0.009	-0.006	-0.001 (Δ 0.001)
r_{33} [-]	1.00	0.990	0.969	1.011	0.991 (Δ 0.009)
x_e [mm]	0.00	0.18	0.49	0.05	0.24 (Δ 0.24)
y_e [mm]	0.00	0.52	0.73	0.68	0.65 (Δ 0.65)
z_e [mm]	66.00	64.09	61.00	66.03	63.73 (Δ 2.27)

Table 5. Results of experiment B for the X vector after identification.

Parameter	Expected [mm]	Experiment B [mm]
x_0 [mm]	611.00	613.88 (Δ 2.88)
y_0 [mm]	29.00	30.94 (Δ 1.94)
z_0 [mm]	-27.00	-25.12 (Δ 1.88)
r_{11} [-]	-	0.506
r_{21} [-]	-	-0.873
r_{31} [-]	-	0.012
r_{12} [-]	-	0.872
r_{22} [-]	-	0.515
r_{32} [-]	-	0.020
r_{13} [-]	-	-0.016
r_{23} [-]	-	-0.013
r_{33} [-]	-	1.043
x_e [mm]	0.00	-0.05 (Δ 0.05)
y_e [mm]	0.00	0.69 (Δ 0.69)
z_e [mm]	66.00	67.59 (Δ 1.59)

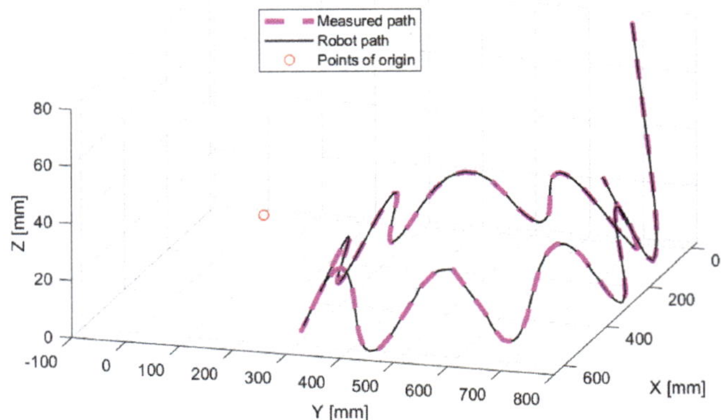

Figure 6. Measured path in comparison with the real robot path of simulation A.

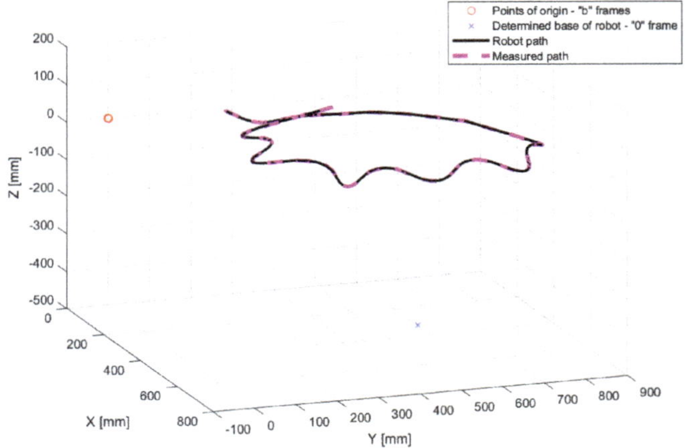

Figure 7. Measured path in comparison with the real robot path of simulation B.

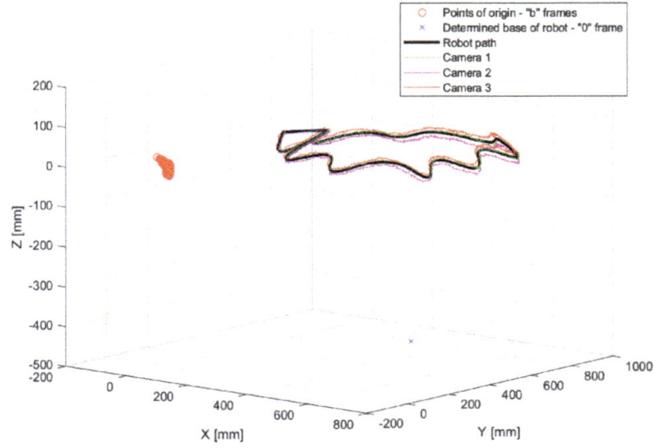

Figure 8. Paths measured by cameras in comparison with the real robot path of experiment A.

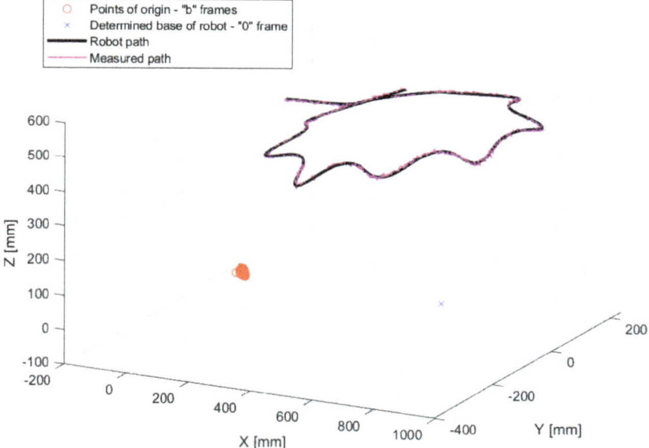

Figure 9. Paths measured by cameras in comparison with the real robot path of experiment A, xy plane.

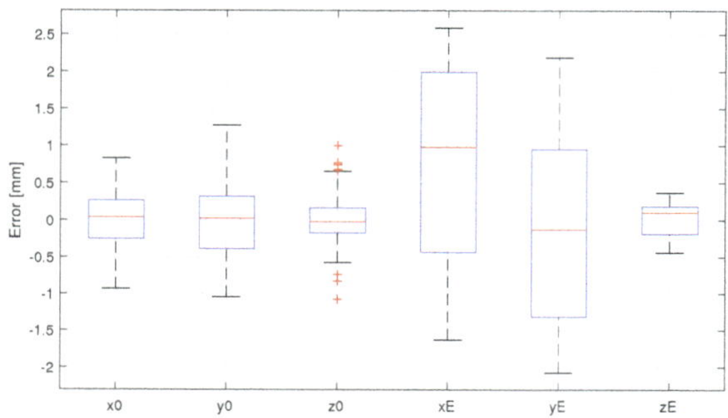

Figure 10. Path measured by a camera in comparison with the real robot path of experiment B.

Figure 11. Errors for the identified values based on simulation B.

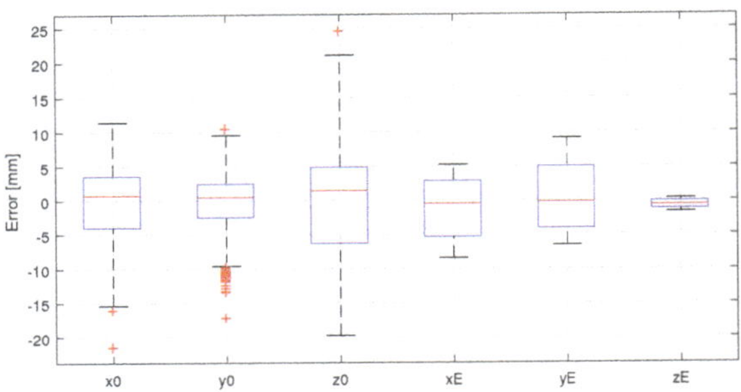

Figure 12. Errors for the identified values based on experiment A.

4. Conclusions

The process of identification of the robot's base frame in a workplace using Aruco markers and OpenCV was presented and verified in this study. This approach may be used for more robots and other components of the workplace at the same time, which brings the main advantage of fast evaluation and later recalibration. The typical scenario is placing all components in their positions, placing markers on them and other points of interest, running a robot's path while measuring end-effector position by a camera and evaluating the results—obtaining the coordinates of robot's base and coordinates of points of interest (manipulated object, pallet, etc.). As the end-effector we used an 3D printed gridboard that might be replaced by a cubic with tags carried by a gripper.

When observing the results provided in the Section 3.5, one can see there is a gap between the accuracy of the simulated workplace and the experiments. As already mentioned above, simulation A demonstrates the possible accuracy of this method when all conditions are close to ideal. Therefore, there are some methods and topics for another study that would help to minimize the errors. The DH parameters of the UR3 were based on its datasheet, but when the robot is manufactured it is calibrated and modified DH parameters are saved in the control unit. This parameters may be retrieved and applied in the identification calculations. In general, UR3 is not as precise as other industrial robots; depending on demands, a different manipulator should provide more accurate results.

Nevertheless, the end-effector was 3D printed and assembled from three parts; more accuracy may be achieved using better manufacturing methods. In some cases, if the end-effector was manufactured precisely beforehand, only the identification of a base frame might be enough (instead of identification of the base frame and end-effector offset, as was presented).

However, the biggest issue seems to be the detection of the tags, as results differed for every camera in the experiment, as shown in Table 4. Positioning of a camera (distance from a tag) seems to have a big influence on the outcome. This topic was researched in the Krajnik's work [25]. Additionally, alternating the OpenCV's software algorithms and filtering leads to better detection; more on this topic is discussed in [23]. A user may seriously consider the use of a self-calibrated camera with higher resolution than 1280 × 720 px, as we used. It is important to note that the presented accuracy in simulation B and real experiment were achieved by cameras that we calibrated. There is no doubt that one with better calibrated hardware may achieve more accurate results.

Another question to focus on is which trajectory and how many measured points are necessary to provide satisfying results; we tested the system with only one path of 250 points.

Once this camera-based method is well optimized for a task depending on certain available equipment, and the accuracy is acceptable, it will stand as sufficient easy-to-deploy and low-cost solution for integrators and researchers. They will be able to quickly place components and robots, tag them and obtain their position coordinates based on prepared universal measurement. In addition, even the current system may serve in the manufacturing process as a continuous safety-check that all required components, including robots, are in the place where they should be, if the detection accuracy is acceptable.

In addition, this method will be used for identification of reference coordinate systems and kinematic parameters of experimental custom manipulators, the design which is a point of interest for Research Centre of Advanced Mechatronic Systems project.

To make this calibration method easier to follow, the Matlab's scripts with the calculations and raw input data obtained by simulation and experiments may be found on the Github page of the Department of Robotics, VSB-Technical University of Ostrava [26]. The calibration methodology and Supplementary Materials provided may serve engineers who have no previous experience with the process; they can use cameras or eventually other sensors, such as laser trackers.

Supplementary Materials: The following are available online at http://www.mdpi.com/2076-3417/10/21/7679/s1.

Author Contributions: Conceptualization, D.H. and Z.B.; methodology, D.H. and A.V.; software, P.O.; validation, D.H., P.O. and T.S.; formal analysis, T.S.; investigation, D.H.; resources, M.V.; data curation, T.S.; writing—original draft preparation, D.H.; writing—review and editing D.H and T.S.; visualization, M.V.; supervision, A.V.; project administration, Z.B.; funding acquisition, Z.B. All authors have read and agreed to the published version of the manuscript.

Funding: This work was supported by the European Regional Development Fund in the Research Centre of Advanced Mechatronic Systems project, project number CZ.02.1.01/0.0/0.0/16_019/0000867 within the Operational Programme Research, Development and Education. This article has been also supported by specific research project SP2020/141 and financed by the state budget of the Czech Republic.

Conflicts of Interest: The authors declare no conflict of interest. The funders had no role in the design of the study; in the collection, analyses or interpretation of data; in the writing of the manuscript, or in the decision to publish the results.

Abbreviations

The following abbreviations are used in this manuscript:

OpenCV	Open Source Computer Vision Library
DH	Denavit–Hartenberg
FOV	Field of View
FPS	Frames per Second
MDH	Modified Denavit–Hartenberg
RGB	Red Green Blue
ROS	Robot Operating System

References

1. Abderrahim, M.; Khamis, A.; Garrido, S.; Moreno, L. Accuracy and calibration issues of industrial manipulators. In *Industrial Robotics Programming, Simulation and Application*; IntechOpen: London, UK, 2004; pp. 131–146.
2. Young, K.; Pickin, C.G. Accuracy assessment of the modern industrial robot. *Ind. Robot. Int. J.* **2000**, *27*, 427–436. [CrossRef]
3. Slamani, M.; Nubiola, A.; Bonev, I. Assessment of the positioning performance of an industrial robot. *Ind. Robot. Int. J.* **2012**, *39*, 57–68. [CrossRef]
4. Nubiola, A.; Bonev, I.A. Absolute calibration of an ABB IRB 1600 robot using a laser tracker. *Robot. Comput. Integr. Manuf.* **2013**, *29*, 236–245. [CrossRef]
5. Wu, Y.; Klimchik, A.; Caro, S.; Furet, B.; Pashkevich, A. Geometric calibration of industrial robots using enhanced partial pose measurements and design of experiments. *Robot. Comput.-Integr. Manuf.* **2015**, *35*, 151–168. [CrossRef]
6. Nguyen, H.N.; Zhou, J.; Kang, H.J. A calibration method for enhancing robot accuracy through integration of an extended Kalman filter algorithm and an artificial neural network. *Neurocomputing* **2015**, *151*, 996–1005. [CrossRef]
7. Joubair, A.; Bonev, I.A. Non-kinematic calibration of a six-axis serial robot using planar constraints. *Precis. Eng.* **2015**, *40*, 325–333. [CrossRef]

8. Filion, A.; Joubair, A.; Tahan, A.S.; Bonev, I.A. Robot calibration using a portable photogrammetry system. *Robot. Comput. Integr. Manuf.* **2018**, *49*, 77–87. [CrossRef]

9. Möller, C.; Schmidt, H.C.; Shah, N.H.; Wollnack, J. Enhanced Absolute Accuracy of an Industrial Milling Robot Using Stereo Camera System. *Procedia Technol.* **2016**, *26*, 389–398. [CrossRef]

10. Lembono, T.S.; Suarez-Ruiz, F.; Pham, Q.C. SCALAR—Simultaneous Calibration of 2D Laser and Robot's Kinematic Parameters Using Three Planar Constraints. In Proceedings of the 2018 IEEE/RSJ International Conference on Intelligent Robots and Systems (IROS), Madrid, Spain, 1–5 October 2018. [CrossRef]

11. Marie, S.; Courteille, E.; Maurine, P. Elasto-geometrical modeling and calibration of robot manipulators: Application to machining and forming applications. *Mech. Mach. Theory* **2013**, *69*, 13–43. [CrossRef]

12. Gan, Y.; Dai, X. Base frame calibration for coordinated industrial robots. *Robot. Auton. Syst.* **2011**, *59*, 563–570. [CrossRef]

13. Arai, T.; Maeda, Y.; Kikuchi, H.; Sugi, M. Automated Calibration of Robot Coordinates for Reconfigurable Assembly Systems. *CIRP Ann.* **2002**, *51*, 5–8. [CrossRef]

14. Motta, J.M.S.; de Carvalho, G.C.; McMaster, R. Robot calibration using a 3D vision-based measurement system with a single camera. *Robot. Comput. Integr. Manuf.* **2001**, *17*, 487–497. [CrossRef]

15. Watanabe, A.; Sakakibara, S.; Ban, K.; Yamada, M.; Shen, G.; Arai, T. A Kinematic Calibration Method for Industrial Robots Using Autonomous Visual Measurement. *CIRP Ann.* **2006**, *55*, 1–6. [CrossRef]

16. Van Albada, G.; Lagerberg, J.; Visser, A.; Hertzberger, L. A low-cost pose-measuring system for robot calibration. *Robot. Auton. Syst.* **1995**, *15*, 207–227. [CrossRef]

17. Santolaria, J.; Brosed, F.J.; Velázquez, J.; Jiménez, R. Self-alignment of on-board measurement sensors for robot kinematic calibration. *Precis. Eng.* **2013**, *37*, 699–710. [CrossRef]

18. Garrido, S.; Nicholson, S. Detection of ArUco Markers. OpenCV: Open Source Computer Vision. 2020. Available online: www.docs.opencv.org/trunk/d5/dae/tutorial_aruco_detection.html (accessed on 23 October 2020).

19. Kroeger, O.; Huegle, J.; Niebuhr, C.A. An automatic calibration approach for a multi-camera-robot system. In Proceedings of the 2019 24th IEEE International Conference on Emerging Technologies and Factory Automation (ETFA), Zaragoza, Spain, 10–13 September 2019; pp. 1515–1518. [CrossRef]

20. Lewis, C. Industrial Calibration. 2020. Available online: https://github.com/ros-industrial/industrial_calibration (accessed on 23 October 2020).

21. Meyer, J. Robot Calibration Tools. 2020. Available online: https://github.com/Jmeyer1292/robot_cal_tools (accessed on 23 October 2020).

22. Craig, J.J. *Introduction to Robotics: Mechanics and Control, 3/E*; Pearson Education India: Bengaluru, India, 2009.

23. Oščádal, P.; Heczko, D.; Vysocký, A.; Mlotek, J.; Novák, P.; Virgala, I.; Sukop, M.; Bobovský, Z. Improved Pose Estimation of Aruco Tags Using a Novel 3D Placement Strategy. *Sensors* **2020**, *20*, 4825. [CrossRef] [PubMed]

24. Intel Corporation. Depth Camera D435i—Intel RealSense. 2020. Available online: https://www.intelrealsense.com/depth-camera-d435i/ (accessed on 23 October 2020).

25. Krajník, T.; Nitsche, M.; Faigl, J.; Vaněk, P.; Saska, M.; Přeučil, L.; Duckett, T.; Mejail, M. A Practical Multirobot Localization System. *J. Intell. Robot. Syst.* **2014**, *76*, 539–562. [CrossRef]

26. Huczala, D. Parameters Identification. 2020. Available online: github.com/robot-vsb-cz/parameters-identification (accessed on 23 October 2020).

Publisher's Note: MDPI stays neutral with regard to jurisdictional claims in published maps and institutional affiliations.

 *applied sciences*

Article

Recognition of Perspective Distorted QR Codes with a Partially Damaged Finder Pattern in Real Scene Images

Ladislav Karrach [1], Elena Pivarčiová [1,*] and Pavol Bozek [2]

[1] Department of Manufacturing and Automation Technology, Faculty of Technology, Technical University in Zvolen, Masarykova 24, 960 01 Zvolen, Slovakia; karrach@zoznam.sk

[2] Institute of Production Technologies, Faculty of Materials Science and Technology, Slovak University of Technology in Bratislava, Vazovova 5, 811 07 Bratislava, Slovakia; pavol.bozek@stuba.sk

* Correspondence: pivarciova@tuzvo.sk

Received: 21 September 2020; Accepted: 30 October 2020; Published: 4 November 2020

Featured Application: Mobile robot navigation, automatic object identification and tracking.

Abstract: QR (Quick Response) codes are one of the most famous types of two-dimensional (2D) matrix barcodes, which are the descendants of well-known 1D barcodes. The mobile robots which move in certain operational space can use information and landmarks from environment for navigation and such information may be provided by QR Codes. We have proposed algorithm, which localizes a QR Code in an image in a few sequential steps. We start with image binarization, then we continue with QR Code localization, where we utilize characteristic Finder Patterns, which are located in three corners of a QR Code, and finally we identify perspective distortion. The presented algorithm is able to deal with a damaged Finder Pattern, works well for low-resolution images and is computationally efficient.

Keywords: QR code detection; adaptive thresholding; finder pattern; perspective transformation

---

1. Introduction

Some of the requirements that are placed on autonomous mobile robotic systems include real-world environments navigation and recognition and identification of objects of interest with which the robotic system have to interact. Computer vision allows machines to obtain a large amount of information from the environment that has a major impact on their behavior. In the surrounding environment there are often numerous different static objects (walls, columns, doors, and production machines) but also moving objects (people, cars, and handling trucks).

Objects that are used to refine navigation, landmarks, can be artificial (usually added by a human) and natural (which naturally occur in the environment) [1,2].

Robotic systems have to work in a real environment and must be able to recognize, for example, people [3], cars [4], product parameters for the purpose of quality control, or objects which are to be handled.

The use of QR (Quick Response) codes (two-dimensional matrix codes) in interaction with robots can be seen in the following areas:

- in the field of navigation as artificial landmarks-analogies of traffic signs that control the movement of the robot in a given area (no entry, driving direction, alternate route, and permitted hours of operation) or as an information board providing context specific information or instructions (such as identification of floor, room, pallets, and working place)

- in the area of object identification, 2D codes are often used to mark products and goods and thus their recognition will provide information about the type of goods (warehouses), the destination of the shipment (sorting lines) or control and tracking during track-and-trace.

QR Codes

QR Codes are classified among 2D matrix codes (similar to Data Matrix codes). QR Codes (Model 1 and Model 2) are squared-shaped 2D matrices of dark and light squares—so called modules. Each module represents the binary 1 or 0. Each QR Code has fixed parts (such as Finder Patterns and Timing Patterns) that are common to each QR Code and variable parts that differ according to the data that is encoded by a QR Code. Finder Patterns, which are located in the three corners of a QR Code, are important for determining the position and rotation of the QR Code. The size of the QR code is determined by the number of modules and can vary from 21 × 21 modules (Version 1) to 177 × 177 (Version 40). Each higher version number comprises four additional modules per side. In the Figure 1 we can see a sample of version 1 QR Code.

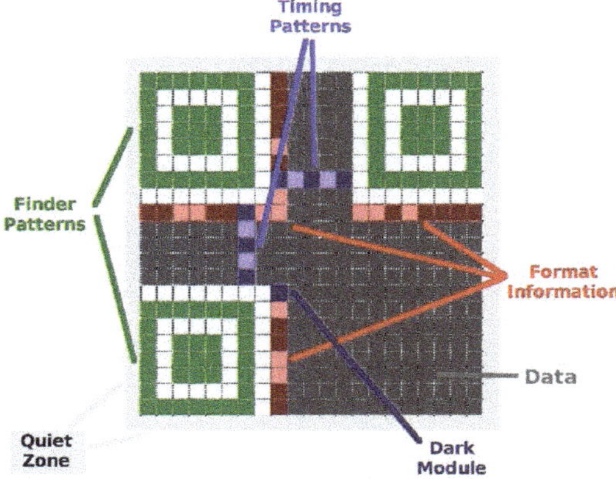

Figure 1. Version 1: 21 × 21 QR Code.

QR Code has error correction capability to restore data if the code is partially damaged. Four error correction levels are available (L–Low, M–Medium, Q–Quartile, and H–High). The error correction level determines how much of the QR Code can be corrupted to keep the data still recoverable (L–7%, M–15%, Q–25%, and H–30%) [5]. The QR Code error correction feature is implemented by adding a Reed–Solomon Code to the original data. The higher the error correction level is the less storage capacity is available for data.

Each QR Code symbol version has the maximum data capacity according to the amount of data, character type and the error correction level. The data capacity ranges from 10 alphanumeric (or 17 numeric) characters for the smallest QR Code up to 1852 alphanumeric (or 3057 numeric) characters for the largest QR Code at highest error correction level [5]. QR Codes support four encoding modes—numeric, alphanumeric, Kanji and binary—to store data efficiently.

QR Code was designed in 1994, in Japan for automotive industry, but currently has a much wider use. QR codes are used to mark a variety of objects (goods, posters, monuments, locations, and business cards) and allow to attach additional information to them, often in the form of a URL to a web page. QR Code is an ISO standard (ISO/IEC 18004:2015) and is freely available without license fees.

In addition to a traditional QR Code Model 1 and 2, there are also variants such as a Micro QR Code (a smaller version of the QR Code standard for applications where a symbol size is limited) or an iQR Code (which can hold a greater amount of information than a traditional QR Code and it supports also rectangular shapes) (Figure 2).

Figure 2. Other types of a QR Code: (**a**) Micro QR code, (**b**) iQR code.

2. Related Work

Prior published approaches for recognizing QR Codes in images can be divided into Finder Pattern based location methods [6–12] and QR Code region based location methods [13–18]. The first group locates a QR Code based on the location of its typical Finder Patterns that are present in its three corners. The second group locates the area of a QR code in the image based on its irregular checkerboard-like structure (a QR Code consists of many small light and dark squares which alternate irregularly and are relatively close to each other).

A shape of the Finder Pattern (Figure 3) was deliberately chosen by the authors of the QR Code, because "it was the pattern least likely to appear on various business forms and the like" [6]. They found out that black and white areas that alternate in a 1:1:3:1:1 ratio are the least common on printed materials.

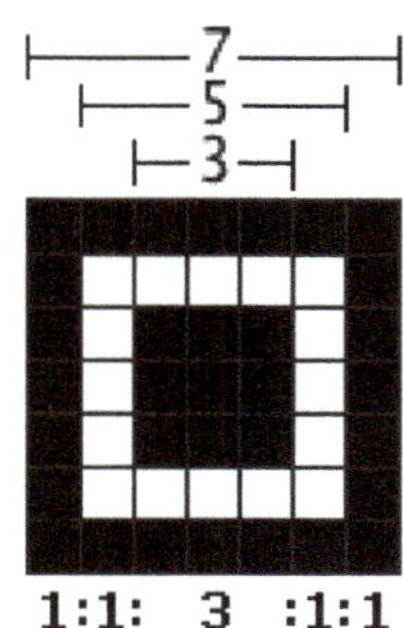

Figure 3. Finder Pattern.

In [7] (Lin and Fuh) all points matching the 1:1:3:1:1 ratio, horizontally and vertically, are collected. Collected points belonging to one Finder Pattern are merged. Inappropriate points are filtered out according to the angle between three of them.

In [8] (Li et al.) minimal containing region is established analyzing five runs in labeled connected components, which are compacted using run-length coding. Second, coordinates of central Finder Pattern in a QR Code are calculated by using run-length coding utilizing modified Knuth–Morris–Pratt algorithm.

In [9] (Belussi and Hirata) two-stage detection approach is proposed. In the first stage Finder Pattern (located at three corners of a QR Code) is detected using a cascaded classifier trained according to the rapid object detection method (Viola–Jones framework). In the second stage geometrical restrictions among detected components are verified to decide whether subsets of three of them correspond to a QR Code or not.

In [10] (Bodnár and Nyúl) Finder Pattern candidate localization is based on the cascade of boosted weak classifiers using Haar-like features, while the decision on a Finder Pattern candidate to be kept or dropped is decided by a geometrical constraint on distances and angles with respect to other probable Finder Patterns. In addition to Haar-like features, local binary patterns (LBP), and histogram of oriented gradients (HOG) based classifiers are used and trained to Finder Patterns and whole code areas as well.

In [11] (Tribak and Zaz) successive horizontal and vertical scans are launched to obtain segments whose structure complies with the ratio 1:1:3:1:1. The intersection between horizontal and vertical segments presents the central pixel of the extracted pattern. All the extracted patterns are transmitted to a filtering process based on principal components analysis, which is used as a pattern feature.

In [12] (Tribak and Zaz) seven Hu invariant moments are applied to the Finder Pattern candidates obtained by initial scanning of an image and using Euclidean metrics they are compared with Hu moments of the samples. If the similarity is less than experimentally determined threshold, then the candidate is accepted.

In [13] (Sun et al.) authors introduce algorithm, which aims to locate a QR Code area by four corners detection of 2D barcode. They combine the Canny edge detector with external contours finding algorithm.

In [14] (Ciążyński and Fabijańska) they use histogram correlation between a reference image of a QR code and an input image divided into a block of size 30×30. Then candidate blocks are joined into regions and morphological erosion and dilation is applied to remove small regions.

In [15] (Gaur and Tiwari) they propose approach which uses Canny edge detection followed by morphological dilation and erosion to connect broken edges in a QR Code into a bigger connected component. They expect that the QR Code is the biggest connected component in the image.

In [16,17] (Szentandrási et al.) they exploit the property of 2D barcodes of having regular distribution of edge gradients. They split a high-resolution image into tiles and for each tile they construct HOG (histogram of oriented gradients) from the orientations of edge points. Then they select two dominant peeks in the histogram, which are apart roughly $90°$. For each tile, a feature vector is computed, which contains a normalized histogram, angles of two main gradient directions, a number of edge pixels and an estimation of probability score of a chessboard-like structure.

In [18] (Sörös and Flörkemeier) areas with high concentration of edge structures as well as areas with high concentration of corner structures are combined to get QR Code regions.

3. The Proposed Method

Our method is primarily based on searching for Finder Patterns and utilizes their characteristic feature—1:1:3:1:1 black and white point ratio in any scanning direction. The basic steps are indicated in the flowchart in Figure 4.

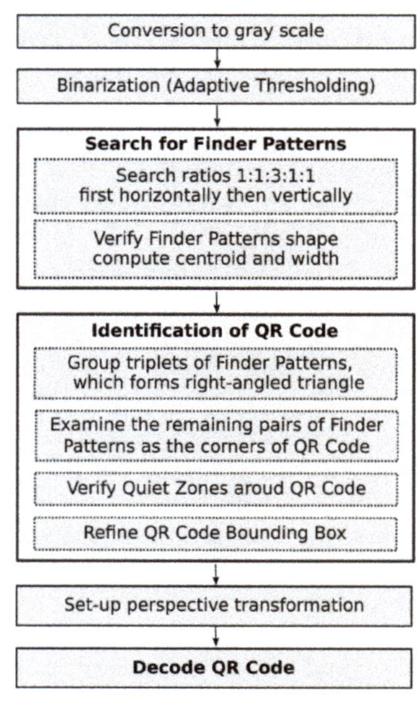

Figure 4. The flow chart of proposed algorithm.

Before searching for a QR Code, the original image (maybe colored) is converted to a gray scale image using Equation (1), because the color information does not bear any significant additional information that might help in QR Code recognition.

$$I = \frac{77R + 151G + 28B}{256} \tag{1}$$

where I stands for gray level and R, G, B for red, green, and blue color intensities of individual pixels in the RGB model, respetively. This RGB to gray scale conversion is integer approximation of widely used luminance calculation as defined in recommendation ITU-R BT.601-7:

$$I = 0.299R + 0.587G + 0.114B \tag{2}$$

Next, the gray scaled image is converted to a binary image using modified adaptive thresholding (with the size of window 35—the window size we choose to be at least five times the size of expected size of QRC module) [19]. We expect that black points, which belong to QRC, will become foreground points.

We use the modification of the well-known adaptive thresholding technique (Equation (3)), which calculates individual threshold for every point in the image. This threshold is calculated using average intensity of points under a sliding window. To speed up the thresholding we pre-calculate the integral sum image and we also use the global threshold value (points with intensity above 180 we always consider as background points). Adaptive thresholding can successfully threshold also uneven illuminated images.

$$B(x,y) = \begin{cases} 0 & \leftarrow & I(x,y) > 180 \\ 0 & \leftarrow & I(x,y) >= T(x,y) \\ 1 & \leftarrow & I(x,y) < T(x,y) \end{cases} \tag{3}$$

$$T(x,y) = m(x,y) - \frac{I(x,y)}{10} - 10$$

$$m(x,y) = \frac{1}{35 \times 35} \sum_{i=-17}^{17} \sum_{j=-17}^{17} I(x+i, y+j)$$

where I is gray scale (input) image, B is binary (output) image, T is threshold value (individual for each pixel at coordinates x, y), and m is average of pixel intensities under sliding window of the size 35×35 pixels.

In order to improve adaptive thresholding results, some of the image pre-processing techniques, such as histogram equalization, contrast stretching or deblurring, are worth to consider.

3.1. Searching for Finder Patterns

First, the binary image is scanned from top to bottom and from left to right, and we look for successive sequences of black and white points in a row matching the ratios 1:1:3:1:1 (W_1:W_2:W_3:W_4:W_5 where W_1, W_3, W_5 indicates the number of consecutive black points which are alternated by W_2, W_4 white points) with small tolerance (tolerance is necessary due to imperfect thresholding and noise in the Finder Pattern area, black and white points in a line do not alternate in ideal ratios 1:1:3:1:1):

$$W_1, W_2, W_4, W_5 \in \langle w - 1.5, w + 2.0 \rangle$$

$$W_3 \in \langle 3w - 2, 3w + 2 \rangle \qquad , \text{ where}$$

$$W_3 \geq \max(W_1 + W_2, W_4 + W_5) \tag{4}$$

$$w = \frac{W_1 + W_2 + W_3 + W_4 + W_5}{7} = \frac{W}{7}$$

For each match in a row coordinates of Centroid (C) and Width ($W = W_1 + W_2 + W_3 + W_4 + W_5$) of the sequence (of black and white points) are stored in a list of Finder Pattern candidates (Figure 5).

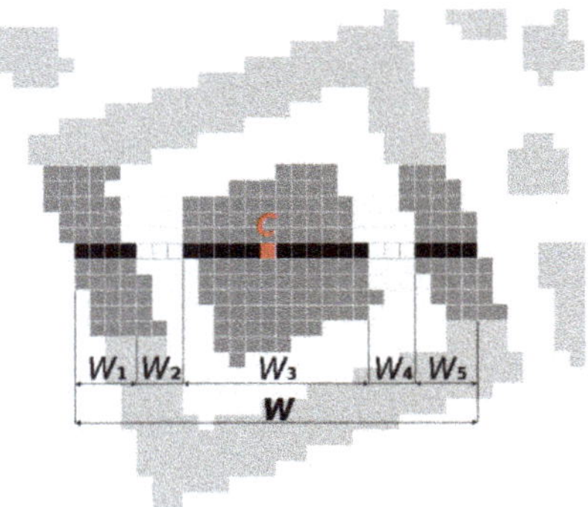

Figure 5. A Finder Pattern candidate matching 1:1:3:1:1 horizontally.

Then, Finder Pattern candidates (from the list of candidates) that satisfy the following criteria are grouped:

- their centroids C are at most 3/7W points vertically and at most 3 points horizontally away from each other,
- their widths W does not differ by more than 2 points.

We expect, that the Finder Pattern candidates in one group belong to the same Finder Pattern and therefore we set the new centroid C and width W of the group as average of x, y coordinates and widths of the nearby Finder Pattern candidates (Figure 6a).

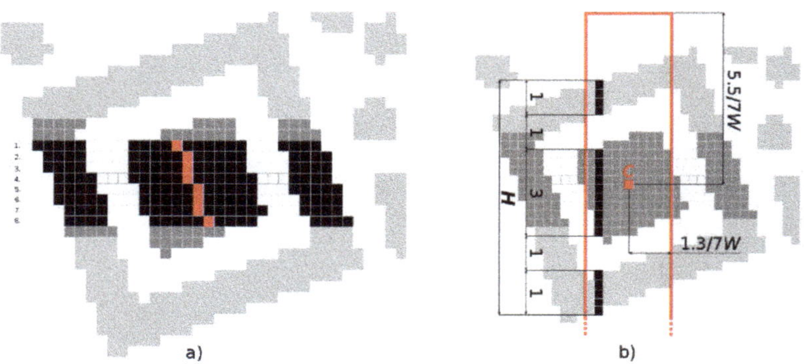

Figure 6. (**a**) Group of 8 Finder Pattern candidates matching 1:1:3:1:1 in rows, (**b**) Finder Pattern candidate matching 1:1:3:1:1 vertically.

After grouping the Finder Patterns, it must be verified whether there are sequences of black and white points also in the vertical direction, which alternate in the ratio 1:1:3:1:1 (Figure 6b). A bounding box around the Finder Pattern candidate, in which vertical sequences are looked for, is defined as

$$x \in \langle C_x \pm 1.3/7W \rangle, \; y \in \langle C_y \pm 5.5/7W \rangle \tag{5}$$

where C (C_x, C_y) is a centroid and W is width of the Finder Pattern candidate. We work with a slightly larger bounding box in case the Finder Pattern is stretched vertically. Candidates, where no vertical match is found or where the ratio $H/W < 0.7$, are rejected. For candidates, where a vertical match is found, the y coordinate of centroid C (C_y) is updated as an average of y coordinates of centers of the vertical sequences.

3.2. Verification of Finder Patterns

Each Finder Pattern consists of a central black square with the side of 3 units (R_1), surrounded by a white frame with the width of 1 unit (R_2), surrounded by a black frame with the width of 1 unit (R_3). In Figure 7 there are colored regions R_1, R_2 and R_3 in red, blue, and green, respectively. For each Finder Pattern candidate Flood Fill algorithm is applied, starting from the centroid C (which lies in the region R_1) and continuing through white frame (region R_2) to black frame (region R_3). As continuous black and white regions are filled, following region descriptors are incrementally computed:

- area ($A = M_{00}$)
- centroid ($C_x = M_{10}/M_{00}$, $C_y = M_{01}/M_{00}$), where M_{00}, M_{10}, M_{00} are raw image moments
- bounding box (Top, Left, Right, Bottom)

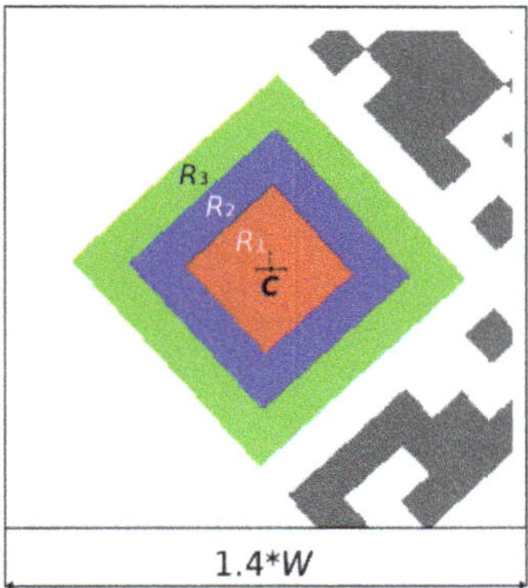

Figure 7. Regions of the Finder Pattern candidate.

The Finder Pattern candidate, which does not meet all of the following conditions, is rejected.

- Area(R_1) < Area(R_2) < Area(R_3) and
  1.1 < Area(R_2)/Area(R_1) < 3.4 and
  1.8 < Area(R_3)/Area(R_1) < 3.9
- 0.7 < AspectRatio(R_2) < 1.5 and
  0.7< AspectRatio(R_3) < 1.5
- |Centroid(R_1), Centroid(R_2)| < 3.7 and
  |Centroid(R_1), Centroid(R_3)| < 4.2

Note: the criteria were set to be invariant to the rotation of the Finder Pattern, and the acceptance ranges were determined experimentally. In an ideal undistorted Finder Pattern, the criteria are met as follows:

- Area(R_2)/Area(R_1) = 16/9 = 1.8 and Area(R_3)/Area(R_1) = 24/9 = 2.7
- AspectRatio(R_1) = 1 and AspectRatio(R_2) = 1 and AspectRatio(R_3) = 1
- Centroid(R_1) = Centroid(R_2) = Centroid(R_3)

In real environments there can be damaged Finder Patterns. The inner black region R_1 can be joined with the outer black region R_3 (Figure 8a) or the outer black region can be interrupted or incomplete (Figure 8b). In the first case the bounding box of the region R_2 is completely contained by bounding box of the region R_1 and in the second case is bounding box of the region R_3 contained in bounding box of the region R_2. These cases are handled individually. If the first case is detected, then the region R_1 is proportionally divided into R_1 and R_3 and if the second case is detected then the region R_2 is instantiated using the region R_1 and R_3.

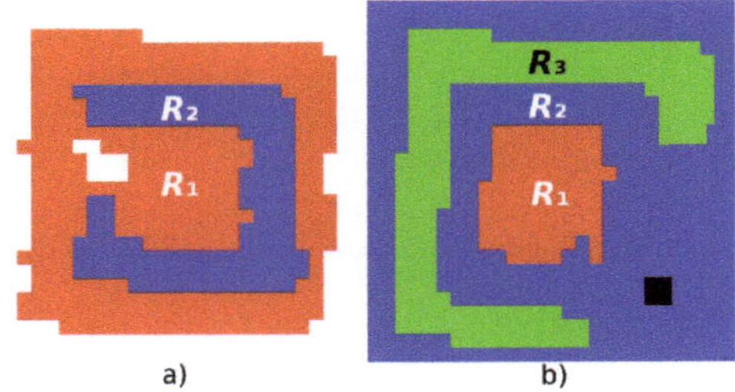

Figure 8. Various damages of Finder Patterns: (**a**) merged inner and outer black regions; (**b**) interrupted outer black region.

The Centroid (*C*) and Module Width (*MW*) of the Finder Pattern candidate are updated using the region descriptors as follows:

$$C = \left(\frac{M_{10}(R_1) + M_{10}(R_2) + M_{10}(R_3)}{M_{00}(R_1) + M_{00}(R_2) + M_{00}(R_3)}, \frac{M_{01}(R_1) + M_{01}(R_2) + M_{01}(R_3)}{M_{00}(R_1) + M_{00}(R_2) + M_{00}(R_3)} \right) \tag{6}$$

$$MW = \sqrt{M_{00}(R_1) + M_{00}(R_2) + M_{00}(R_3)}/7$$

3.3. Grouping of Finder Patterns

3.3.1. Grouping Triplets of Finder Patterns

In the previous steps, Finder Patterns in the image were identified and now such triplets (from the list of all Finder Patterns) must be selected, which can represent 3 corners of a QR Code. Matrix of the distances between the centroid of all Finder Patterns is build and all 3-element combinations of all Finder Patterns are examined. For each triplet it is checked whether it is possible to construct a right-angled triangle from it so that the following conditions are met:

- the size of each triangle side must be in predefined range
- the difference in sizes of two legs must be less than 21
- the difference in size of the real and theoretical hypotenuse must be less than 12

In this way, a list of QR Code candidates (defined by a triplet FP_1, FP_2, FP_3) is built. However, such a candidate for a QR Code is selected only on the basis of the mutual position of the 3 FP candidates. As is shown in Figure 9 not all QR Code candidates are valid (dotted red $FP_{3'}$-$FP_{3''}$-$FP_{2''}$ is false positive). These false positive QR Code candidates will be eliminated in the next steps.

Figure 9. QR Code candidates.

Finally, Bottom-Left and Top-Right Finder Pattern from the triplet (FP$_1$, FP$_2$, FP$_3$) is determined by using formula:

If (FP$_3$.x − FP$_2$.x)(FP$_1$.y − FP$_2$.y) − (FP$_3$.y − FP$_2$.y)(FP$_1$.x − FP$_2$.x) < 0 then Bottom-Left is FP$_3$ and Top-Right is FP$_1$ else vice versa.

3.3.2. Grouping Pairs of Finder Patterns

If any QR Code has one of the 3 Finder Patterns significantly damaged, then this Finder Pattern might not be identified and there remains two Finder Patterns in the Finder Patterns list, that were not selected (as the vertices of a right-angled triangle) in the previous step (Figure 10a). The goal of this step is to identify these pairs and determine the position of the third missing Finder Pattern. A square shape of the QR Code is assumed.

Figure 10. QR Code with one damaged Finder Pattern: (**a**) two known Finder Patterns; (**b**) two possible regions of QR Code; (**c**) two possible positions of 3rd Finder Pattern.

All two element combinations of remaining Finder Patterns, whose distance is in a predefined interval, are evaluated. A pair of Finder Patterns can represent Finder Patterns that are adjacent corners of a QR Code square (Figure 10a) or are in the opposite corners. If they are adjacent corners, then there are two possible positions of the QR Code (in Figure 10b depicted as a red square and green square). If they are opposite corners, then there are other two possible positions of the QR Code.

All four possible positions of the potential QR Code are evaluated against the following criteria:

- Is there a Quiet Zone around the bounding square at least 1 *MW* wide?
- Is there a density of white points inside bounding square in interval (0.4; 0.65)?
- Is there a density of edge points inside bounding square in interval (0.4; 0.6)?

Density of edge points is computed as the ratio of the number of edge points to area*2/*MW*.

Region of a QR Code is expected to have relative balanced density of white and black points and relative balanced ratio of edges to area.

A square region that meets all the above conditions is considered a candidate for a QR Code. There are two possible corners in the QR Code bounding square where 3rd Finder Pattern can be located (Figure 10c). For both possible corners Finder Pattern match score is computed and one with better score is selected (in other words question, "In which corner is the structure that more closely resembles the ideal Finder Pattern?" must be answered). Match score is computed as

$$MS = \mathrm{argmin}(OS + \min(BS, WS)) \tag{7}$$

where *MS* is match score (lower is better), *OS* is overall pattern match score, *BS* is black module match score and *WS* is white module match score. *BS* stores matches only for expected black points and *WS* stores matches only for expected white points between the mask and the image. *BS* and *WS* was introduced to handle situations when over the area of Finder Pattern is placed black or white spot, which would cause a low match score if only a simple pattern matching technique were used.

The match score is computed for several Finder Pattern mask positions by moving the mask in a spiral from its initial position up to a radius of *MW* with a step of *MW*/2 (for the case of small geometric deformations of the QR Code).

3.4. Verification of Quiet Zone

According to ISO standard a Quiet Zone is defined as "a region 4X wide which shall be free of all other markings, surrounding the symbol on all four sides". So, it must be checked if there are only white points in the image in the rectangular areas wide $1MW$ which is parallel to line segments defined by FP_1–FP_2 and FP_2–FP_3 (Figure 11). For fast scanning of the rectangle points Bresenham's line algorithm is utilized [20].

Figure 11. Quiet Zones.

QR Code candidates which do not have quiet zones around are rejected. Rejected are also QR Code candidates whose outer bounding box (larger) contains outer bounding box (smaller) of another QR Code candidate.

3.5. QR Code Bounding Box

Centroids of the 3 Finder Patterns (which represent the QR Code candidate) are the vertices of the triangle FP_1–FP_2–FP_3. This inner triangle must be expanded to outer triangle P_1–P_2–P_3, where the arms of the triangle pass through boundary modules of the QR Code (Figure 12). For instance, the shift of FP_3 to P_3 may be expressed as

$$P_3 = FP_3 + \frac{FP_3 - FP_1}{|FP_3, FP_1|} MW \sqrt{18} \tag{8}$$

where MW is module width (Equation (6))

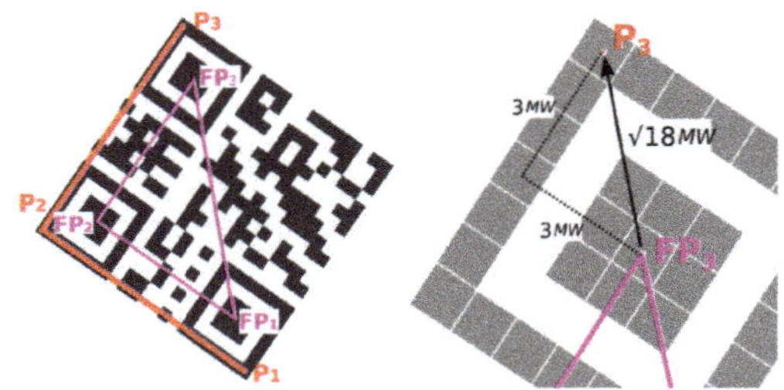

Figure 12. Bounding Box.

3.6. Perspective Distortion

For perspective undistorted QR Codes (only shifted, scaled, rotated, or sheared) it is sufficient to have only 3 points to set-up the affine transformation from a square to destination parallelogram. However, for perspective (projective) distorted QR Codes 4 points are required to set-up perspective transformation from a square to destination quadrilateral [21].

Some authors (for example [7,22]) search for Alignment Pattern to obtain 4th point. However, version 1 QR Codes does not have Alignment Pattern, so we have decided not to rely on Alignment Patterns.

Instead of that we use an iterative approach to find the opposite sides to P_2–P_1 and P_2–P_3, which aligns to QR Code borders.

1. We start from initial estimate of P_4 as an intersection of the line L_1 and L_3, where L_1 is parallel to P_2–P_3 and L_3 is parallel to P_2–P_1 (Figure 13a).
2. We count the number of pixels which are common to the line L_3 and the QR Code for each third of the line L_3 (Figure 13a).
3. We shift the line L_3 by width of one module away from the QR Code and again we count the number of pixels which are common to the shifted line $L_{3'}$ and the QR Code. The module width is estimated as *MW* (from Equation (6)) (Figure 13b).
4. We compare overlaps of the line L_3 from the step 2 and 3, and

    a. If L_3 was whole in the QR Code and the shifted $L_{3'}$ is out of the QR Code, then initial estimation of P_4 is good and we end.
    b. If L3 was whole in the QR Code and 3rd third of the shifted $L_{3'}$ is again in the QR Code, then we continue by step 5.
    c. If 3rd third of L_3 was in quiet zone and 2nd and 3rd third of the shifted $L_{3'}$ is in the quiet zone or if 2nd third of L_3 was in the quiet zone and 1st and 2nd third of the shifted $L_{3'}$ is in the quiet zone then we continue by step 6.

5. We start to move P_4 end of line segment P_3–P_4 away from the QR Code until 3rd third of L_3 touches the quiet zone (Figure 13c).
6. We start to move P_4 end of line segment P_3–P_4 towards the QR Code until 3rd third of L_3 touches the QR Code.
7. We apply the same procedure also to the line L_1 like for the line L_3.
8. The intersection of the shifted lines L_1 and L_3 is a new P_4 position.

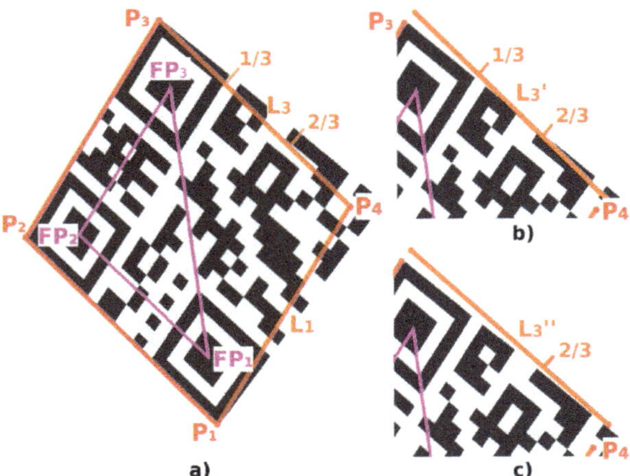

Figure 13. Perspective distortion: (**a**) initial estimate of the point P_4 and lines L_1, L_3; (**b**) first shift of the line L_3; (**c**) second shift of the line L_3.

Once the position of 4th point, P_4, is obtained, perspective transformation from the source square representing the ideal QR Code to destination quadrilateral representing the real QR Code in the image can be set-up (Figure 14).

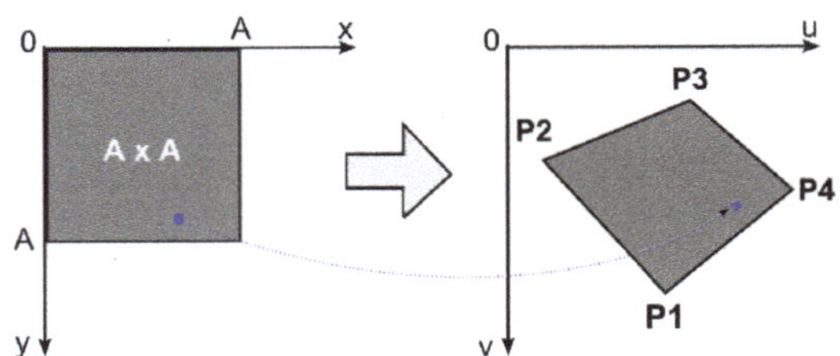

Figure 14. Perspective transformation.

Using equations [21]:

$u = \frac{ax+by+c}{gx+hy+1}$, $v = \frac{dx+ey+f}{gx+hy+1}$, where the transformation coefficients can be calculated from the coordinates of the points $P_1(u_1, v_1)$, $P_2(u_2, v_2)$, $P_3(u_3, v_3)$, $P_4(u_4, v_4)$ as

$$a = (u_3 - u_2)/A + gu_3, \ b = (u_1 - u_2)/A + hu_1, \ c = u_2$$
$$d = (v_3 - v_2)/A + gv_3, \ e = (v_1 - v_2)/A + hv_1, \ f = v_2$$
$$g = \frac{\begin{vmatrix} du_3 & du_2 \\ dv_3 & dv_2 \end{vmatrix}}{\begin{vmatrix} du_1 & du_2 \\ dv_1 & dv_2 \end{vmatrix}}, \ h = \frac{\begin{vmatrix} du_1 & du_3 \\ dv_1 & dv_3 \end{vmatrix}}{\begin{vmatrix} du_1 & du_2 \\ dv_1 & dv_2 \end{vmatrix}}$$
$$du_1 = (u_3 - u_2)A, \ du_2 = (u_1 - u_4)A, \ du_3 = u_2 - u_3 + u_4 - u_1$$
$$dv_1 = (v_3 - v_2)A, \ dv_2 = (v_1 - v_4)A, \ dv_3 = v_2 - v_3 + v_4 - v_1$$

It sometimes happens, that the estimate of P_4 position is not quite accurate so we move P_4 in spiral from its initial position (obtained in previous step) and we calculate match score of bottom-right Alignment Pattern (Alignment Pattern exists only in QR codes version 2 and above). For each shift we recalculate coefficients for the perspective transformation, and we recalculate also match score. For the given version of the QR Code we know the expected position and size of bottom-right Alignment Pattern so we can calculate match between expected and real state. The final position P_4 is the one with the highest match score.

Another alternative method how to handle perspective distorted QR Codes and how to determine position of the P_4 point is based on edge directions and edge projections analysis [23].

3.7. Decoding of a QR Code

A QR Code is 2D square matrix in which the dark and light squares (modules) represent bits 1 and 0. In fact, each such a module on the pixel level is usually made up of cluster of adjacent pixels. In a QR Code decoding process, we have to build a 2D matrix that has elements with a value of 1 or 0 (Figure 15).

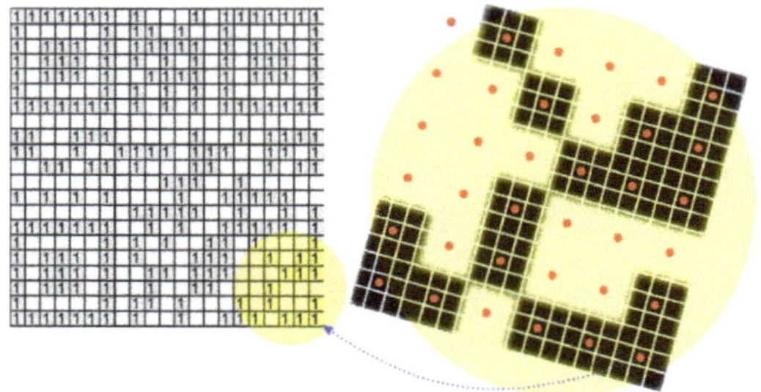

Figure 15. Transformation of the QR Code into the binary grid.

The pixels, where the brightness is lower than the threshold, are declared as 1 and the others are declared as 0. When analyzing such a cluster of pixels (module), the central pixel of the module plays a decisive role. If the calculated position of the central pixel does not align to integer coordinates in the image, the brightness is determined by bilinear interpolation.

Once the binary matrix of 1 and 0 is created, the open-source ZBar library [24] can be used to start the final decoding of the binary matrix and to receive the original text encoded by the QR Code.

4. Results

We used a test dataset of 595 QR Code samples to verify the method described in this paper. The testing dataset contained 25 artificial QR codes of different sizes and rotations, 90 QR Codes from Internet images, and 480 QR Codes from a specific industrial process. Several examples of the testing samples are in Figure 16.

Figure 16. Testing samples: (**a**) artificial, (**b**) Internet, (**c**) industrial.

In Table 1 and Figure 17 there are our results compared to competing QR Code decoding solutions (commercial and also open-source). In the table are the numbers of correctly decoded QR Codes from the total number of 595 QR Codes.

Table 1. Comparison of competing solutions with our proposed method.

Solution	Artificial Samples	Internet Samples	Industrial Samples
ZXing (open-source) [25]	2	72	48
Quirc (open-source) [22]	12	69	45
LEADTOOLS QR Code SDK [26]	9	72	147
ZBar (open-source) [24]	23	84	398
Inlite Barcode Reader SDK [27]	25	80	421
DataSymbol Barcode Reader SDK [28]	25	89	471
Dynamsoft Barcode Reader SDK [29]	25	88	478
Our solution	25	87	480

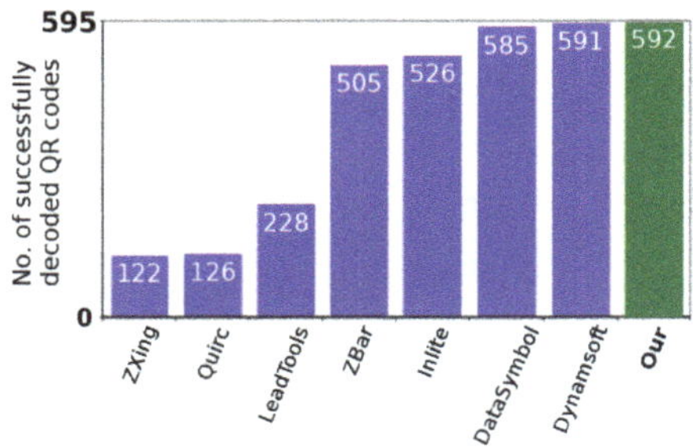

Figure 17. Comparison of the recognition rate of our method against competing solutions.

Our method successfully detected all QR Codes but failed to decode three samples. Two samples were QR Codes placed on a bottle, where perspective and cylindrical distortions were combined. How to deal with this type of combined distortion is a challenge for future research.

As the commercial solutions have closed source code, we performed the testing using the black-box method. We have compiled our own QR Code test dataset (published together with this article under "Supplementary Material") to evaluate and compare our method, as a standardized universal dataset is not publicly available.

In Table 2 the computational complexity of our algorithm is compared to competing open-source solutions (commercial solutions were tested online). Our algorithm was implemented in Free Pascal and tests were run on an i5-4590 3.3GHz CPU (Intel Corporation, Santa Clara, CA, USA).

Table 2. Dependence of computational complexity on image resolution and number of QR codes in an image.

Solution	1296×960		2592×1920		
	1 Code	10 Codes	1 Code	10 Codes	50 Codes
ZBar (open-source)	85 ms	185 ms	347 ms	372 ms	1744 ms
Quirc (open-source)	13 ms	26 ms	45 ms	130 ms	493 ms
Our solution	18 ms	21 ms	73 ms	77 ms	107 ms

We see the main contribution of our method in the way the broken Finder Pattern is dealt with (Section 3.3.2) and how the QR Code bounding box is determined (Section 3.6), especially for perspective distorted QR Codes. The presented method can still locate a QR Code if one of the three Finder Patterns is significantly damaged. Consecutive tests in the real manufacturing process showed that this situation occurs much more often than a situation where two or three opposite Finder Patterns are damaged. In order to detect a QR Code with multiple damage Finder Patterns, it will be necessary to combine Finder Pattern based localization with the region based localization.

5. Conclusions

We have designed and tested a computationally efficient method for precise location of 2D QR Codes in arbitrary images under various illumination conditions. The proposed method is suitable for low-resolution images as well as for real time processing. The designed Finder Pattern based localization method uses three typical patterns of QR Codes to identify three corners of QR Codes in an image. If one of the three Finder Patterns is so destroyed that it cannot be localized, we have suggested a way to deal with it. The input image is binarized and scanned horizontally to localize the Finder Pattern candidates, which are subsequently verified in order to localize the raw QR Code region. For distorted QR Codes, the perspective transformation is set-up by gradually approaching the boundary of the QR Code.

This method was validated on the testing dataset consisting of a wide variety of samples (synthetic, real world, and specific industrial samples) and it was compared to competing software. The experimental results show that our method has a high detection rate.

The application of QR Codes and their optical recognition has wide use in identification, tracing or monitoring of items in production [30], storing and distribution processes, to aid visually impaired and blind people, to let autonomous robots [31] to acquire context-relevant information, to support authorization during log-in process, to support electronic payments, to increase industrial production surety factor [32], etc.

In cases where is required to place the 2D matrix codes on a very small area it may be preferable to use Data Matrix codes [33].

Supplementary Materials: The following are available online at http://www.mdpi.com/2076-3417/10/21/7814/s1, one ZIP file contains images of QR Codes testing dataset used to evaluate the competing solutions.

Author Contributions: Conceptualization, methodology, software, writing—original draft preparation, L.K.; validation, writing—review and editing, visualization, E.P.; supervision, project administration, funding acquisition, P.B.; All authors have read and agreed to the published version of the manuscript.

Funding: The article is funded of the research project KEGA 013TUKE-4/2019 "Modern educational tools and methods for shaping creativity and increasing the practical skills and habits of graduates of technical departments of universities".

Acknowledgments: The contribution is sponsored by the project.

Conflicts of Interest: The authors declare no conflict of interest.

References

1. Zhang, X.; Zhu, S.; Wang, Z.; Li, Y. Hybrid visual natural landmark–based localization for indoor mobile robots. *Int. J. Adv. Robot. Syst.* **2018**, *15*. [CrossRef]
2. Bozek, P.; Pokorný, P.; Svetlik, J.; Lozhkin, A.; Arkhipov, I. The calculations of Jordan curves trajectory of the robot movement. *Int. J. Adv. Robot. Syst.* **2016**, *13*, 1–7. [CrossRef]
3. Wang, K.; Zhou, W. Pedestrian and cyclist detection based on deep neural network fast R-CNN. *Int. J. Adv. Robot. Syst.* **2018**, *16*. [CrossRef]
4. Zhang, X.; Gao, H.; Xue, C.; Zhao, J.; Liu, Y. Real-time vehicle detection and tracking using improved histogram of gradient features and Kalman filters. *Int. J. Adv. Robot. Syst.* **2018**, *15*. [CrossRef]
5. Denso Wave Incorporated. What is a QR Code? 2018. Available online: http://www.qrcode.com/en/about/ (accessed on 6 September 2018).
6. Denso Wave Incorporated. History of QR Code. 2018. Available online: http://www.qrcode.com/en/history/ (accessed on 6 September 2018).
7. Lin, J.-A.; Fuh, C.-S. 2D Barcode Image Decoding. *Math. Probl. Eng.* **2013**, *2013*, 848276. [CrossRef]
8. Li, S.; Shang, J.; Duan, Z.; Huang, J. Fast detection method of quick response code based on run-length coding. *IET Image Process.* **2018**, *12*, 546–551. [CrossRef]
9. Belussi, L.F.F.; Hirata, N.S. Fast Component-Based QR Code Detection in Arbitrarily Acquired Images. *J. Math. Imaging Vis.* **2012**, *45*, 277–292. [CrossRef]
10. Bodnár, P.; Nyúl, L.G. Improved QR Code Localization Using Boosted Cascade of Weak Classifiers. *Acta Cybern.* **2015**, *22*, 21–33. [CrossRef]
11. Tribak, H.; Zaz, Y. QR Code Recognition based on Principal Components Analysis Method. *Int. J. Adv. Comput. Sci. Appl.* **2017**, *8*, 241–248. [CrossRef]
12. Tribak, H.; Zaz, Y. QR Code Patterns Localization based on Hu Invariant Moments. *Int. J. Adv. Comput. Sci. Appl.* **2017**, *8*, 162–172. [CrossRef]
13. Sun, A.; Sun, Y.; Liu, C. The QR-code reorganization in illegible snapshots taken by mobile phones. In Proceedings of the 2007 International Conference on Computational Science and its Applications (ICCSA 2007), Kuala Lumpur, Malaysia, 26–29 August 2007; pp. 532–538.
14. Ciążyński, K.; Fabijańska, A. Detection of QR-Codes in Digital Images Based on Histogram Similarity. *Image Process. Commun.* **2015**, *20*, 41–48. [CrossRef]
15. Gaur, P.; Tiwari, S. Recognition of 2D Barcode Images Using Edge Detection and Morphological Operation. *Int. J. Comput. Sci. Mob. Comput. IJCSMC* **2014**, *3*, 1277–1282.
16. Szentandrási, I.; Herout, A.; Dubská, M. Fast detection and recognition of QR codes in high-resolution images. In Proceedings of the 28th Spring Conference on Computer Graphics, Budmerice, Slovakia, 2–4 May 2012; ACM: New York, NY, USA, 2012.
17. Dubská, M.; Herout, A.; Havel, J. Real-time precise detection of regular grids and matrix codes. *J. Real Time Image Process.* **2016**, *11*, 193–200. [CrossRef]
18. Sörös, G.; Flörkemeier, C. Blur-resistant joint 1D and 2D barcode localization for smartphones. In Proceedings of the 12th International Conference on Mobile and Ubiquitous Multimedia, MUM, Lulea, Sweden, 2–5 December 2013; pp. 1–8.
19. Bradley, D.; Roth, G. Adaptive Thresholding using the Integral Image. *J. Graph. Tools* **2007**, *12*, 13–21. [CrossRef]
20. Bresenham, J.E. Algorithm for computer control of a digital plotter. *IBM Syst. J.* **1965**, *4*, 25–30. [CrossRef]
21. Heckbert, P. Fundamentals of Texture Mapping and Image Warping. Master's Thesis, University of California, Berkeley, CA, USA, June 1989.

22. Beer, D. Quirc—QR Decoder Library. 2018. Available online: https://github.com/dlbeer/quirc (accessed on 22 September 2018).
23. Karrach, L.; Pivarčiová, E.; Božek, P. Identification of QR Code Perspective Distortion Based on Edge Directions and Edge Projections Analysis. *J. Imaging* **2020**, *6*, 67. [CrossRef]
24. Terriberry, T.B. ZBar Barcode Reader. 2018. Available online: http://zbar.sourceforge.net (accessed on 22 September 2018).
25. Google. ZXing (Zebra Crossing) Barcode Scanning Library for Java, Android. 2018. Available online: https://github.com/zxing (accessed on 22 September 2018).
26. Leadtools. QR Code SDK Technology. 2018. Available online: http://demo.leadtools.com/JavaScript/Barcode/index.html (accessed on 22 September 2018).
27. Inlite Research Inc. Barcode Reader SDK. 2018. Available online: https://online-barcode-reader.inliteresearch.com (accessed on 22 September 2018).
28. DataSymbol. Barcode Reader SDK. 2018. Available online: http://www.datasymbol.com/barcode-reader-sdk/barcode-reader-sdk-for-windows/online-barcode-decoder.html (accessed on 22 September 2018).
29. Dynamsoft. Barcode Reader SDK. 2018. Available online: https://www.dynamsoft.com/Products/Dynamic-Barcode-Reader.aspx (accessed on 22 September 2018).
30. Bako, B.; Božek, P. Trends in Simulation and Planning of Manufacturing Companies. *Procedia Eng.* **2016**, *149*, 571–575. [CrossRef]
31. Frankovsky, P.; Pastor, M.; Dominik, L.; Kicko, M.; Trebuna, P.; Hroncova, D.; Kelemen, M. Wheeled mobile robot in structured environment. In Proceedings of the 12th International Conference ELEKTRO 2018, Mikulov, Czech Republic, 21–23 May 2018; pp. 1–5.
32. Pivarčiová, E.; Božek, P. Industrial production surety factor increasing by a system of fingerprint verification. In Proceedings of the 2014 International Conference on Information Science, Electronics and Electrical Engineering (ISEEE 2014), Sapporo, Japan, 26–28 April 2014; pp. 493–497.
33. Karrach, L.; Pivarčiová, E.; Nikitin, Y.R. Comparing the impact of different cameras and image resolution to recognize the data matrix codes. *J. Electr. Eng.* **2018**, *69*, 286–292. [CrossRef]

Publisher's Note: MDPI stays neutral with regard to jurisdictional claims in published maps and institutional affiliations.

Letter

Smart Building Surveillance System as Shared Sensory System for Localization of AGVs

Petr Oščádal, Daniel Huczala, Jan Bém, Václav Krys and Zdenko Bobovský *

Department of Robotics, Faculty of Mechanical Engineering, VSB-Technical University of Ostrava, 708 00 Ostrava, Czech Republic; petr.oscadal@vsb.cz (P.O.); daniel.huczala@vsb.cz (D.H.); jan.bem.st@vsb.cz (J.B.); vaclav.krys@vsb.cz (V.K.)
* Correspondence: zdenko.bobovsky@vsb.cz

Received: 3 November 2020; Accepted: 25 November 2020; Published: 27 November 2020

Featured Application: Existing camera systems may provide localization and tracking of everything marked, while AGVs do not have to carry sensors for localization and navigation themselves. This method provides centralized supervision of deployed robotic systems.

Abstract: The objective of this study is to extend the possibilities of robot localization in a known environment by using the pre-deployed infrastructure of a smart building. The proposed method demonstrates a concept of a Shared Sensory System for the automated guided vehicles (AGVs), when already existing camera hardware of a building can be utilized for position detection of marked devices. This approach extends surveillance cameras capabilities creating a general sensory system for localization of active (automated) or passive devices in a smart building. The application is presented using both simulations and experiments for a common corridor of a building. The advantages and disadvantages are stated. We analyze the impact of the captured frame's resolution on the processing speed while also using multiple cameras to improve the accuracy of localization. The proposed methodology in which we use the surveillance cameras in a stand-alone way or in a support role for the AGVs to be localized in the environment has a huge potential utilization in the future smart buildings and cities. The available infrastructure is used to provide additional features for the building control unit, such as awareness of the position of the robots without the need to obtain this data directly from the robots, which would lower the cost of the robots themselves. On the other hand, the information about the location of a robot may be transferred bidirectionally between robots and the building control system to improve the overall safety and reliability of the system.

Keywords: smart buildings; shared sensory system; surveillance system; localization; marker detection

---

1. Introduction

Imagine that in the future every building will look like a living organism comparable to a hive where both humans, robots, and other devices are going to cooperate on their tasks. For example, mobile robotic systems will assist people with basic repetitive tasks, such as package and food delivery, human transportation, security, cleaning, and maintenance. Besides fulfilling the purpose, robots need to be safe for any person that comes in contact with one, so any possible harm is avoided. A robot position tracking and collision avoidance plays a key role to ensure such capabilities.

There are many methodologies for motion planning and their theoretical approaches are described in a survey by Mohanan and Salgoankar [1]. One of the most used methods is Simultaneous Localization and Mapping (SLAM) [2] where the environment is not known, so a vision based system is placed on a robot to observe the area and create a map of it, eventually to perform trajectory planning on top of it. There were many enhancements of the SLAM methods proposed, some of them are compared by Santos [3] and the most common sensors used there are lidars [4–7] and depth cameras [4,8–10]. However, when the environment is known (indoor areas frequently), the advanced SLAM systems may become redundant and if more robots are deployed, it becomes costly when every single robot has to carry all sensors.

Therefore, some alternatives come to mind, such as real-time locating systems (RTLS), when devices working on the radio frequency communication principle are placed in the environment to determine the position of a robot. Under this technology, one can imagine WiFi [11,12], when time-difference measurements between the towers and the monitored object are obtained, or RFID [13–15], when tags shaped in a matrix on the floor or walls are detected by a transmitter on the robot. However, deployment of RTLS is not always possible, when WiFi needs uninterrupted visual contact and reliable communication between the towers. The RFID tags on the floor can be damaged during a time and the accuracy is based on the density (amount) of the tags. In general, RTLS is suitable only for indoor areas. On the other hand, the most common GNSS systems for outdoor navigation cannot provide a reliable solution in the case of indoor environments.

Another approach for robot navigation is odometry, which analyzes the data from onboard sensors and estimates the position change over time. Mostly the wheel encoders are used [16], or a camera for visual odometry [17]. Although, the odometry measurements have a disadvantage when they accumulate an error with time, as investigated by Ganganath [18], where he proposed a solution using a Kinect sensor that is detecting tag markers placed in the environment. Similar methods of robot localization when the map is known and visual markers are deployed are presented in [19,20]. The other possible solution of an object localization is a marker-based motion capture system, in which the functioning with multiple robots was demonstrated in [21], but only a single camera was used and all paper-printed markers were the same, so it is not possible to use it for device identification. Infra-red markers have a similar disadvantage, they were used for tracking AGVs with either a number of high-speed cameras in [22–24] or the Kinect system in [25]. When many robots at the same place are used, a field Swarm Robotics [26] comes to mind. Some algorithms tested with real robots were introduced [27,28], but they were not applied on different types of robots and have not been tested in cooperation with humans or in an environment where people randomly appear and disappear.

Except for the RTLS and motion capture, the previously mentioned methods are based on sensor systems carried by a robot. Furthermore, only the robot determines its pose and in case of an unexpected situation, it might lead to a fault state and connection lost without providing any reliable feedback to the control system of a building, as shown in Figure 1a. The RTLS avoid this situation, but only because of added devices such as communication towers. The marker-based motion capture system, if used in common areas like rooms and corridors of a building, needs many cameras deployed and a specific pattern of IR markers on every device. If there are tens or hundreds of devices, to define and to recognize the markers becomes a difficult task. However, smart buildings may already have their sensor systems, mostly cameras and related devices for security purposes. Nowadays, face-recognition and object detection software has been used in airports [29], governmental buildings, other public places, and even in smart phones. These capabilities lead us to an approach when the surveillance cameras would be involved in the facial recognition and tracking of any robot or device at the same time, meaning that every device has its unique easily printable marker that characterizes it, so the control system of the whole building has an independent feedback on a device's pose continuously. This also extends the possibility of tracking things in an environment even though they are not electronic devices, for example important packages and boxes. Such a solution acts, as we call it, as a Shared Sensory System. This approach simplifies the tracking of new devices, as only their ID is needed, as well as a new marker to be placed on a device or thing to the control system.

Furthermore, the hardware costs are minimized when the devices do not need to carry any computational units or sensors. The image analysis, object detection, localization, and trajectory planning of AGVs may run in the cloud. The scheme of the system is shown in Figure 1b.

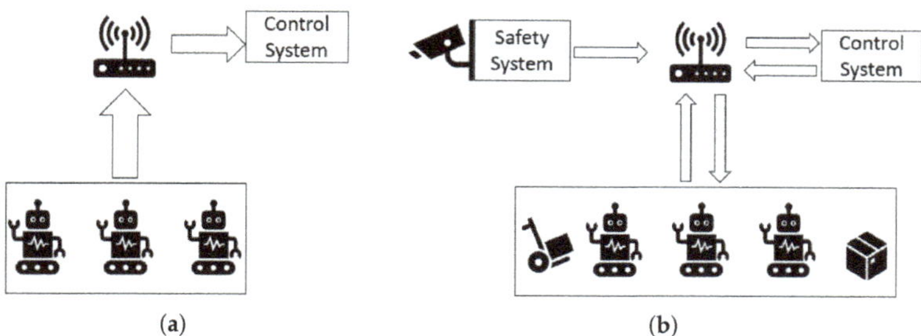

(a)   (b)

Figure 1. A robot pose determination: (**a**) Nowadays robots inform control systems about their pose; (**b**) Proposed method of determination of any device pose, robots may provide their own pose estimation so two-way detection is performed.

2. Test Benches and Methods

There are more options of detecting an object in an image as compared to a survey of Parekh [30]. For image processing and tag marker detection we used OpenCV library [31] and methodology described in [32] for detecting an Aruco 3D gridboard, which improves the reliability and accuracy of the detection in comparison with basic 2D tags. The gridboard (Figure 2) represents a coordinate frame and may be placed on a robot.

Figure 2. Aruco 3D gridboard with detected coordinate frame XYZ—red, green, blue, respectively.

The OpenCV Aruco module analyzes an image from a camera and calculates the transformation from a tag to the camera. This transformation is represented by a homogenous transformation matrix, including the position and rotation of the detected coordinate frame. Every unique marker may represent a unique robot or device. When there are more of them placed in the field of view of a camera, the transformation between them is calculated directly, or in reference to the base coordinate frame. The homogenous transformation matrix from a global coordinate frame (base frame B) to the robot frame is calculated by Equation (1). The T_{BC} is a known transformation from the global coordinate frame to the camera frame. The T_{CR} is the detected transformation from the camera frame to the robot frame, obtained from image processing. The transformation matrices, in terms of robot kinematics, are described by Legnani [33].

$$T_{BR} = T_{BC} T_{CR} \tag{1}$$

We assume that the placements of all cameras are known, the transformations between the cameras may be determined by their layout related to a world coordinate frame. With this information, it is possible to track and calculate the positions between tagged objects even if they are in different parts of a building. Number of cameras has an impact on the accuracy of the detection, as it is described in Section 3. Results show, however, that even a single camera provides reliable data on the location of an AGV. To demonstrate the difference between the number of cameras used, we designed a camera system with 3 cameras of the same type, which is shown in Figure 3 schematically and in Figure 4 as captures of simulation. Finally, the same system was deployed in the experiment, which is shown in

Figure 5. Fields of view of those cameras intersect and are described in Figure 3. Such a system might serve, for example, as a face-recognition every time a human wants to enter a door. If this person is allowed to enter, the door can automatically unlock and open, which may be appreciated in clean rooms, such as biomedical labs, where, for example, samples of dangerous diseases are analyzed. Our methodology utilizes these kinds of cameras for AGVs localization and makes them multipurpose.

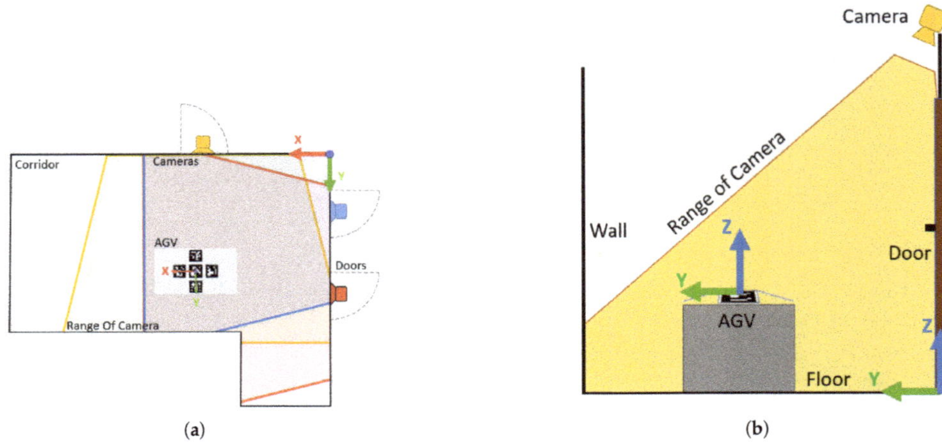

Figure 3. Schematics of cameras placement: (**a**) top view; (**b**) side view.

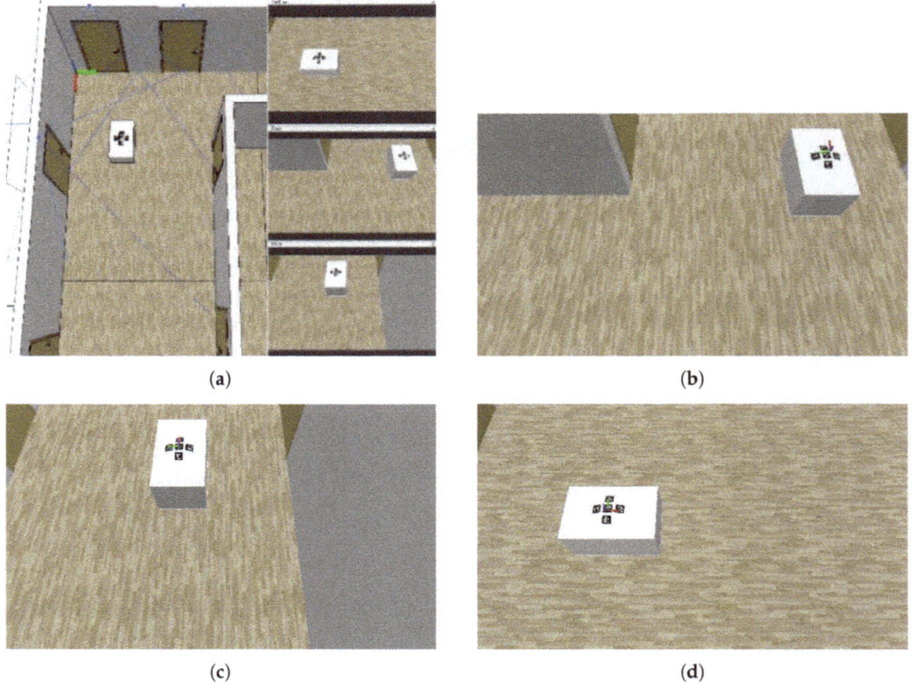

Figure 4. Simulation environment: (**a**) top view; (**b**) red camera view; (**c**) blue camera view; (**d**) yellow camera view. Resolution for every camera was set to 1280 × 720 px this time.

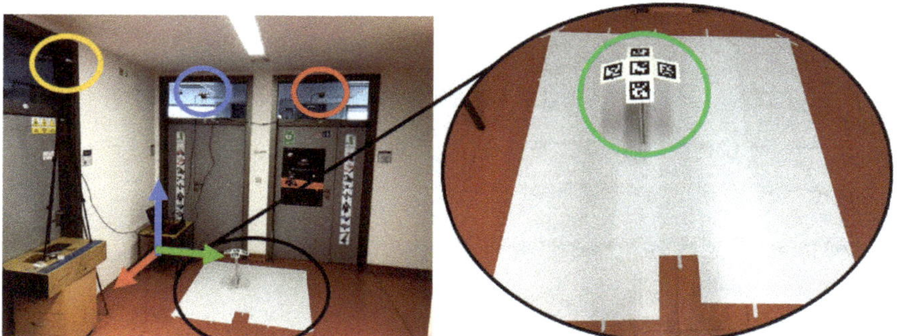

Figure 5. Real test: the cameras are highlighted with yellow, blue, and red circles. The gridboard is in a green circle representing the position of an automated guided vehicles (AGV).

2.1. Simulation with Single Robot

The concept was simulated and tested in an environment similar to that described in the previous chapter and as shown in Figure 4a. The cameras were deployed above the doors in a corridor. In the simulation, an AGV position was represented as a box with a 3D gridboard on top of it. The robot was moving along a trajectory loop under the cameras. The results of this simulation are supposed to provide an idea about how accurate the detection can be if the environmental conditions (sun light, reflections, artificial lightning, etc.) are ideal, secondly what impact the cameras resolution has on the detection. The crucial outcome is the comparison between accuracy and processing time when resolution differs. As the processing time, we understand the period when the OpenCV algorithm is analyzing an image that was previously captured and stored in a buffer as the most recent. The average processing time depends on the number of pixels and stays the same no matter if an analyzed image was captured in the simulation or real test.

In Figure 4b–d, the images captured by all three cameras at the same time can be seen. The images were already processed, so the coordinate frame of the detected tag is visible. The resolution has an impact on a tag detection—higher resolution may detect a smaller (more distant) marker. The simulated resolutions for the chosen tags with dimensions of 70 × 70 mm were:

- 1920 × 1080 px
- 1280 × 720 px
- 640 × 480 px

2.2. Experiment

Based on the simulation result we performed the real test, as shown in Figure 5. The objective of the experiment was to determine the absolute accuracy of the detection. Therefore, the gridboard was placed on a static rod, which allowed us to check the detected position with its absolute value that was measured using a precisely placed matrix of points on the floor (white canvas in Figure 5). The rod was moved into 14 different positions, which were 250 mm apart from each other. In every position, 40 measurements were taken for a dataset that was analyzed later. The height of the gridboad above the floor on the simulated robot was the same as the height of the gridboard on the rod in the experiment, 460 mm exactly. The cameras were 2500 mm above the floor. Resolution was set to 1280 × 720 px for the cameras in the experiment when it provided the best performance in the simulation, as described in the Results Section. The control PC was running on Ubuntu 16.04 with a CPU of 4 cores at 2.80 GHz, the USB 3 was providing connection to the Intel D435i [34] depth cameras; however, only their RGB image was analyzed. It is important to note that camera calibration process is crucial for the accuracy of the detection, as it is described in [35]. The cameras were calibrated following these instructions.

In both simulation and experiment, the relation between the number of cameras and the detected position accuracy was examined.

For presenting the accuracy results of this method, we chose to determine the absolute distance from the point where a tag's coordinate frame was expected to the point where it was measured. Only X and Y coordinates were considered, the Z coordinate is known for every robot—the height of a tag in our case. In Equation (2), the calculated accuracy error is presented. The X_0 and Y_0 represent the expected values, the X_1 and Y_1 represent the measured coordinates by a camera.

$$error = \sqrt{(X_0 - X_1)^2 + (Y_0 - Y_1)^2} \tag{2}$$

In the simulation and experiment, only the position values X and Y of the gridboard were later analyzed and presented in the Results Section. However, during the image processing, all data as X, Y, Z position values and R_x, R_y, R_z values for orientation were acquired.

2.3. Simulation with Multiple Robots

The purpose of the second simulation is to present the capabilities of our algorithm to detect multiple robots using multiple cameras. The trajectories and lookout of the robots are shown in Figure 6. The tags on the black robot were placed at different positions in comparison with the white robot, but the system can handle this. In comparison with the first simulation, the size of the tags was changed to 170×170 mm to highlight the scalability of the system. At some point of the simulation the robots are detected by one or more cameras and at another point both robots are detected by all cameras. The environment and camera position stayed the same as in the previous simulation with a single robot. The 1280×720 px camera resolution was used. The simulation step was set to $\Delta t = 0.0333$ s, which corresponds to 30 fps, which is a general value for most of the cameras. Based on this setup, the velocity v_i of the robot was calculated as the change of the position in two captured frames in time, as shown in Equation (3). The X_i, Y_i represent the position of a robot in i_{th} frame, the X_{i-1}, Y_{i-1} represent the position of the robot in the previous frame.

$$v_i = \frac{1}{\Delta t}\sqrt{(X_i - X_{i-1})^2 + (Y_i - Y_{i-1})^2} \tag{3}$$

The default speed of the robots was set to 1.5 m/s in the simulation, the error between this value and the calculated velocity v based on the marker detection is presented in the Results Section.

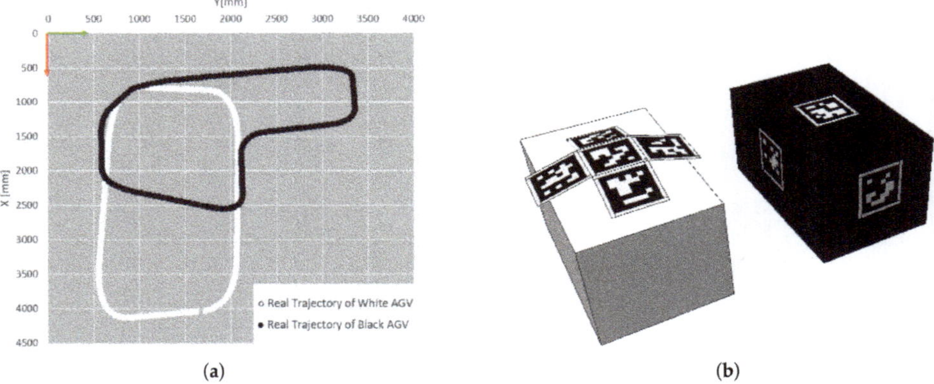

Figure 6. Multiple robots simulation: (a) trajectories of the robots in the previously described environment; (b) marked robots, the size of the tags was set to 170×170 mm.

A video with an additional description of the simulation of multiple robots was uploaded on a YouTube channel [36] of the Department of Robotics, VSB-TU of Ostrava. A Python source code of the detection is available as a Github repository [37].

3. Results

In the case of the simulation with a single robot, we present data on the detection accuracy error for the three resolutions in Figure 7a, when the image from a single camera was used. The zero value of the graph represents the real position of the robot, measured by the CoppeliaSim simulation engine, where it is possible to read the absolute position data of every object. As expected, with a smaller resolution of a camera the accuracy drops down. However, the difference between Full HD and HD image resolution is only 4.8 mm on average. In addition, when observing the processing time for an image in Figure 7b the HD resolution may provide twice as fast detection as Full HD.

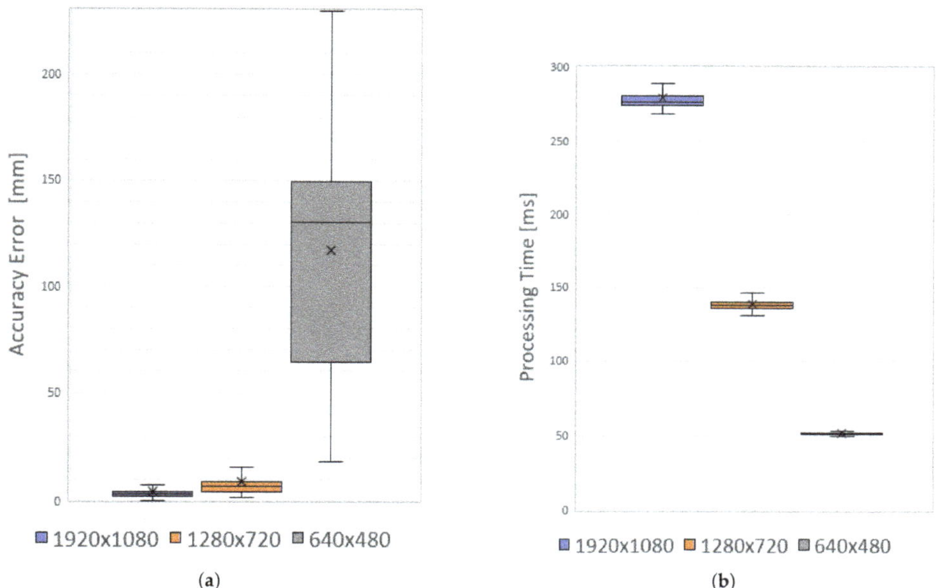

Figure 7. Simulation results: (**a**) detection accuracy error; (**b**) processing time.

The simulation results with the single robot are presented in Table 1. The data show the relation between the resolution of a camera and its impact on detection accuracy and processing time. The accuracy can be influenced by many aspects, but the processing time of an image is very closely related to the number of pixels. Therefore, in the following real cameras experiment, 1280×720 px resolution was used when it provides (observing Table 1) good performance between accuracy and processing time.

Table 1. Average accuracy and processing time of detection in simulation.

Cameras Resolution (px)	Average Accuracy (mm)	Processing Time (ms)
1920×1080	4.1	278.9
1280×720	8.9	138.5
640×480	117.5	51.6

More cameras can detect a gridboard all at once. In this simulation and experiment, we detected it with one, two or three cameras to compare the results. An average value of the position vector for two or three cameras was determined. In Figure 8 there is a comparison of the position accuracy error for single or multiple cameras. As shown in Figure 8b for the experiment, when more cameras were used, the absolute error was not improved significantly; however, the variance of the error was lowered. The reason why the error does not go down substantially when more cameras are used is that at a random position the robot appears in different areas of the images of every camera. In the center of an image, the localization is the most precise, in the sides of an image, the precision of detection is decreasing. This effect is described later in Figure 9.

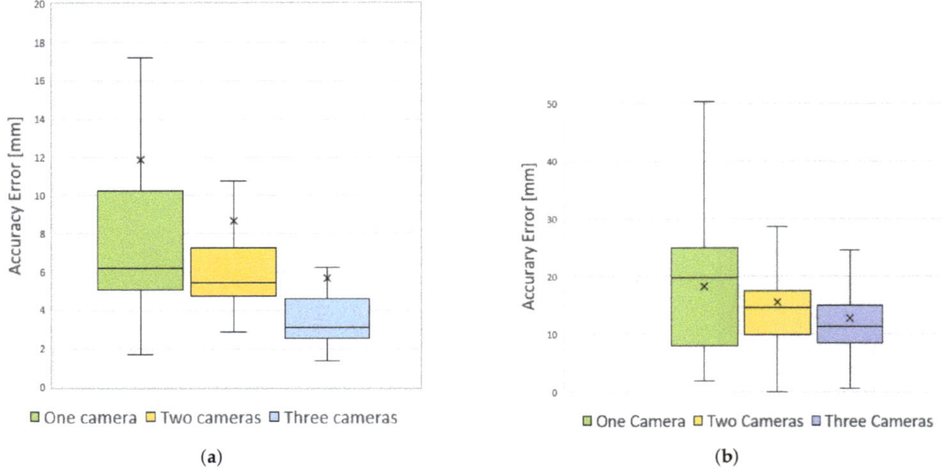

(**a**)   (**b**)

Figure 8. Comparison of detection when one, two or three cameras were used: (**a**) simulation; (**b**) experiment.

The average values of the accuracy error in the simulation with the single robot and experiment presented in Figure 8 are stated in Table 2.

Table 2. Comparison of average accuracy error of detection in simulation with the single robot vs. real test.

Environment	Average Accuracy of 1 Camera (mm)	Average Accuracy of 2 Cameras (mm)	Average Accuracy of 3 Cameras (mm)
Simulation	11.8	8.6	5.7
Experiment	18.3	15.5	12.7

Figure 9 represents the distortion [35] of nonperfect calibration of a camera. Therefore, more cameras cannot guarantee the absolute accuracy of the detection; on the other hand, the error can be significantly lowered when there two or more cameras are used. In Figure 9, the 14 different positions were measured 40 times by a camera highlighted by the red circle in Figure 5. The set of points A with X, Y mm was very close to the center of the camera's field of view. The set of points B was the most distant and the dispersion of the 40 measurements is the most significant.

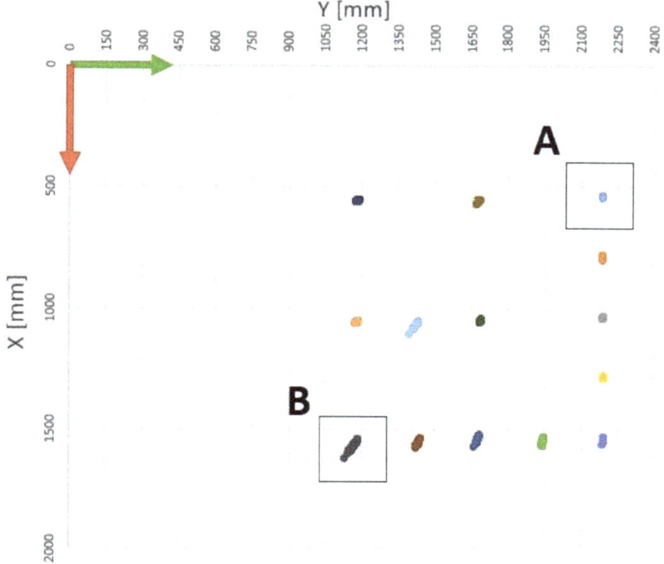

Figure 9. Detection accuracy of a single camera in a real experiment. Every color represents a different set of measured positions. It is the red camera in Figure 5, specifically.

The following graphs are the results of the simulation with multiple robots. In Figure 10, the accuracy of the algorithm is presented, comparing the defined path of the robots and the detected path.

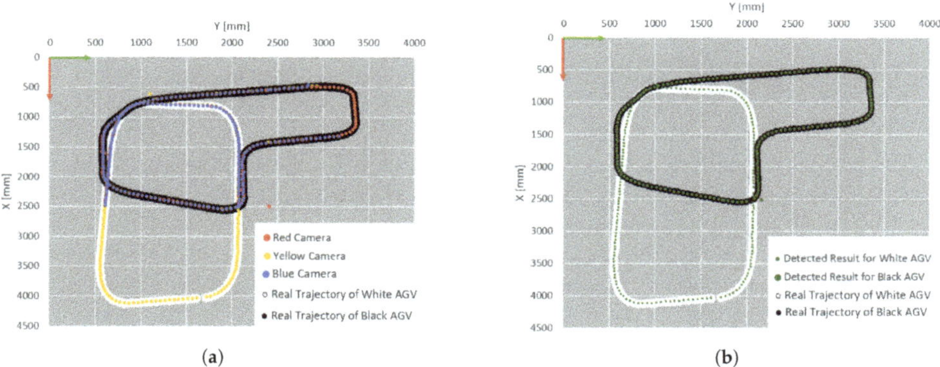

Figure 10. Comparison of detected and real positions during the simulation with multiple robot: (**a**) results for every camera; (**b**) average value.

The graphs of Figure 10 prove excellent general accuracy of the system. The results are overlapping with the absolute position values measured by the simulation engine and the difference is not clearly visible in this scale. Therefore, in Figure 11 the error values of the detected X and Y coordinates are presented. Zero error means that the detected position is equal to the real position of the robot. The colors of lines represent the cameras and the dashed line is the average value. There is no smoothing filter applied on the data, some peak values are visible. The application of filtering is one of the subjects for the future research.

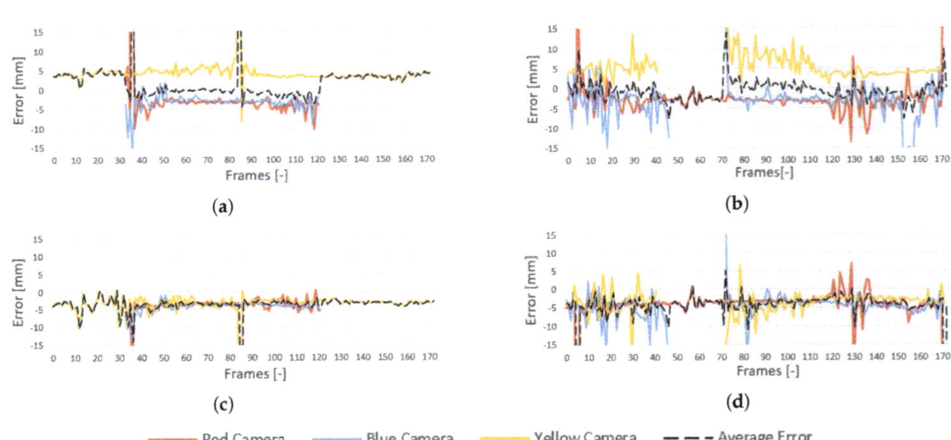

Figure 11. Error between detected and real position of the robots: (**a**) X direction of white robot; (**b**) X direction of black robot; (**c**) Y direction of white robot; (**d**) Y direction of black robot.

The obtained data may be analyzed for other purposes, for example, in the graphs of Figure 12 the speed of the robots was calculated based on the difference between the positions in time, comparing i_{th} and i_{th-1} frames. In the simulation for multiple robots, 175 frames were obtained in total.

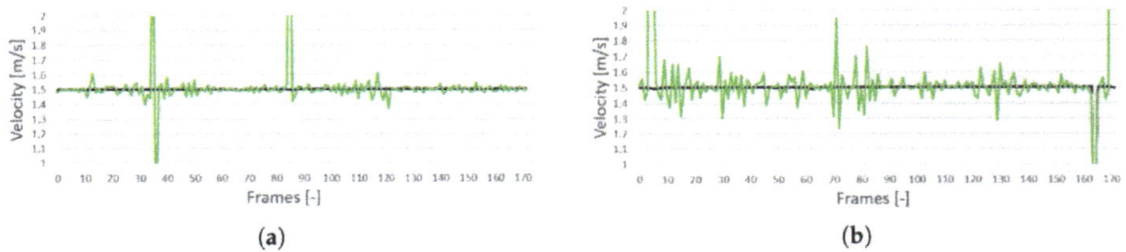

Figure 12. Calculated velocity of robots: (**a**) white robot; (**b**) black robot.

4. Conclusions

In addition to the previous chapter where the results were presented, some other ideas and relations to the research of others may be discussed here. This study shows how the hardware that is already deployed may be used for another purpose in the case of smart cities and buildings, when only a software module with a control unit is added. The only prerequisite is having access to real-time image data of the installed camera system and computing capabilities for image processing, which can be run locally or in the cloud. In general, the price of a smart building is already high, and it increases with every added technology. This approach may lead to lower cost and create a Shared Sensory System for all AGVs at the level of a whole building and surrounding area.

We presented a possible use case when images of face-recognition security cameras, which were placed above the doors to provide automatic unlocking and opening, were also used for tracking of robots and other devices. This method can be applied to a robot navigation adding algorithm [38] for human and object detection of autonomous cars. Such capabilities would allow to stop or slow down a robot when a person is approaching while hidden behind a corner, so the human is not in the robot's field of view. Alternatively, the methodology may be used along with SLAM to add another safety feature to the control system.

The choice of camera resolution has an influence on the reliability and processing time of the system. In our measurements, for the real-time tracking application, the 640×480 px resolution would be necessary, but the drop in the case of accuracy might be a problem. However, the processing time may be solved if more powerful control PC is used. In the experiment, we used resolution 1280×720 px, which provided the best ratio between accuracy and speed of localization for us. When more cameras are deployed, a higher computing power is demanded. Therefore, more processing units or cloud computing for analyzing the images should be considered. Another disadvantage of the system is that a device position might be lost if it is not detected by any camera. In this case, the control system should be prepared for dealing with such a situation. For example, the robot shall be stopped until allowed to continue in its trajectory or reset.

For the most accurate detection of a tag, every camera should be calibrated. On the other hand, if many cameras of the same type are deployed, the calibration constants can be determined statistically performing the calibration with a few of these cameras and distributing the calibration file with constants of average values. If a new camera is added to the system, its precise position can be determined using an additional device with tag markers placed at a certain distance from the camera using the device's known dimensions.

In multiple camera detection, we use a simple distance average for the detected objects. Another approach to determine the position more accurately is to apprise the distance of every camera to the detected object. Therefore, the nearest camera that should provide the most precise detection would obtain a higher weight than the others. This is a topic for further research.

Our algorithm can deal with multiple marker detection. This can be deployed whenever in a building with coverage of the camera's field of view and used for localization of any device with the marker, not only robots. In addition to that, this system could be applied in the outdoor environment using the surveillance camera systems deployed by governments. For example, in the case of the possibility of using the image data, car-sharing or autonomous taxi companies could place unique markers on the roofs of their vehicles fleet, and the position of a vehicle could be inquired in the governmental system, which might be an additional benefit for a public city.

Author Contributions: Conceptualization, P.O. and D.H.; methodology, Z.B. and P.O.; software, P.O.; validation, J.B. and Z.B.; formal analysis,V.K., D.H. and P.O.; investigation, Z.B. and J.B.; resources, D.H., V.K. and J.B.; data curation, P.O. and J.B.; writing—original draft preparation, D.H. and P.O.; writing—review and editing, D.H.,V.K. and Z.B.; visualization, P.O., D.H.,V.K. and J.B.; supervision, Z.B.; project administration,V.K. and Z.B.; funding acquisition, V.K. and Z.B. All authors have read and agreed to the published version of the manuscript.

Funding: This work was supported by the Research Platform focused on Industry 4.0 and Robotics in Ostrava Agglomeration project, project number CZ.02.1.01/0.0/0.0/17_049/0008425 within the Operational Programme Research, Development and Education. This article has also been supported by a specific research project SP2020/141 and financed by the state budget of the Czech Republic.

Conflicts of Interest: The authors declare no conflict of interest. The funders had no role in the design of the study; in the collection, analyses, or interpretation of data; in the writing of the manuscript, or in the decision to publish the results.

Abbreviations

The following abbreviations are used in this manuscript:

AGV	Automated Guided Vehicle
FPS	Frames per Second
GNSS	Global Navigation Satellite System
OpenCV	Open Source Computer Vision Library
RFID	Radio Frequency Identification
RGB	Red Green Blue
RTLS	Real Time Locating System
SLAM	Simultaneous Localization and Mapping

References

1. Mohanan, M.; Salgoankar, A. A survey of robotic motion planning in dynamic environments. *Robot. Auton. Syst.* **2018**, *100*, 171–185, [CrossRef]

2. Montemerlo, M.; Thrun, S.; Koller, D.; Wegbreit, B. FastSLAM: A factored solution to the simultaneous localization and mapping problem. *Aaai/iaai* **2002**, 593-598.

3. Santos, J.M.; Portugal, D.; Rocha, R.P. An evaluation of 2D SLAM techniques available in Robot Operating System. In Proceedings of the 2013 IEEE International Symposium on Safety, Security, and Rescue Robotics (SSRR), Linkoping, Sweden, 21–26 October 2013; [CrossRef]

4. Pandey, G.; McBride, J.R.; Eustice, R.M. Ford Campus vision and lidar data set. *Int. J. Robot. Res.* **2011**, *30*, 1543–1552, [CrossRef]

5. Blanco-Claraco, J.L.; Moreno-Dueñas, F.Á.; González-Jiménez, J. The Málaga urban dataset: High-rate stereo and LiDAR in a realistic urban scenario. *Int. J. Robot. Res.* **2013**, *33*, 207–214, [CrossRef]

6. Li, L.; Liu, J.; Zuo, X.; Zhu, H. An Improved MbICP Algorithm for Mobile Robot Pose Estimation. *Appl. Sci.* **2018**, *8*, 272, [CrossRef]

7. Olivka, P.; Mihola, M.; Novák, P.; Kot, T.; Babjak, J. The Design of 3D Laser Range Finder for Robot Navigation and Mapping in Industrial Environment with Point Clouds Preprocessing. In *Modelling and Simulation for Autonomous Systems*; Springer International Publishing: Berlin/Heidelberg, Germany, 2016; pp. 371–383, [CrossRef]

8. Biswas, J.; Veloso, M. Depth camera based indoor mobile robot localization and navigation. In Proceedings of the 2012 IEEE International Conference on Robotics and Automation, Saint Paul, MN, USA, 14–18 May 2012; [CrossRef]

9. Cunha, J.; Pedrosa, E.; Cruz, C.; Neves, A.J.; Lau, N. *Using a Depth Camera for Indoor Robot Localization and Navigation*; DETI/IEETA-University of Aveiro: Aveiro, Portugal, 2011.

10. Yao, E.; Zhang, H.; Xu, H.; Song, H.; Zhang, G. Robust RGB-D visual odometry based on edges and points. *Robot. Auton. Syst.* **2018**, *107*, 209–220, [CrossRef]

11. Jachimczyk, B.; Dziak, D.; Kulesza, W. Using the Fingerprinting Method to Customize RTLS Based on the AoA Ranging Technique. *Sensors* **2016**, *16*, 876, [CrossRef] [PubMed]

12. Denis, T.; Weyn, M.; Williame, K.; Schrooyen, F. Real Time Location System Using WiFi. Available online: https://www.researchgate.net/profile/Maarten_Weyn/publication/265275067_Real_Time_Location_System_using_WiFi/links/54883ed60cf2ef3447903ced.pdf (accessed on 11 November 2020).

13. Kulyukin, V.; Gharpure, C.; Nicholson, J.; Pavithran, S. RFID in robot-assisted indoor navigation for the visually impaired. In Proceedings of the 2004 IEEE/RSJ International Conference on Intelligent Robots and Systems (IROS) (IEEE Cat. No.04CH37566), Sendai, Japan, 28 September–2 October 2004; [CrossRef]

14. Chae, H.; Han, K. Combination of RFID and Vision for Mobile Robot Localization. In Proceedings of the 2005 International Conference on Intelligent Sensors, Sensor Networks and Information Processing, Melbourne, Australia, 5–8 December 2005; [CrossRef]

15. Su, Z.; Zhou, X.; Cheng, T.; Zhang, H.; Xu, B.; Chen, W. Global localization of a mobile robot using lidar and visual features. In Proceedings of the 2017 IEEE International Conference on Robotics and Biomimetics (ROBIO), Macau, China, 5–8 December 2017; [CrossRef]

16. Chong, K.S.; Kleeman, L. Accurate odometry and error modelling for a mobile robot. In Proceedings of the International Conference on Robotics and Automation, Albuquerque, NM, USA, 25 April 1997; [CrossRef]

17. Scaramuzza, D.; Fraundorfer, F. Visual Odometry [Tutorial]. *IEEE Robot. Autom. Mag.* **2011**, *18*, 80–92, [CrossRef]

18. Ganganath, N.; Leung, H. Mobile robot localization using odometry and kinect sensor. In Proceedings of the 2012 IEEE International Conference on Emerging Signal Processing Applications, Las Vegas, NV, USA, 12–14 January 2012; [CrossRef]

19. Ozkil, A.G.; Fan, Z.; Xiao, J.; Kristensen, J.K.; Dawids, S.; Christensen, K.H.; Aanaes, H. Practical indoor mobile robot navigation using hybrid maps. In Proceedings of the 2011 IEEE International Conference on Mechatronics, Istanbul, Turkey, 13–15 April 2011; [CrossRef]

20. Farkas, Z.V.; Szekeres, K.; Korondi, P. Aesthetic marker decoding system for indoor robot navigation. In Proceedings of the IECON 2014-40th Annual Conference of the IEEE Industrial Electronics Society, Dallas, TX, USA, 29 October–1 November 2014; [CrossRef]

21. Krajník, T.; Nitsche, M.; Faigl, J.; Vaněk, P.; Saska, M.; Přeučil, L.; Duckett, T.; Mejail, M. A Practical Multirobot Localization System. *J. Intell. Robot. Syst.* **2014**, *76*, 539–562, [CrossRef]

22. Goldberg, B.; Doshi, N.; Jayaram, K.; Koh, J.S.; Wood, R.J. A high speed motion capture method and performance metrics for studying gaits on an insect-scale legged robot. In Proceedings of the 2017

IEEE/RSJ International Conference on Intelligent Robots and Systems (IROS), Vancouver, BC, Canada, 24–28 September 2017; [CrossRef]

23. Qin, Y.; Frye, M.T.; Wu, H.; Nair, S.; Sun, K. Study of Robot Localization and Control Based on Motion Capture in Indoor Environment. *Integr. Ferroelectr.* **2019**, *201*, 1–11. [CrossRef]

24. Bostelman, R.; Falco, J.; Hong, T. Performance Measurements of Motion Capture Systems Used for AGV and Robot Arm Evaluation. Available online: https://hal.archives-ouvertes.fr/hal-01401480/document (accessed on 11 November 2020).

25. Bilesan, A.; Owlia, M.; Behzadipour, S.; Ogawa, S.; Tsujita, T.; Komizunai, S.; Konno, A. Marker-based motion tracking using Microsoft Kinect. *IFAC-PapersOnLine* **2018**, *51*, 399–404, [CrossRef]

26. Bayındır, L. A review of swarm robotics tasks. *Neurocomputing* **2016**, *172*, 292–321, [CrossRef]

27. Hayes, A.T.; Martinoli, A.; Goodman, R.M. Swarm robotic odor localization: Off-line optimization and validation with real robots. *Robotica* **2003**, *21*, 427–441, [CrossRef]

28. Rothermich, J.A.; Ecemiş, M.İ.; Gaudiano, P. Distributed Localization and Mapping with a Robotic Swarm. In *Swarm Robotics*; Springer: Berlin/Heidelberg, Germany, 2005; pp. 58–69, [CrossRef]

29. Zhang, N.; Jeong, H.Y. A retrieval algorithm for specific face images in airport surveillance multimedia videos on cloud computing platform. *Multimed. Tools Appl.* **2016**, *76*, 17129–17143, [CrossRef]

30. Parekh, H.S.; Thakore, D.G.; Jaliya, U.K. A survey on object detection and tracking methods. *Int. J. Innov. Res. Comput. Commun. Eng.* **2014**, *2*, 2970–2979.

31. Garrido, S.; Nicholson, S. Detection of ArUco Markers. OpenCV: Open Source Computer Vision. Available online: www.docs.opencv.org/trunk/d5/dae/tutorial_aruco_detection.html (accessed on 1 October 2020).

32. Oščádal, P.; Heczko, D.; Vysocký, A.; Mlotek, J.; Novák, P.; Virgala, I.; Sukop, M.; Bobovský, Z. Improved Pose Estimation of Aruco Tags Using a Novel 3D Placement Strategy. *Sensors* **2020**, *20*, 4825, [CrossRef] [PubMed]

33. Legnani, G.; Casolo, F.; Righettini, P.; Zappa, B. A homogeneous matrix approach to 3D kinematics and dynamics—I. Theory. *Mech. Mach. Theory* **1996**, *31*, 573–587, [CrossRef]

34. Intel RealSense Depth Camera D435i. Intel RealSense Technology. Available online: www.intelrealsense.com/depth-camera-d435i (accessed on 1 October 2020).

35. Camera Calibration With OpenCV. OpenCV: Open Source Computer Vision. Available online: docs.opencv.org/2.4/doc/tutorials/calib3d/camera_calibration/camera_calibration.html (accessed on 1 October 2020).

36. Oščádal, P. Smart Building Surveillance Security System as Localization System for AGV. Available online: www.youtube.com/watch?v=4oLhoYI5BSc (accessed on 1 October 2020).

37. Oščádal, P. 3D Gridboard Pose Estimation, GitHub Repository. Available online: https://github.com/robot-vsb-cz/3D-gridboard-pose-estimation (accessed on 1 October 2020).

38. Asvadi, A.; Garrote, L.; Premebida, C.; Peixoto, P.; Nunes, U.J. Multimodal vehicle detection: fusing 3D-LIDAR and color camera data. *Pattern Recognit. Lett.* **2018**, *115*, 20–29, [CrossRef]

Publisher's Note: MDPI stays neutral with regard to jurisdictional claims in published maps and institutional affiliations.

applied sciences

Article

CNN Training Using 3D Virtual Models for Assisted Assembly with Mixed Reality and Collaborative Robots

Kamil Židek, Ján Piteľ *, Michal Balog, Alexander Hošovský, Vratislav Hladký, Peter Lazorík, Angelina Iakovets and Jakub Demčák

Department of Industrial Engineering and Informatics, Faculty of Manufacturing Technologies with a Seat in Presov, Technical University of Kosice, Bayerova 1, 08001 Presov, Slovakia; kamil.zidek@tuke.sk (K.Ž.); michal.balog@tuke.sk (M.B.); alexander.hosovsky@tuke.sk (A.H.); vratislav.hladky@tuke.sk (V.H.); peter.lazorik@tuke.sk (P.L.); angelina.iakovets@tuke.sk (A.I.); jakub.demcak@tuke.sk (J.D.)
* Correspondence: jan.pitel@tuke.sk; Tel.: +421-90-524-1605 or +421-55-602-6455

Abstract: The assisted assembly of customized products supported by collaborative robots combined with mixed reality devices is the current trend in the Industry 4.0 concept. This article introduces an experimental work cell with the implementation of the assisted assembly process for customized cam switches as a case study. The research is aimed to design a methodology for this complex task with full digitalization and transformation data to digital twin models from all vision systems. Recognition of position and orientation of assembled parts during manual assembly are marked and checked by convolutional neural network (CNN) model. Training of CNN was based on a new approach using virtual training samples with single shot detection and instance segmentation. The trained CNN model was transferred to an embedded artificial processing unit with a high-resolution camera sensor. The embedded device redistributes data with parts detected position and orientation into mixed reality devices and collaborative robot. This approach to assisted assembly using mixed reality, collaborative robot, vision systems, and CNN models can significantly decrease assembly and training time in real production.

Keywords: assisted assembly; mixed reality; collaborative robot; digital twin; convolutional neural networks

Citation: Židek, K.; Piteľ, J.; Balog, M.; Hošovský, A.; Hladký, V.; Lazorík, P.; Iakovets, A.; Demčák, J. CNN Training Using 3D Virtual Models for Assisted Assembly with Mixed Reality and Collaborative Robots. *Appl. Sci.* **2021**, *11*, 4269. https://doi.org/10.3390/app11094269

Academic Editor: Pavol Božek

Received: 13 April 2021
Accepted: 6 May 2021
Published: 8 May 2021

Publisher's Note: MDPI stays neutral with regard to jurisdictional claims in published maps and institutional affiliations.

1. Introduction and Related Works

Collaborative robots and their implementation in the assisted assembly process is an important part of the Industry 4.0 concept. They can work in the same workspace as human workers and perform basic manipulations or simple monotonous assembly tasks. This area is open to new research, methodology development and definition of basic requirements, because real applications in production processes are currently still limited.

The main advantage of using collaborative robots in the assembly process is a minimal transport delay of assembly parts between manual and automated operation. Other benefits are, for example, an integrated vision system for additional inspection of manual operation, the possibility to provide the interface for digital data collection from sensors and communication with external cloud platforms.

Appropriate human-robot cooperation can significantly improve assembly time, but both must have exactly defined methods of communication between them. For example, the collaborative robot can check the success of a worker assembly operation by the integrated camera and the worker can get information about this status by mixed reality devices. The mixed reality device also can shortcut the time of staff training. Configuration principles of a collaborative robot in an assembly task were introduced in [1], a framework to implement collaborative robots in the manual assembly in [2] and a human-robot collaboration framework for improving ergonomics in [3]. The important condition for the assisted assembly process is the synchronization of augmented (AR), virtual (VR), or mixed

(MR) reality devices with the digital twin for full digitalization of the used technology. A nice review of virtual, mixed, and augmented reality for immersive systems research is presented in [4]. Some other research results of the mixed assembly process between human and collaborative robots are described in [5–7]. An AR-based worker support system for human-robot collaboration using AR libraries was proposed in [8] and an anchoring support system using the AR toolkit was developed in [9]. A novel approach for end-user-oriented no-code process modeling in IoT domains using mixed reality technology is presented in [10] and a holistic analysis towards understanding consumer perceptions of virtual reality devices in the post-adoption phase in [11]. Technologies of virtual, augmented and mixed reality have an important role also in educational and training purposes as it was, for example, stated in [12,13].

Our research in the field of a digital twin implementation into assembly processes started with the digitalization of the experimental manufacturing assembly system described in [14]. A digital twin can visualize the real status of a manufacturing system as a 3D simulation model with real-time actualization, so there is a need to have an appropriate simulation model, for example a generic simulation model of the flexible manufacturing system developed in [15]. An automatic generation of a simulation-based digital twin of an industrial process plant is described in [16]. The basic overview about the usability of a digital twin can be found in [17], the possibility of improving the efficiency of the production by the implementation of digital twins is presented in [18]. Some learning experiences after establishing digital twins are described in [19] and using a digital twin to enhance the integration of ergonomics in the workplace design in [20]. Recommendations for future research and practice in the use of digital twins in the field of collaborative robotics are given in [21]. Also, this brief overview shows that a digital twin is an important element in the implementation of an assisted assembly into the production process.

The next condition of successful implementation of an assisted assembly into the production process is synchronization with the recognition system mainly based on vision devices with some SMART features like object detection, assembly parts identification and localization with their actual orientation in a workspace. Based on the obtained knowledge from the research on diagnostics of errors at the component surface by vision recognition systems using machine learning algorithms [22] we have started to use a convolutional neural network (CNN) for the recognition of standardized industrial parts (hexagon screw/nut and circular hole assembly elements) trained by real image input [23]. CNN with deep learning can work reliably for parts recognition, but the problem is manual preparation of input data for the learning process, because a very large quantity of input samples need to be prepared, usually several hundred for one part. The solution is to replace real image samples with 3D virtual models [24], but an important task is an automated sample generation from 3D models, which can be simplified using a web interface [25]. Besides automated input data preparation for CNN training, automated image analysis is also important. Some principles of this process were described in [26]. An interesting case study on the recognition of mark images using deep CNN was published in [27]. A methodology for synthesizing novel 3D image classifiers by generative enhancement of existing non-generative classifiers was proposed and verified in [28]. A viewpoint estimation in images using CNNs trained with rendered 3D model views was published in [29] and an accurate and fast CNN-based 6DoF object pose estimation using synthetic training in [30].

The sections of this article are structured in the following manner: following the introduction and related works in this Section a methodology of CNN training using 3D virtual models is introduced in Section 2. Section 3 describes the experimental assisted assembly work cell and the assembled product, in Section 4 the principles of the 3D virtual model preparation and 2D sample generation for CNN training are presented. Section 5 contains results and discussion, including implementation of parts recognition into the collaborative work cell. Finally, Section 6 presents a summary of the article along with some ideas for future work.

The main novelty and the innovation contribution of the article is a complex methodology for CNN training by virtual 3D models and design of a communication framework for assisted assembly devices like collaborative robot and mixed reality device.

2. Methodology of Deep Learning Implementation into the Assisted Assembly Process

A methodology of CNN training using 3D virtual models for deep learning implementation into the assisted assembly process is based on an automated generation of input sample data for learning without any monotonous manual work. All tasks, such as an object detection position, background and material change can be automated by the scripting language. This methodology can be divided into eight steps:

(1) Creation of 3D virtual models from the experimental assembly by any 3D design software or point cloud creation by laser scanning technology with conversion to some standard 3D format (OBJ, FBX, STL, IGES, etc.);

(2) Import 3D models into the software with cinematic rendering and some simulation of dynamics;

(3) Algorithms design of an automatic data queue of parts positioning, rotating, and camera setup by parts size;

(4) Rendering two sets of images: the first for CNN teaching and the second for an automated annotation algorithm;

(5) Creating of XML file for single shot detection and JSON format for instance segmentation;

(6) Automated ratio sorting to training and testing samples and moving to separate folder;

(7) Training of convolutional neural network for parts classification and localization (using single shot detection and instance segmentation);

(8) Transformation of CNN models into some type of embedded devices for inference of the trained model and results distribution of the detected position data to assisted assembly systems: a collaborative robot internal Cartesian system and mixed reality device anchoring system;

The detected objects are placed on the floor and they can be rotated only around one axis with a chosen increment from 20° to 360° by step from 1 to 18. The translation and rotation of virtual parts in scene are computed by standard translation and rotation matrixes for placement in 3D environment [24].

The detected object can be too small, for example as nuts or washers, so it is necessary to get magnification to change the field of view for the camera, according to Equation (1):

$$\text{if } \frac{FOV_{H,V}}{D_{X,Y}} > 3; \text{ then } H = Mf_L, \tag{1}$$

where $FOV_{H,V}$ is field of view (horizontal or vertical), $D_{X,Y}$ is object dimension in X or Y axis, H is distance to object [mm], M is magnification setup to 0.5x or 0.25x and f_L is focal length [mm].

The number of generated 2D images from all imported parts is counted by simple Equation (2):

$$N = pfn_{\alpha}, \tag{2}$$

where p is the number of imported parts, f is the number of used floors and n_{α} is the number of rotations for every part (in the range from 1 to 18).

Figure 1 presents a diagram of the proposed methodology of automatic data preparation for CNN automated training using experimental values, evaluation, and execution in the embedded device.

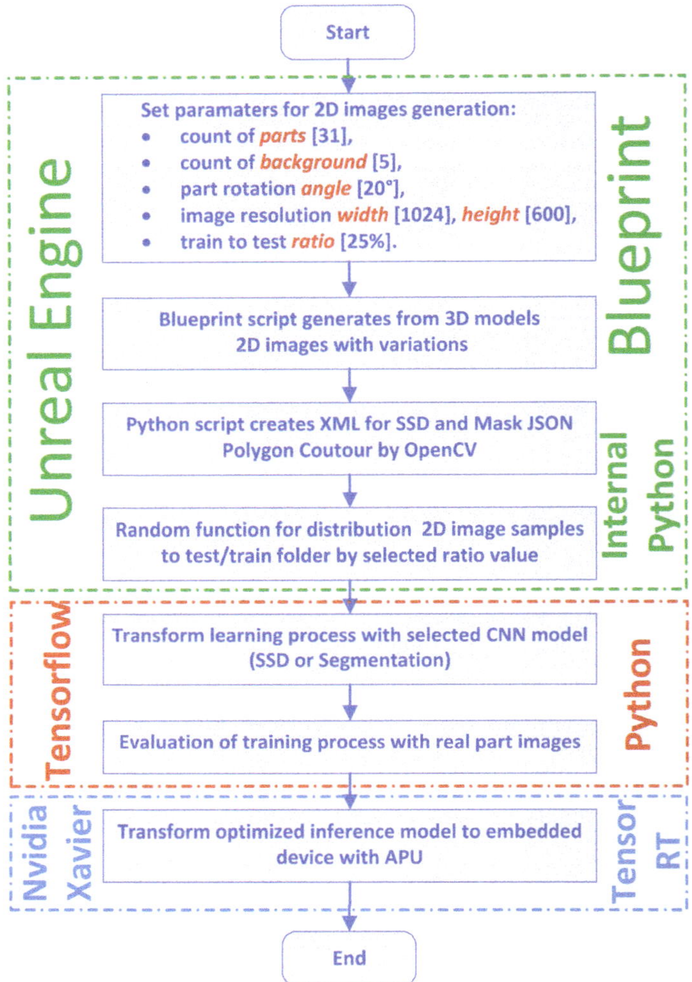

Figure 1. The simplified algorithm for samples generation from 3D virtual models of assembly parts.

3. Experimental Platform

The research has been provided in the SmartTechLab for Industry 4.0 at the Faculty of Manufacturing Technologies of Technical University of Kosice. There is installed an experimental SMART manufacturing system established primarily for research purposes, but also for collaboration with companies and teaching purposes. An important part of this system is a work cell for assisted assembly with incorporated technologies for parts recognition, mixed reality and collaborative robotics (Figures 2 and 3).

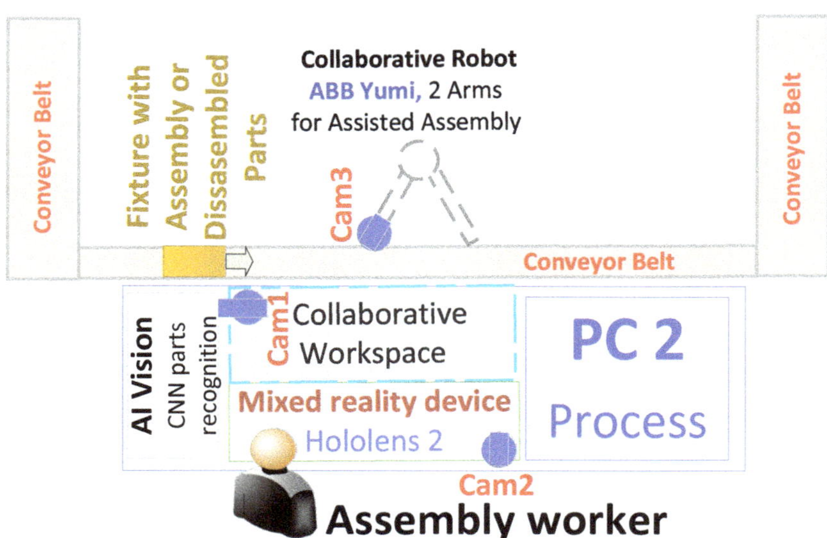

Figure 2. A scheme of the experimental assisted assembly work cell with CNN processing unit, mixed reality device and collaborative robot.

Figure 3. The experimental SMART manufacturing system; red frame: the workplace with an assisted assembly work cell with collaborative robot ABB Yumi and Microsoft Hololens 2 mixed reality device.

The point of interest for experimental assembly is a cam switch consisting of 31 parts made from different materials: plastic, rubber, stainless steel, and brass. The disassembled parts are shown in Figure 4a, and the assembled product is shown in Figure 4b.

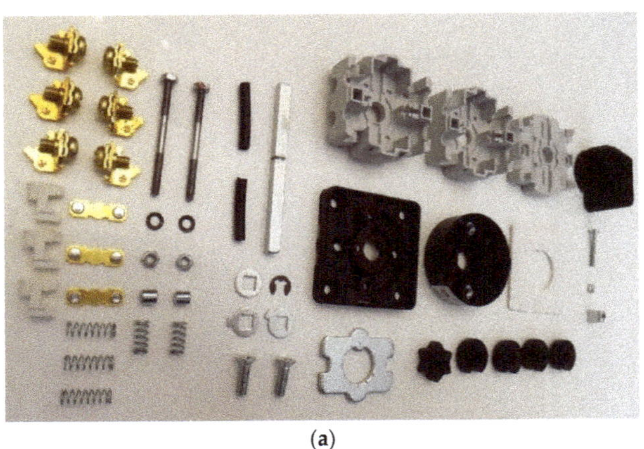 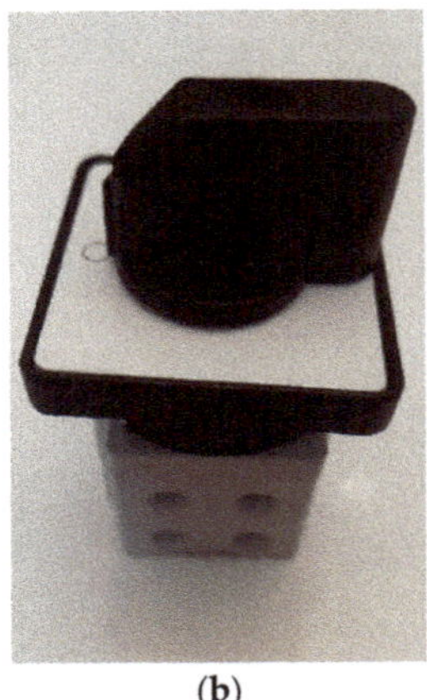

(a) (b)

Figure 4. The experimental product used in research: (**a**) The disassembled product parts for identification by CNN; (**b**) The assembled cam switch.

4. Input Data Preparation for CNN Training

CNN can work reliably for assembly parts recognition, but a problem is the preparation of input data for their training. A very large quantity of input samples need to be prepared, usually several hundred for one assembly part, because it has to be captured with different angular/translation variations and also with different backgrounds and materials. The replacement of real images of assembly parts with their 3D virtual models can significantly accelerate this process. Applying virtual models is also a trend of the Industry 4.0 concept and they can represent the real production process or product. Such virtual models digitally replicate all aspects of real products and they are called digital twins. They consist of 3D models of parts grouped into assemblies with the possibility of data synchronization with a real product, so the first step in the methodology of deep learning implementation into the assisted assembly process is a digital twin creation of the assembled product, in our case a cam switch (Figure 5). This digital twin will serve also as an input model into a mixed reality device for staff training.

Figure 5. An exploded view of a cam switch digital twin used in the assisted assembly process for implementation into mixed reality devices.

4.1. *An Automated 2D Images Generation by the Unreal Engine*

In the previous research [24] there was combined Blender 3D software with Python scripting language for an automated generation of the CNN training set, but the Blender rendering engine does not provide cinematic quality of the generated samples. This disadvantage significantly decreases the classification precision of trained convolutional networks by about 20 to 30%. The new approach is based on cinematic rendering from the Unreal Engine combined with Blueprint and Python scripting language. The next new feature is a dynamic collision used for realistic shadow rendering placed under generated samples on a different surface. An example of part position in the 3D inside of Unreal Engine editor setup before dynamic simulation (left) and the part with shadow after simulation (right) is shown in Figure 6.

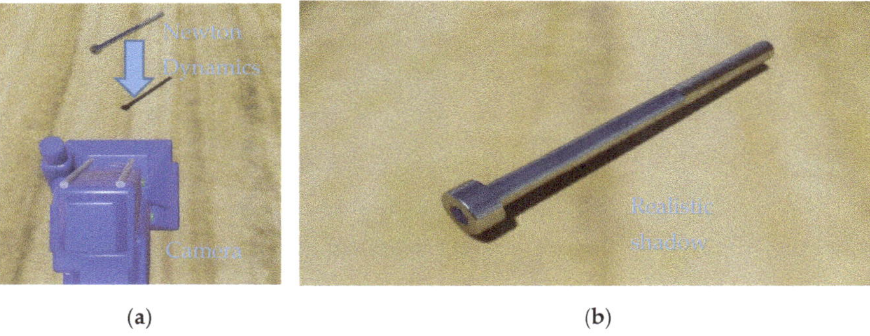

(a) (b)

Figure 6. An example project in the Unreal Engine: (**a**) 3D virtual model of a screw in the Unreal Engine editor; (**b**) 3D after dynamic simulation with cinematics texture and shadow.

The basic algorithm is coded in the Blueprint Scripting Language, an example of subprogram for 3D virtual part rotation in Z axis is shown in Figure 7.

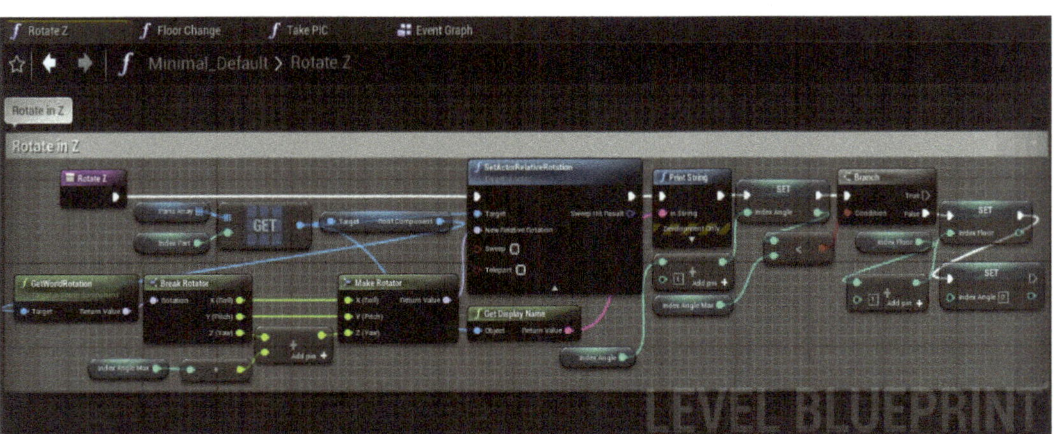

Figure 7. An example of a visual script for automated part rotation in Z-axis using the algorithm coded in the Blueprint Scripting Language.

The initial parameters for 2D sample generation are rotation angle, type of CNN model which provides basic image resolution, number of generated backgrounds as floors, and annotation file type. Basic parameters selection in the Unreal HUD menu before automated generation start is shown in Figure 8. Full assembly of the cam switch consists of 31 different parts, the basic setup uses 5 different floor textures and the angle of rotation can be set up from 20° to 360° in Z-axis.

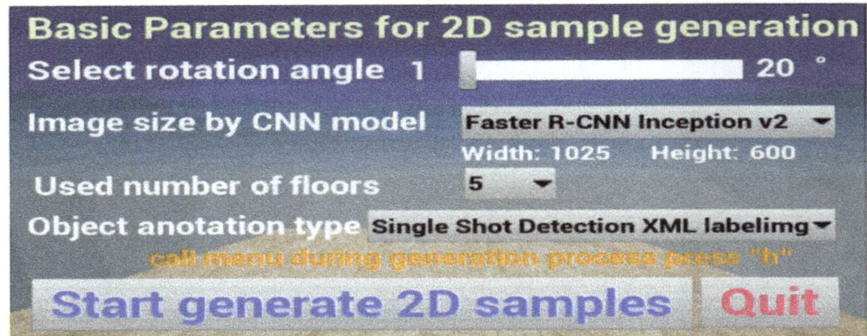

Figure 8. The Unreal Engine application setup of basic parameters for automated generation of 2D samples training and testing set.

4.2. An Automated Annotation by the OpenCV Algorithms

The input condition for automated annotation is a black background and binary threshold to get clear object edges. Two basic annotation methods were selected for generated 2D images from 3D virtual models:

- Single Shot Detection (SSD) annotation by the basic unrotated bounding box in XML format for LabelImg;
- Instance segmentation annotation by a polygon with variable approximation in JSON format for LabelMe.

The process of automated SSD annotation and evaluation is shown in Figure 9. The resolution of generated images can be changed exactly for the used CNN model. The first tested model was Faster R-CNN with Inception v2 with default resolution $600 \times 1024 \times 3$. Test samples are separated randomly from the train set for every generated part by a default value of 25%.

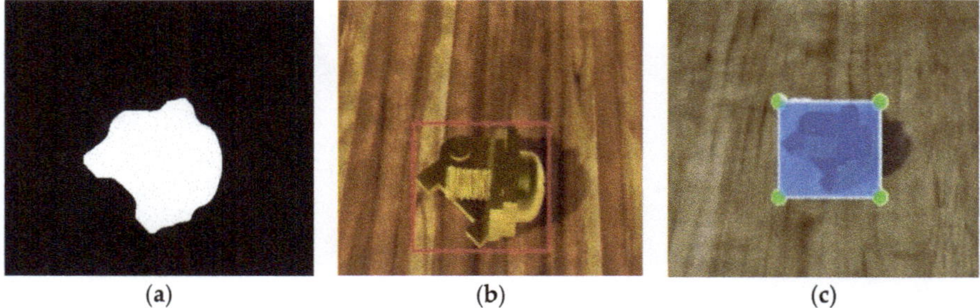

(a)   (b)   (c)

Figure 9. An autogeneration of the part localization by OpenCV for SSD: (**a**) binary threshold; (**b**) contour detection with bounding box generation; (**c**) LabelImg check of XML data generation.

Segmentation needs much more precise thresholding like single shot detection. The closing algorithm was used to get a precise contour of the object. An example of thresholding with object contour closing is shown in Figure 10.

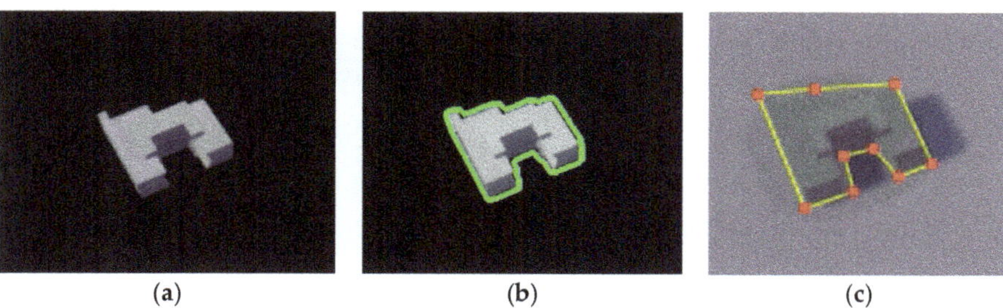

(a) (b) (c)

Figure 10. An autogeneration of the part localization by OpenCV for Segmentation: (**a**) the input image with black background from the Unreal Engine; (**b**) the result of OpenCV algorithm for precise contour generation; (**c**) an example of manual contour selection in VGG Image Annotator.

XML format for SSD is accepted as standard, but instance segmentation has many formats, COCO JSON, CSV, LabelMe JSON, RLE, etc. The simplest JSON structure has LabelMe format with polygon shape, which can be easily implemented to the automated process of contour annotation by OpenCV. An automated contour detection by OpenCV can provide a better contour as a manual process and it can significantly improve CNN instance segmentation after the training process, as can be seen in Figure 11.

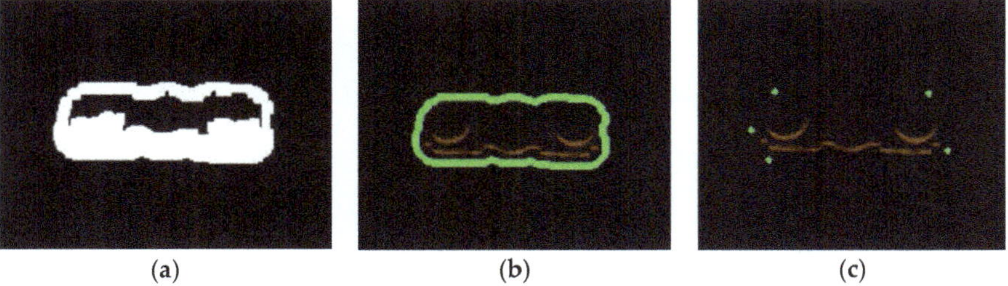

(a) (b) (c)

Figure 11. OpenCV detected contour representation: (**a**) binarized image; (**b**) all detected points; (**c**) reduced number of points optimized by Douglas-Peucker algorithm.

4.3. The Generated Training and Testing Sample Set

An example of results from an automated sample generation with XML annotation converted to CSV files is shown in Figure 12.

To start the CNN model training process is only necessary to copy folder *train*, *test*, and two cumulated annotations to *CSV* files to the TensorFlow folder and create TF_record files for training and testing.

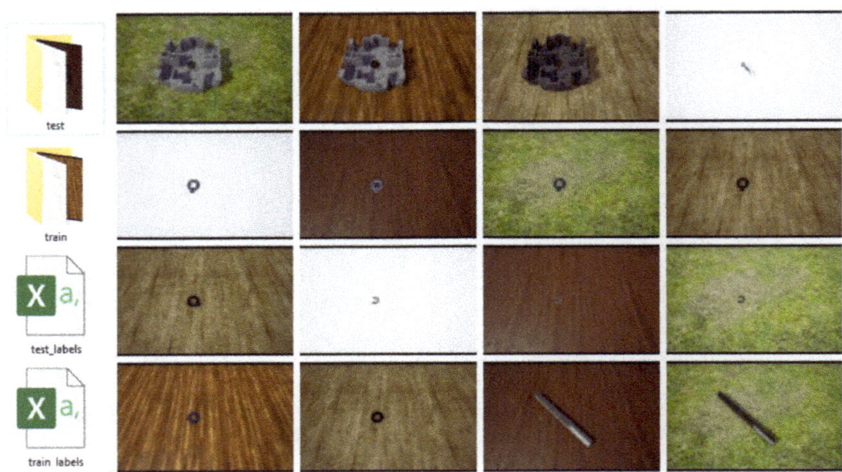

Figure 12. An example of the autogenerated training/tests set with XML bounding box annotation.

All necessary information to identify the sample is encoded in the sample image name:

00007 3**plastic_button** pic or 00007 3**plastic_button** picCV**.png**
where:
00007—number of the generated image [image 7]
3—part identification in the assembly [part 3]
plastic_button—part name
pic—image type
CV—OpenCV generated binary image with black background
.png/.xml/.json—the type of file necessary for TF record algorithm

5. Experimental Results and Implementation into the Assembly Process

An initial experiment of the cam switch parts recognition was executed using a small set of training samples (five per part, 155 samples altogether) with the different floor. Considering this small teaching set the obtained results are acceptable (see Table 1). The training process for Inception V2 is shown in Figure 13, where the unit on the X-axis is the number of cycles and the unit on the Y-axis is mAP.

Table 1. The acquired results by autogenerated training samples based on virtual 3D models.

Pretrained CNN Model	Testing Set	Number of Samples	Classification Results [%]	Training Time [h]
Faster RCNN Inception V2 SSD [1]	virtual	155	67–86	1.35
Mask RCNN Resnet101 [1]	virtual	155	91–95	5.20

[1] Both CNN models were used with a pretrained COCO dataset.

CNN models with single shot detection can be retrained very fast by transfer learning with accepted results within less than 2 h of training without dedicated GPU (for example the used pretrained Faster RCNN Inception V2 SSD reached the required accuracy within 1.35 h). CNN models with segmentation for the same input samples need 4–5 times more time for successful training (for example the used pretrained Mask RCNN Resnet101 reached required accuracy up to 5.20 h). But in contrast to SSD where a model is saved after unpredictable numbers of iteration, the training of models with Segmentation is stored after each epoch. The results of recognition of other images (not training set) of some parts for both tested CNN models (SSD and instance segmentation) are shown in Figure 14.

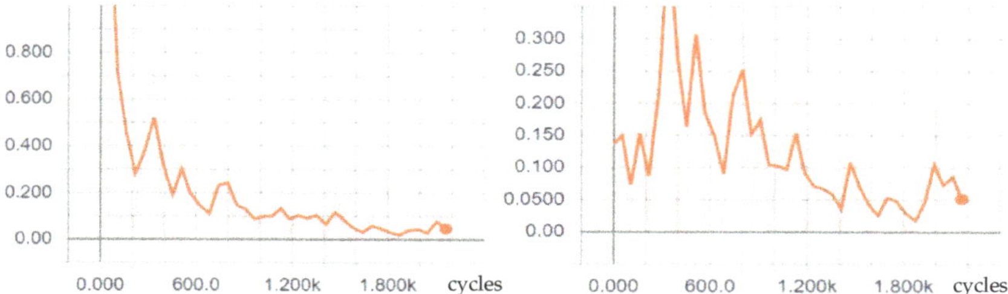

Figure 13. Training graphs for classification (**left**) and localization (**right**).

Figure 14. Inference experiments with the SSD trained model (**top**) and instance segmentation (**bottom**).

The inference time experiments with trained CNN models (SSD and Mask) have been performed on many different platforms. The delay results presented in Tables 2 and 3 for inference time is average value from a multiple test in loop for recognition of 40 sample images. The obtained results for CNN model SSD Inception V2 and TensorFlow 1 are in Table 2, for CNN Segmentation model Resnet101 and TensorFlow 2 with Pixelib in Table 3.

Table 2. The acquired inference results for the SSD CNN model.

Testing Environment	Platform Type	Platform Parameters	Delay [ms]
Google Colab	CPU	Intel Xeon 2.30GHz Gen. 6	2726.18
Google Colab	GPU	Tesla P4	358.37
Google Colab	TPU	Cloud TPU	2474.20
NVIDIA APU	Xavier AGX	Saved model native	885.18
NVIDIA APU	AGX	FP32 Tensor RT	445.43
NVIDIA APU	AGX	FP16 Tensor RT	339.41

Table 3. The acquired inference results for the Mask CNN model.

Testing Environment	Platform Type	Platform Parameters	Delay [ms]
NVIDIA APU	Xavier AGX	Maximum power 30W	690.71
Conda GPU	GPU	Nvidia RTX 2060	166.00
Conda CPU	CPU	Intel i9-10900KF 3,7 Ghz	701.12
WSL2 CPU	CPU	i7-9750H 2,7 Ghz	1522.88
WSL2 GPU	AGX	NVIDIA GTX 1660 Ti DirectML	5807.61

The FP16 SSD Inception V2 CNN model can reach about 3 FPS, which is an acceptable parts identification delay for checking worker assembly tasks and collaborative robot assembly status. The experiment with the Mask RCNN segmentation model reached in AGX device about 700 ms delay, which is acceptable in comparison to Desktop PC with high-performance CPU.

The mixed reality device is based on ARM64 architecture which does provide enough power for the execution of the trained inference model. The new approach is to stream video data to NVIDIA Xavier APU which runs an inference model and sends only extracted data: bounding box, contour polygon, and a result of classification as feedback.

The collaborative work cell contains a SMART vision system consisting of three cameras. The primary camera is connected to NVIDIA Xavier AGX (Figure 15a), where is uploaded trained CNN model. The second camera is integrated into mixed reality devices (Figure 15b) and the third camera is integrated into the right hand of the collaborative robot (Figure 15c). The principle of parts detection can be described in these steps:

(1)   Get images from the mixed reality device and collaborative robot hand;
(2)   Send images to the front of data;
(3)   Get images by NVIDIA Xavier embedded device;
(4)   Inference images by selected CNN model and acquire part position data;
(5)   Send bounding box (contour) and classification value by TCP communication to both devices (mixed reality and collaborative robot).

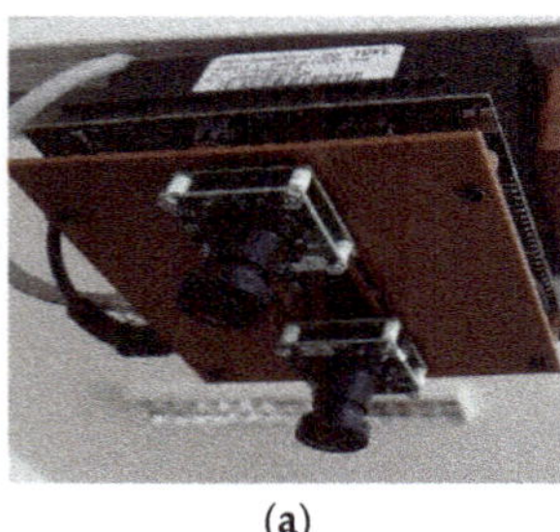  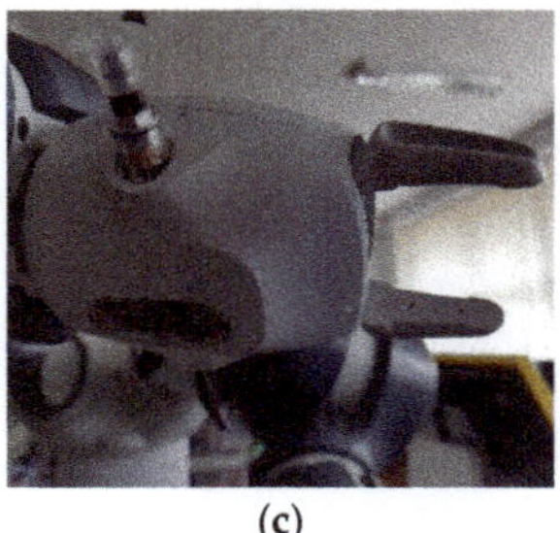

(a)   (b)   (c)

Figure 15. SMART vision system of the collaborative work cell: (a) CNN processing unit NVIDIA Xavier AGX; (b) Mixed reality device camera; (c) Collaborative robot ABB Yumi Vision system integrated into the right hand.

The implementation principle of parts recognition into the collaborative work cell is shown in Figure 16.

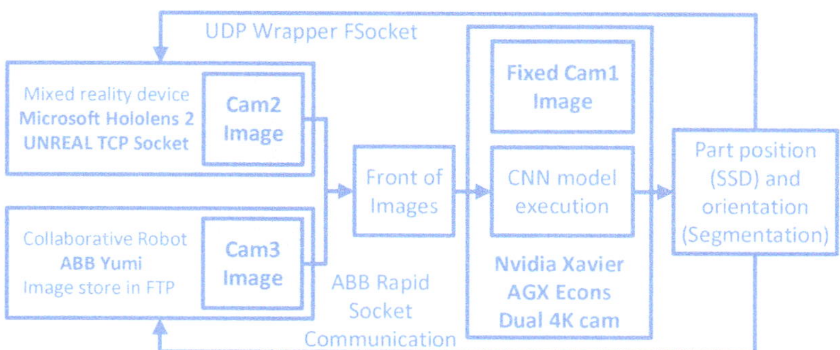

Figure 16. The principle of an experimental assisted assembly work cell with CNN processing unit, mixed reality device, and collaborative robot.

Images captured by all vision systems (static dual 4K e-con cameras connected to NVIDIA Xavier AGX and JetPack 4.4, integrated Cognex 7200 camera in ABB Yumi right hand and Microsoft Hololens 2 internal head camera) are shown in Figure 17.

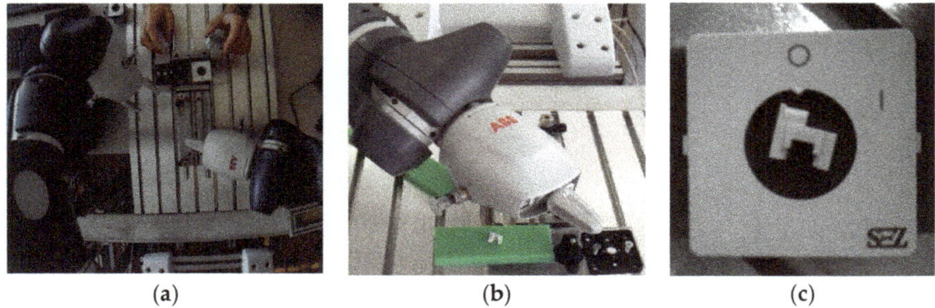

(a)	(b)	(c)

Figure 17. Images captured in the assisted assembly work cell by: (**a**) Nvidia Xavier AGX e-con camera; (**b**) Mixed reality Microsoft Hololens 2 head camera; (**c**) ABB Yumi Cognex Vision system.

The application for the mixed reality device Hololens 2 is coded in the same software (Unreal Engine) as a sample generation software but does not use Python programming language and is coded only by Blueprint programming language with UXTool library.

The designed calibration principle of all used vision systems to one Cartesian coordinate is shown in Figure 18 and their synchronization is realized in these steps:

- NVIDIA Xavier AGX system with e-con dual 4K camera is static and default zero position is set as fixed X, Y offset in pixels;
- ABB Yumi collaborative robot position is realized by offset from home point to the axis of the left-hand vision system in Rapid programming language;
- The mixed reality device Microsoft Hololens 2 is synchronized by QR code placed on the lower-left corner of the assembly table and rotation measured by integrated MEMS sensors.
- The information about part position adjusted by QR code is shared between all devices as Part identification, where position is X, Y; size W, H, and contour of the detected part is represented by array of points C.

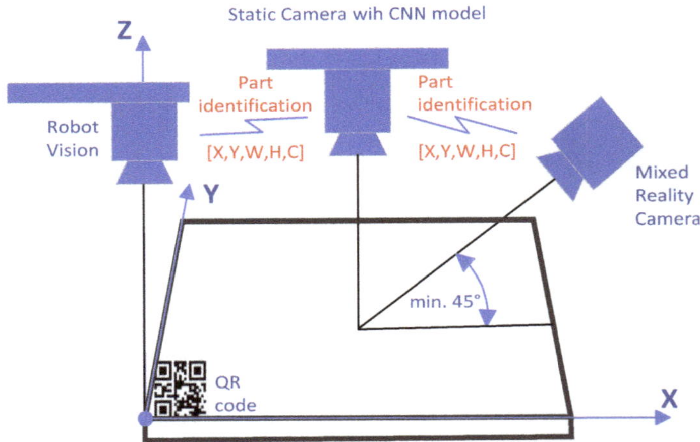

Figure 18. The principle of vision systems synchronization to one Cartesian coordinate.

Current deep learning frameworks provide only image augmentation, which only reduces the number of images that are needed to be prepared. It means, the most monotonous works in deep learning implementation to real application still exist. This is the main reason why is not profitable to use deep learning in small series assembly tasks. On the other hand, an automated generation of training samples from CAD models, which are available before production starts, can help to implement more assisted assembly solutions into practice.

The research field in an automated generation of training samples for CNN models from 3D virtual models has a big potential to expand. Current progress in GPU with real-time raytracing can provide new rendering possibilities to reach cinematic quality in object visualization and fast preparation of virtual samples. An interesting project is Kaolin from NVIDIA, a modular differentiable rendering for applications like high-resolution simulation environments, though it is still only as a library under research.

The early research progress is improving the presented tested solution with parts overlay recognition implemented into the Unreal Engine. An example of the first testing implementation with overlays is shown in Figure 19.

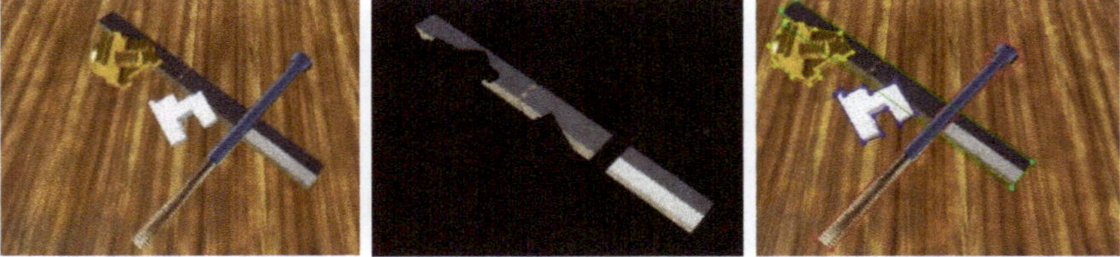

Figure 19. Parts overlay early research implemented into the Unreal Engine.

6. Conclusions

An automated generation of training samples based on 3D virtual models is a new approach in the field of deep learning that can save many hours of manual work. The presented research in the article introduces a methodology of CNN training for deep learning implementation into the assisted assembly process. This methodology was evaluated in an experimental SMART manufacturing system with assisted assembly work cell using cam switch as chosen assembly product from real production where is still used fully manual assembly process [31].

To summarize, those experiments have been performed and these main research results have been acquired in the field of CNN training for parts recognition in the assisted assembly process:

- A Blueprint program for an automated generation of 2D images from 3D virtual models as CNN training set in the Unreal Engine 4 software has been created;
- A Python algorithm with OpenCV library has been implemented for an automated image annotation for single shot detection as XML and instance segmentation as JSON;
- Two separate CNN models (SSD and Mask) have been trained in TensorFlow 2 framework for evaluation of the proposed methodology of deep learning implementation into the assisted assembly process;
- Inference experiments with trained CNN models on different platforms including an embedded APU have been performed and evaluated;
- Parts recognition data transfer among mixed reality devices, a collaborative robot and an embedded APU device has been designed and implemented.

The future works can be divided into more steps because there are further plans mainly in extending the current software:

- To increase rendered image quality to cinematic by real-time raytracing implemented in the new generation of GPUs;
- To implement an automated parts overlay recognition, which can be simply solved using Unreal Newton Physics and automatic switching to black texture for objects overlapping as input for OpenCV annotation;
- Transform current experimental project for free use as a plugin into the Unreal Engine software.

Author Contributions: Conceptualization, K.Ž. and J.P.; methodology, M.B.; software, P.L.; validation, V.H. and A.I.; formal analysis, M.B.; investigation, A.H.; resources, J.P.; data curation, K.Ž.; writing—original draft preparation, K.Ž.; writing—review and editing, J.P.; visualization, J.D.; supervision, A.H.; project administration, K.Ž.; funding acquisition, J.P. All authors have read and agreed to the published version of the manuscript.

Funding: This work was supported by the Slovak Research and Development Agency under the contract No. APVV-19-0590 and also by the projects VEGA 1/0700/20, 055TUKE-4/2020 granted by the Ministry of Education, Science, Research and Sport of the Slovak Republic.

Institutional Review Board Statement: Not applicable.

Informed Consent Statement: Not applicable.

Acknowledgments: The article was written as a result of the successful solving of the Project of the Structural Funds of the EU, ITMS code: 26220220103.

Conflicts of Interest: The authors declare no conflict of interest. The funders had no role in the design of the study; in the collection, analyses, or interpretation of data; in the writing of the manuscript, or in the decision to publish the results.

References

1. Realyvásquez-Vargas, A.; Arredondo-Soto, K.C.; García-Alcaraz, J.L.; Márquez-Lobato, B.Y.; Cruz-García, J. Introduction and configuration of a collaborative robot in an assembly task as a means to decrease occupational risks and increase efficiency in a manufacturing company. *Robot. Comput. Manuf.* **2019**, *57*, 315–328. [CrossRef]
2. Malik, A.A.; Bilberg, A.; Katalinic, B. Framework to Implement Collaborative Robots in Manual Assembly: A Lean Automation Approach. In Proceedings of the 29th International DAAAM Symposium 2018, Zadar, Croatia, 8–11 November 2017; Volume 1, pp. 1151–1160.
3. Kim, W.; Peternel, L.; Lorenzini, M.; Babič, J.; Ajoudani, A. A Human-Robot Collaboration Framework for Improving Ergonomics During Dexterous Operation of Power Tools. *Robot. Comput. Manuf.* **2021**, *68*, 102084. [CrossRef]
4. Liberatore, M.J.; Wagner, W.P. Virtual, mixed, and augmented reality: A systematic review for immersive systems research. *Virtual Real.* **2021**, 1–27. [CrossRef]
5. Khatib, M.; Al Khudir, K.; De Luca, A. Human-robot contactless collaboration with mixed reality interface. *Robot. Comput. Manuf.* **2021**, *67*, 102030. [CrossRef]

6.   Akkaladevi, S.C.; Plasch, M.; Maddukuri, S.; Eitzinger, C.; Pichler, A.; Rinner, B. Toward an Interactive Reinforcement Based Learning Framework for Human Robot Collaborative Assembly Processes. *Front. Robot. AI* **2018**, *5*, 126. [CrossRef] [PubMed]
7.   Ghadirzadeh, A.; Chen, X.; Yin, W.; Yi, Z.; Bjorkman, M.; Kragic, D. Human-Centered Collaborative Robots With Deep Reinforcement Learning. *IEEE Robot. Autom. Lett.* **2021**, *6*, 566–571. [CrossRef]
8.   Liu, H.; Wang, L. An AR-based Worker Support System for Human-Robot Collaboration. *Procedia Manuf.* **2017**, *11*, 22–30. [CrossRef]
9.   Takaseki, R.; Nagashima, R.; Kashima, H.; Okazaki, T. Development of Anchoring Support System Using with AR Toolkit. In Proceedings of the 2015 7th International Conference on Emerging Trends in Engineering & Technology (ICETET), Kobe, Japan, 18–20 November 2015; pp. 123–127.
10.   Dehghani, M.; Acikgoz, F.; Mashatan, A.; Lee, S.H. (Mark) A holistic analysis towards understanding consumer perceptions of virtual reality devices in the post-adoption phase. *Behav. Inf. Technol.* **2021**, 1–19. [CrossRef]
11.   Seiger, R.; Kühn, R.; Korzetz, M.; Aßmann, U. HoloFlows: Modelling of processes for the Internet of Things in mixed reality. *Softw. Syst. Model.* **2021**, 1–25. [CrossRef]
12.   Allcoat, D.; Hatchard, T.; Azmat, F.; Stansfield, K.; Watson, D.; Von Mühlenen, A. Education in the Digital Age: Learning Experience in Virtual and Mixed Realities. *J. Educ. Comput. Res.* **2021**. [CrossRef]
13.   Kesim, M.; Ozarslan, Y. Augmented Reality in Education: Current Technologies and the Potential for Education. *Procedia Soc. Behav. Sci.* **2012**, *47*, 297–302. [CrossRef]
14.   Židek, K.; Piteľ, J.; Adámek, M.; Lazorík, P.; Hošovský, A. Digital Twin of Experimental Smart Manufacturing Assembly System for Industry 4.0 Concept. *Sustainability* **2020**, *12*, 3658. [CrossRef]
15.   Luscinski, S.; Ivanov, V. A simulation study of Industry 4.0 factories based on the ontology on flexibility with using Flexsim®software. *Manag. Prod. Eng. Rev.* **2020**, *11*, 74–83. [CrossRef]
16.   Martinez, G.S.; Sierla, S.; Karhela, T.; Vyatkin, V. Automatic Generation of a Simulation-Based Digital Twin of an Industrial Process Plant. In Proceedings of the IECON 2018—44th Annual Conference of the IEEE Industrial Electronics Society, Washington, DC, USA, 21–23 October 2018; pp. 3084–3089.
17.   Tomko, M.; Winter, S. Beyond digital twins—A commentary. *Environ. Plan. B Urban Anal. City Sci.* **2019**, *46*, 395–399. [CrossRef]
18.   Shubenkova, K.; Valiev, A.; Shepelev, V.; Tsiulin, S.; Reinau, K.H. Possibility of Digital Twins Technology for Improving Efficiency of the Branded Service System. In Proceedings of the 2018 Global Smart Industry Conference (GloSIC), Chelyabinsk, Russian, 13–15 November 2018; pp. 1–7.
19.   David, J.; Lobov, A.; Lanz, M. Learning Experiences Involving Digital Twins. In Proceedings of the IECON 2018—44th Annual Conference of the IEEE Industrial Electronics Society, Washington, DC, USA, 21–23 October 2018; pp. 3681–3686.
20.   Caputo, F.; Greco, A.; Fera, M.; Macchiaroli, R. Digital twins to enhance the integration of ergonomics in the workplace design. *Int. J. Ind. Ergon.* **2019**, *71*, 20–31. [CrossRef]
21.   Malik, A.A.; Brem, A. Digital twins for collaborative robots: A case study in human-robot interaction. *Robot. Comput. Manuf.* **2021**, *68*, 102092. [CrossRef]
22.   Židek, K.; Piteľ, J.; Hošovský, A. Machine learning algorithms implementation into embedded systems with web application user interface. In Proceedings of the IEEE 21st International Conference on Intelligent Engineering Systems 2017 (INES 2017); IEEE. 2017; pp. 77–81.
23.   Židek, K.; Hosovsky, A.; Piteľ, J.; Bednár, S. Recognition of assembly parts by convolutional neural networks. In *Advances in Manufacturing Engineering and Materials; Lecture Notes in Mechanical Engineering*; Springer: Cham, Germany, 2019; pp. 281–289.
24.   Židek, K.; Lazorík, P.; Piteľ, J.; Hošovský, A. An Automated Training of Deep Learning Networks by 3D Virtual Models for Object Recognition. *Symmetry* **2019**, *11*, 496. [CrossRef]
25.   Pollák, M.; Baron, P.; Telišková, M.; Kočiško, M.; Török, J. Design of the web interface to manage automatically generated production documentation. *Tech. Technol. Educ. Manag. TTEM* **2018**, *7*, 703–707.
26.   Gopalakrishnan, K. Deep Learning in Data-Driven Pavement Image Analysis and Automated Distress Detection: A Review. *Data* **2018**, *3*, 28. [CrossRef]
27.   Mao, K.; Lu, D.; E, D.; Tan, Z. A Case Study on Attribute Recognition of Heated Metal Mark Image Using Deep Convolutional Neural Networks. *Sensors* **2018**, *18*, 1871. [CrossRef] [PubMed]
28.   Varga, M.; Jadlovský, J.; Jadlovská, S. Generative Enhancement of 3D Image Classifiers. *Appl. Sci.* **2020**, *10*, 7433. [CrossRef]
29.   Su, H.; Qi, C.R.; Li, Y.; Guibas, L.J. Render for CNN: Viewpoint Estimation in Images Using CNNs Trained with Rendered 3D Model Views. In Proceedings of the 2015 IEEE International Conference on Computer Vision (ICCV), Santiago, Chile, 7 December 2015; pp. 2686–2694.
30.   Su, Y.; Rambach, J.; Pagani, A.; Stricker, D. SynPo-Net—Accurate and Fast CNN-Based 6DoF Object Pose Estimation Using Synthetic Training. *Sensors* **2021**, *21*, 300. [CrossRef] [PubMed]
31.   Lazár, I.; Husár, J. Validation of the serviceability of the manufacturing system using simulation. *J. Effic. Responsib. Educ. Sci.* **2012**, *5*, 252–261. [CrossRef]

 *applied sciences*

Article

Design of a Mechanical Part of an Automated Platform for Oblique Manipulation

Miroslav Blatnický [1], Ján Dižo [1,*], Milan Sága [2], Juraj Gerlici [1] and Erik Kuba [1]

[1] Department of Transport and Handling Machines, Faculty of Mechanical Engineering, University of Žilina, Univerzitná 8215/1, 01026 Žilina, Slovakia; miroslav.blatnicky@fstroj.uniza.sk (M.B.); juraj.gerlici@fstroj.uniza.sk (J.G.); erik.kuba@fstroj.uniza.sk (E.K.)

[2] Department of Applied Mechanics, Faculty of Mechanical Engineering, University of Žilina, Univerzitná 8215/1, 01026 Žilina, Slovakia; milan.saga@fstroj.uniza.sk

* Correspondence: jan.dizo@fstroj.uniza.sk; Tel.: +421-41-513-2560

Received: 24 September 2020; Accepted: 26 November 2020; Published: 27 November 2020

Abstract: Handling machines are increasingly being used in all sectors of the industry. Knowledge of the theory of transport and handling machines are basic prerequisites for their further technical development. Development in the field of manipulators is reflected not only in their high technical level, but also in increasing safety and economy. The article presents results of research focused on the complete engineering design of a manipulator, which will serve as a mean of the oblique transport of pelletised goods. The manipulator takes the form of a platform moving between two destinations by means of an electromotor. The engineering design of the platform including the track and a working principle is described. The design includes analytical and numerical calculations of main loaded components of the platform. Extensive functional and dimensional calculations serve as the base for preparation of the technical documentation. An important step will be the creation of a parametric model of the force and moment load acting on a platform drivetrain. Based on this, optimal parameters of an electromotor and its dimensional calculation are performed.

Keywords: manipulation; design; analytical calculation; numerical analysis; manipulation platform

---

1. Introduction

Recently, it has become possible to meet handling machines of various forms almost at every turn. There are mainly transport means, robotics, lifts, cranes, trolleys etc. Their use is not only in transport, construction or health care services, but their applications are implemented also in other branches of industry, such as storage. Storage is an inseparable part of material flow in every sphere of the economy. The need for material storage of every kind rises due to different timings of production and consumption, different flow of material in an individual part of a logistic chain at every level. Storage is a necessary part of production technology as well.

The problem of warehouses can be studied from different points of view, e.g., from the building point of view, warehouse organisation, used technology etc. In term of logistic objects, the attention is mainly paid to three kind of warehouses, and these are warehouses of bulk materials, warehouses of palletised good and metallurgical materials [1,2]. When one wants to decide which particular kind of warehouse has to be chosen, it is necessary to draw up a thorough analysis of the problem. All requirements of a new device have to be taken into account, because the use of the state of art of technology may not bring the highest functional and economical effects. Handling and transport operations are connected with a relatively large number of workers, and therefore modern engineering production does not function without precisely solved manipulation [3–6].

The objective of this article is to achieve the described requirements by means of implementation of a design of a platform manipulator (Figure 1), which is intended to be mounted in a warehouse. It will transport palletised material on standardised pallets with dimensions 1.2 × 0.8 m with the maximal weight up to 225 kg (up to 300 kg exceptionally) and up to the maximal height of 5 m.

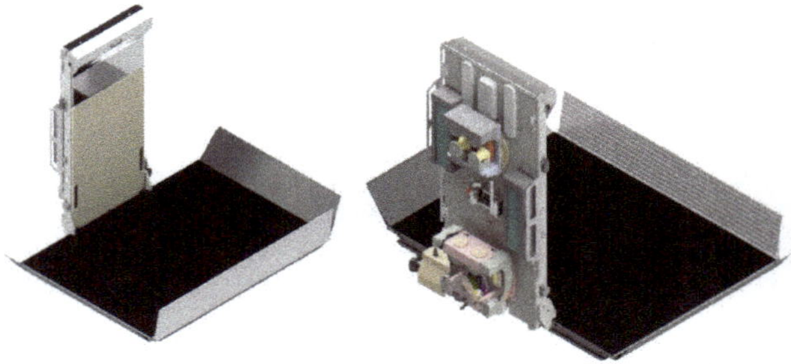

Figure 1. A three-dimensional model of the design of the created platform manipulator.

A modern warehouse is not only for material storage. In principle, it is the set of warehouse devices, transport, lifting, sorting and packing machines etc. which can create whole transport lines. The choice of a transport system in a warehouse is a decisive phase of warehouse establishment. The proper choice of the system influences technical, functional, economical and ergonomic aspects of processes during its entire lifetime [1,7].

2. Analysis of Theoretical Aspects Influencing the System Choice

Basic data of a warehouse, for which a manipulator has to be designed, are quantified by technical factors and storage possibilities of the given plant. The kind and amount of stored material and warehouse capacity are the main input quantities. In this research, main attention is focused on the infrastructure in term of communications, building layout as well as fire protection. Besides that, we have to take into account existing storage facilities, such as racks, conveyors, further auxiliary means, i.e., pallets, binding and gripping means [8–10], technical means, i.e., electric and electronic elements, compressed air distribution and others. The main criteria, which were taken into account in the process of the proper system choice, are technical and ecological safety, economy, reliability, life cycle, demands on operation, and affordability. When all the described parameters were accepted, the use of a forklift truck seemed to be the easiest form of manipulation of material. However, due to the character of the transported material, it must not be exposed to exhalants produced by a conventional combustion engine regardless of whether they are gasoline, gas [11,12], diesel [13,14] engines or hybrids [15]. Although there are forklift trucks produced with an electric drivetrain [16–18], in consideration of its acquisition costs and its minimal other use, the purchase of an electric forklift truck was refused. Moreover, the parameters of the transported material would lead to a dangerous situation from the dynamic properties point of view.

Thus, if we do not consider warehouses in which pallets are transported by means of forklift trucks, we have to focus on such warehouses in which material is handled by other devices. These devices are generally automatised and controlled by computers [19–21]. However, such a solution suits a warehouse with large capacity and mainly for large assortment, where it is possible to easily reach every one of them. Very narrow aisle trucks, which move in the aisle of a warehouse fully meet that requirement [22,23]. The use of a transelevator [24] was assessed as unprofitable, and therefore it was not considered. An implementation of a shelf stacker was not rejected, because this solution satisfies demands on one hand in terms of the geometrical parameters of the warehouse and on the other hand in terms of the special environment of the warehouse (humidity, temperature and purity) as well as in terms of the price. Figure 2 shows a partial three-dimensional view of the workplace, for which the manipulator is designed. It is obvious, that neither a common elevator cannot be installed in the warehouse, because its structure would cause as an obstacle to the emergency exit. After consideration and acceptance of all facts described above, the solution of an oblique transport of material by means of the platform was suggested. Such a solution includes all needed properties and meets all operational conditions.

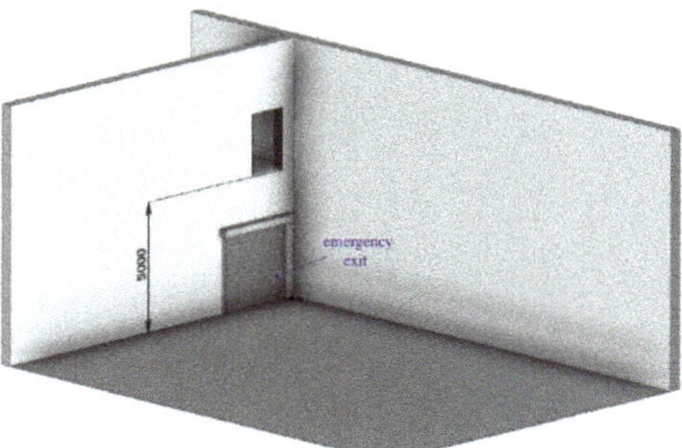

Figure 2. A partial three-dimensional view of the workplace for placement of the designed manipulator.

3. Design of the Accepted Solution

The design of the handling machine consists of two individual, but interconnected parts.

First, it was necessary to define all external loads acting on the device. Among them, the total weight of the platform and the weight of transported material are the most important input parameters. This mutual load acts on the whole bearing structure and it imposes requirements on parameters of the used drivetrain, i.e., to the drive motor power.

The other part of the design consists of the design of a track, on which the platform will move. It is necessary to take into account its structure, mounting onto a wall, drivetrain mechanism, reliability and safety. The design methodology of such a specific kind of a manipulation machine comes from the STN EN 81-40 standard [25]. As structural designs have to be performed in accordance with the industrial property office requirements, a survey of current technical solutions was carried out. Based on the results, the engineering design of the platform can be used for the intended operational conditions. The platform will move on two steel tubes, which represent the guidance of the platform. The steel tubes with a circular cross-section are mounted on supporting columns. The guidance including columns can be fixed on the wall; the mechanism drivetrain consists of an electromotor and a gearbox, which are built in a platform skeleton. The power transmission is ensured by means of a pinion and a gear rack, which are the part of the guiding track.

The steel structure loaded is during operation by gravitational forces, which cause the strain change. If the distribution of forces, which act on the structure, are improperly analysed, the platform structure would be dimensioned insufficiently and it could result in dangerous operation. The STN EN 81-40 standard [25] defines the minimal loads of platforms of 250 kg·m^{-2}. It is necessary to prove that a platform is able to transmit such loads. As the designed device will transport palletised material on the pallet with dimensions of 1.2 × 0.8 m, the carrying capacity of the platform must be at least of 240 kg. Therefore, the suggested carrying capacity meets the requirements of the minimum carrying capacity. A conceptual design of the solved manipulator is shown in Figure 3. The solved platform will move in two trolleys placed one above the other and which are rotationally connected with the platform. The trolleys tilt during the platform operation, which results in the change of the forces distribution in the structure. The distance between rotating points of both trolleys is of 0.594 m (Figure 4).

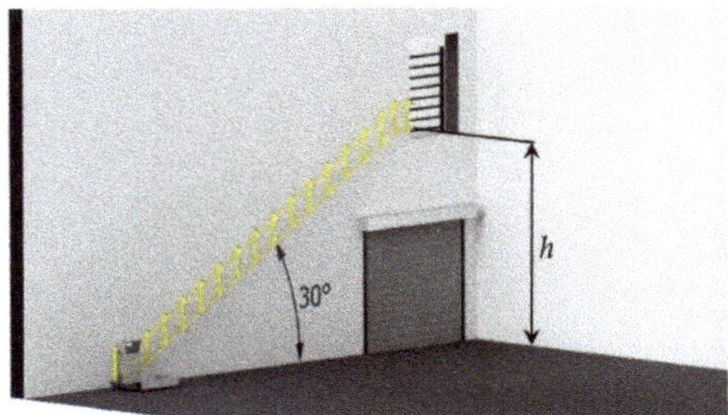

Figure 3. A conceptual design of the manipulator.

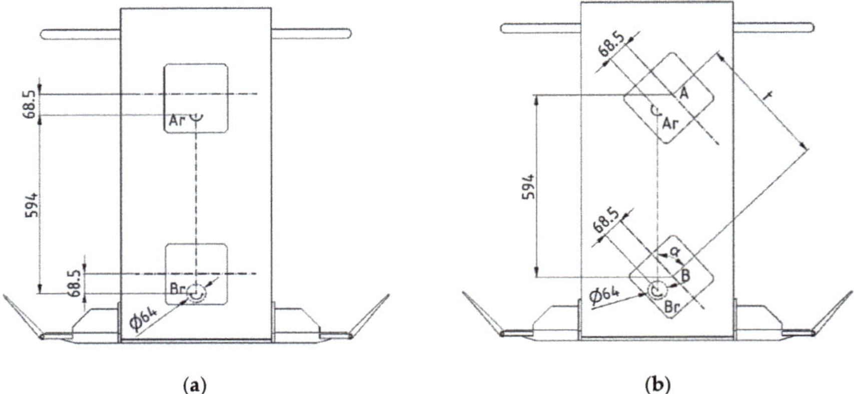

(a)   (b)

Figure 4. The main dimensions of the travel mechanism of the platform: (a) when it moves in a straight line; (b) when it moves on a slope.

This distance is constant despite the varying track slope. The distance between the two rotating points of the trolleys and the longitudinal axes of both track tubes is constant as well. It is 0.0685 m. Because the points of rotation of both trolleys (Figure 4, points A_r and B_r) are on the mutual vector line of the gravitational acceleration, the vertical distance of resulted reaction effects must be 0.594 m. It is necessary to realise, that the track slope change influences the perpendicular distance of individual reaction forces, in Figure 4a marked as t.

It is possible to determine parameter t analytically by the following equation:

$$t = 0.594 \cdot \sin \alpha, \tag{1}$$

Where α (°) is the track slope. The constant vertical distance of the trolleys' reactions also influences the calculation of the horizontal reactions. These reactions compensate the moment of the platform gravitational force (Figure 5a—G_1 force) as well as the moment of the transported material gravitational force (Figure 5b—G_2 force). They act on arms marked as g_1 and g_2. The free body diagram is shown in Figure 5. From it, we can determine the equations of equilibrium as the following:

$$\sum_i M_{iA} = 0 \rightarrow R_{bz} \cdot 0.594 - G_1 \cdot g_1 - G_2 \cdot g_2 = 0, \tag{2}$$

$$\sum_i F_{iz} = 0 \rightarrow R_{bz} - R_{az} = 0. \tag{3}$$

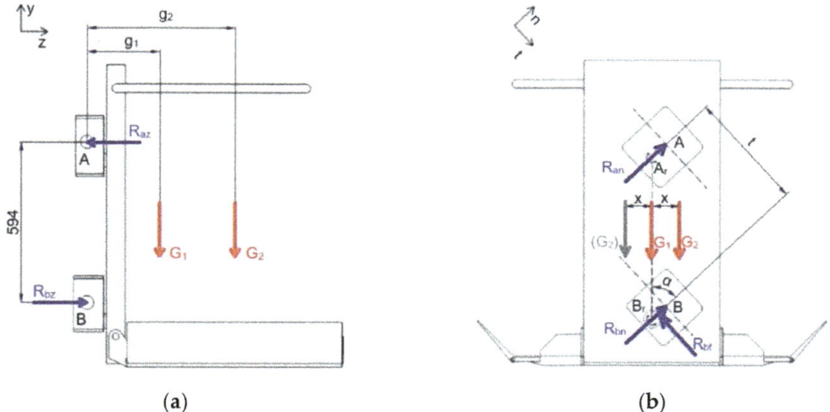

Figure 5. Application of the free body diagram method for the platform: (**a**) when it moves in a straight line; (**b**) when it moves on a slope.

Solving Equations (2) and (3) we obtain:

$$R_{bz} = \frac{G_1 \cdot g_1 + G_2 \cdot g_2}{0.594}, \tag{4}$$

and

$$R_{az} = R_{bz}, \tag{5}$$

$$\sum_i F_{it} = 0 \rightarrow G_1 \cdot \sin\alpha + G_2 \cdot \sin\alpha - R_{bt} = 0, \tag{6}$$

$$\sum_i M_{iBr} = 0 \rightarrow R_{an} \cdot t + G_2 \cdot x - R_{bt} \cdot 32 = 0. \tag{7}$$

The solution of Equations (8)–(10) gives

$$R_{bz} = (G_1 + G_2) \cdot \sin\alpha, \tag{8}$$

$$R_{bz} = \frac{R_{bt} \cdot 32 - G_2 \cdot x}{t} \rightarrow R_{an} = \frac{(G_1 + G_2) \cdot \sin\alpha \cdot 32 - G_2 \cdot x}{t}, \tag{9}$$

$$\sum_i F_{in} = 0 \rightarrow R_{an} + R_{bn} - G_1 \cdot \cos\alpha - G_2 \cdot \cos\alpha = 0. \tag{10}$$

From Equation (10) we obtain:

$$R_{bn} = (G_1 + G_2) \cdot \cos\alpha - R_{an} \rightarrow R_{bn} = (G_1 + G_2) \cdot \cos\alpha - \frac{(G_1 + G_2) \cdot \sin\alpha \cdot 32 - G_2 \cdot x}{t}. \tag{11}$$

The validity of the Equations (11) and (12) will be for any value of the angle α, if the parameter x (the distance of the gravitational force G_2 from the vertical axis between points A_r and B_r, Figure 5b) in the right direction will gain a positive value and in the left direction it will gain a negative value. Considering the dimensions of the platform of 1.2×0.8 m and the centre of gravity location, the parameter x is in the interval from −0.100 to +0.100 m.

Table 1 contains all geometrical parameters and Table 2 includes all force effects, which are needed for the calculation. The maximal slope (angle α) of the track including safety conditions is determined to the value of $\alpha = 47°$.

Table 1. Geometrical parameters for calculation.

Parameter	Value (m)
g_1	0.2730
g_2	0.5630
t	0.4344
x	±0.1000

Table 2. Determined values of external loads of the platform.

Considered Weight (kg)	Gravitational Force (N)
154	1510.74
225	2207.25
300	2943.00
375	3678.75

The engineering design of the platform considers the maximal carrying load capacity of 300 kg, when this load is transported up to the end station. Within the safety conditions, the platform can be loaded up to the weight of 375 kg, but, in such a case, the platform must not move. Otherwise, it could lead to the platform damage. The kerb weight of the platform is 154 kg.

Substituting of values from Table 2 into Equations (1)–(10) gives results of reaction effects introduced in Table 3, where:

R_{az}—tangential reaction of wheels of the upper trolley;

R_{bz}—tangential reaction of wheels of the bottom trolley;

R_{an}—normal reaction of wheels of the upper trolley;

R_{bn}— normal reaction of wheels of the bottom trolley;

R_{bt}—slope resistance.

Table 3. Calculated values of reaction effects considering the geometry of the device and values of loads.

$X = +0.1$ m	225 kg	300 kg	375 kg	$X = -0.1$ m	225 kg	300 kg	375 kg
R_{az} (N)	2786.39	3483.74	4171.09	R_{az} (N)	2786.39	3483.74	4181.09
R_{bz} (N)	2786.39	3483.74	4181.09	R_{bz} (N)	2786.39	3483.74	4181.09
R_{bt} (N)	2719.17	3257.26	3795.35	R_{bt} (N)	2719.17	3257.26	3795.35
R_{an} (N)	−307.81	−437.54	−567.27	R_{an} (N)	708.42	917.43	1126.44
R_{bn} (N)	2843.47	3474.98	4106.5	R_{bn} (N)	1827.24	2120.01	2412.78
$\sum|R_{an}| + |R_{bn}|$ (N)	3151.28	3912.52	4673.77	$\sum R_{an} + R_{bn}$ (N)	2535.66	3037.44	3539.22

Based on the performed calculation, the greatest load of the platform is in the moment, when the force G_2 is shifted by the value of $x = +0.1$ m. Just this position of the load is chosen for the calculation of resistance forces. The change of the reaction forces due to the platform geometry is calculated by means of the Matlab 2017a program (MathWorks Inc., Natick, MA, USA) [26–28] and their graphs are shown in Figure 6, where a blue line represents the waveform of the reaction force R_{an}, a green line depicts the reaction force R_{bn} and a red curve R_{sum} is the waveform of the sum of reaction forces.

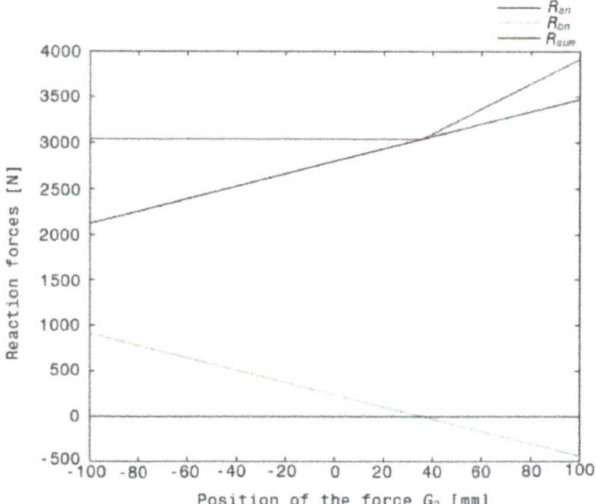

Figure 6. Dependence of values of the reaction forces R_{an}, R_{bn} and R_{sum} on the platform geometry (position of the centre of gravity G_2).

Furthermore, these results are used for the calculation of resistance forces, which arise during the platform operation on the track. Because of the rolling of guiding wheels of the trolleys (Figure 7), rolling resistance arises. This rolling resistance occurs between wheels and pins. Movement of the platform causes elastic deformation of the track as well as the platform wheels. This elastic deformation results in resistance moment arising. As we can see in Figure 7, the upper trolley transmits the reactions R_{az} and R_{an}. Four guiding wheels are mounted on the trolley, however they do not ensure the transmitting of the force R_{az}, therefore, it is assumed, that a sliding force between the guiding wheels and the spacer rings side surfaces will arise. On the contrary, the force R_{an} causes rolling resistance and pin friction during the operation. The total resistance of the guiding wheel is given by the following formulations:

$$M_{a1} = R_{an} \cdot (e_{a1} + r_{ca1} \cdot f_{ca1}), \tag{12}$$

$$M_{a1} = \frac{(G_1 + G_2) \cdot \sin \alpha \cdot 32 - G_2 \cdot x}{t} \cdot (e_{a1} + r_{ca1} \cdot f_{ca1}). \tag{13}$$

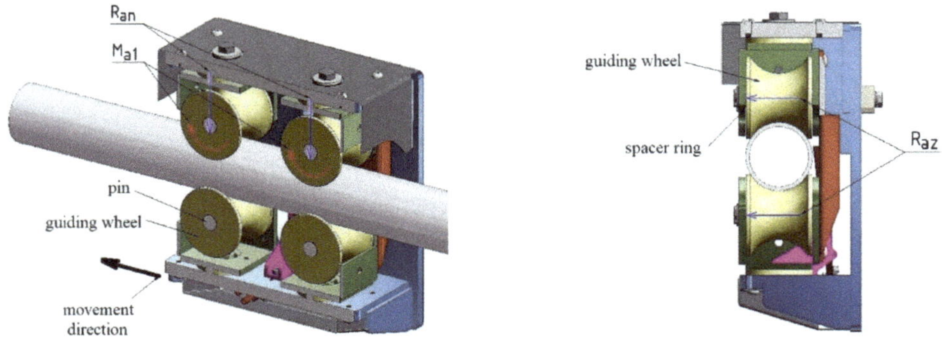

Figure 7. Transmission of reaction forces on the upper trolley of the platform.

The friction resistance moment of the spacer ring is calculated as follows:

$$M_{a2} = f_k \cdot \int_r^R x \cdot dR_{az}. \tag{14}$$

The area (Figure 8), on which the force R_{az} acts, is determined:

$$S = \pi \cdot \left(R^2 - r^2\right).$$ (15)

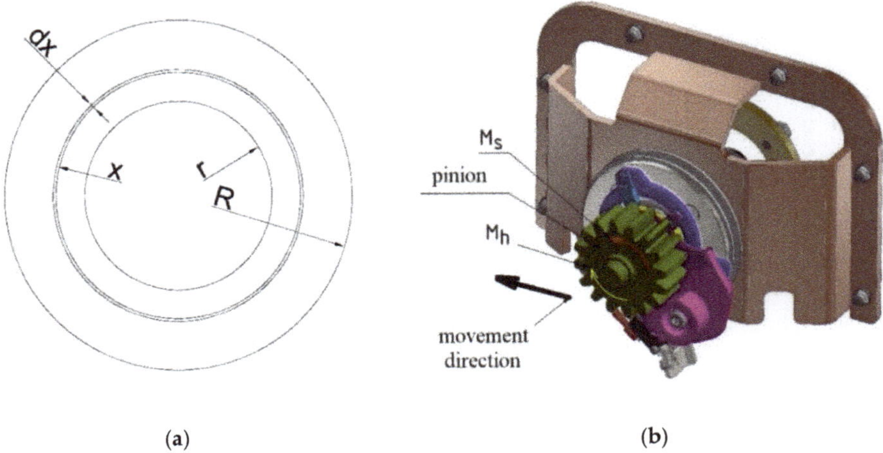

(a)   (b)

Figure 8. (a) The elementary area and integration boundaries for determining the resistance moment; (b) the reduction in the slope resistance to the output shaft of the platform drivetrain.

The value of the pressure acting on the spacer ring is calculated:

$$p = \frac{R_{az}}{S} = \frac{R_{az}}{\pi \cdot (R^2 - r^2)}.$$ (16)

Substituting the Equations (20) and (21) into the Equation (18) we obtain:

$$M_{a2} = f_k \cdot \int_r^R x \cdot \frac{R_{az}}{\pi \cdot (R^2 - r^2)} \cdot dA,$$ (17)

where the elementary surface dA is shown in Figure 8 and it is given by the formulation:

$$dA = 2 \cdot \pi \cdot x \cdot dxM.$$ (18)

By application of the Equation (16) into the Equation (14) we obtain:

$$M_{a2} = f_k \cdot \int_r^R x \cdot \frac{R_{az}}{\pi \cdot (R^2 - r^2)} \cdot 2 \cdot \pi \cdot x \cdot dx,$$ (19)

$$M_{a2} = f_k \cdot \int_r^R x^2 \cdot \frac{2 \cdot R_{az}}{(R^2 - r^2)} \cdot dx = f_k \cdot \frac{2 \cdot R_{az}}{(R^2 - r^2)} \cdot \left[\frac{x^3}{3}\right]_r^R,$$ (20)

$$M_{a2} = f_k \cdot \frac{2 \cdot R_{az}}{(R^2 - r^2)} \cdot \frac{R^3 - r^3}{3}.$$ (21)

As we can see in Figure 9, the bottom trolley transmits the force reactions R_{bn}, R_{bz} and R_{bt}. The first reaction R_{bn} is transmitted by means of guiding wheels, which are mounted on rolling bearings.

The force R_{bz} is transmitted to the platform mechanism by means of side rollers, which are mounted on steel pins. The last and at the same time the greatest R_{bt} value is caused due to the slope resistance. The total resistance of the upper trolley wheels is given:

$$M_{b1} = R_{bn} \cdot e_{b1} + 4 \cdot M_{BR}. \tag{22}$$

where M_{BR} (N·m) is the friction moment in a bearing and its value is calculated by means of SKF Bearing Calculator [29]. Substituting appropriate relations into the Equation (17) we obtain the following relation:

$$M_{b1} = \left[(G_1 + G_2) \cdot \left(\cos \alpha - \sin \alpha \cdot \frac{32}{t} \right) + G_2 \cdot \frac{x}{t} \right] \cdot e_{b1} + 4 \cdot M_{BR}. \tag{23}$$

Figure 9. Transmission of reaction forces on the bottom trolley of the platform.

The total resistance of the rollers of the bottom trolley is:

$$M_{b2} = R_{bz} \cdot (e_{b2} + r_{cb2} \cdot f_{cb2}), \tag{24}$$

$$M_{b2} = \frac{G_1 \cdot g_1 + G_2 \cdot g_2}{594} \cdot (e_{b2} + r_{cb2} \cdot f_{cb2}). \tag{25}$$

and the slope resistance is:

$$M_S = \frac{d}{2} \cdot R_{bt}, \tag{26}$$

$$M_S = \frac{d}{2} \cdot (G_1 + G_2) \cdot \sin \alpha. \tag{27}$$

where d (mm) is the thread effective diameter of the pinion (Figure 8b). All resistance effects are reduced on this diameter.

Using Equations (16)–(33) and values listed in Tables 3 and 4, we obtain:

$$M_{a1} = R_{an} \cdot (e_{a1} + r_{ra1} \cdot f_{ra1}) = 437.54 \cdot (0.00004 + 7.5 \cdot 0.00005) = 0.182 \, \text{N} \cdot \text{m}, \tag{28}$$

$$M_{a2} = f_k \cdot \frac{2 \cdot R_{az}}{R^2 - r^2} \cdot \frac{R^3 - r^3}{3} = 0.1 \cdot \frac{2 \cdot 3483.74}{13^2 - 7^2} \cdot \frac{(0.013)^3 - (0.007)^3}{3} = 3.588 \, \text{N} \cdot \text{m}, \tag{29}$$

$$M_{b1} = R_{bn} \cdot e_{b1} + 4 \cdot M_{BR} = 3474.98 \cdot 0.00004 + 4 \cdot 69.7 = 0.418 \, \text{N} \cdot \text{m}, \tag{30}$$

$$M_{b2} = R_{bz} \cdot (e_{b1} + r_{rb2} \cdot f_{rb2}) = 3483.74 \cdot (0.00003 + 0.004 \cdot 0.05) = 0.800 \, \text{N} \cdot \text{m}, \tag{31}$$

$$M_s = \frac{d}{2} \cdot R_{bt} = \frac{0.064}{2} \cdot 3257.26 = 104.23 \, \text{N} \cdot \text{m}, \tag{32}$$

$$\sum_i M_{oi} = M_{a1} + M_{a2} + M_{b1} + M_{b2} + M_s =$$
$$= 0.182 + 3.588 + 0.418 + 0.800 + 104.23 = 109.22 \ \text{N} \cdot \text{m} \tag{33}$$

Table 4. Input parameters for calculation of resistance forces and moments.

Parameter	Value	Parameter	Value
f_{p1}	0.05	n_m	46.666 s^{-1}
f_{p2}	0.05	g	9.81 m·s^{-2}
r_{p1}	0.0075 m	α	max. 47°
r_{p2}	0.004 m	v	0.1 m·s^{-1}
e_{a1}	0.00004 m	i	100
e_{b1}	0.00004 m	m	0.004 m
e_{b2}	0.00003 m	z	16
η_1	0.65	M_{BR}	69.7 N·m
η_2	0.98	R	0.013 m
η_3	0.99	r	0.007 m
m_1	154 kg	d	0.064 m
m_2	300 kg	ω	2.932 rad·s^{-1}

The electromotor of the platform, which serves as a drive source, must evolve the torque M_h (N·m) to be greater than the sum of all resistance torques (Equation (34)). We consider the angular velocity of the drive shaft $\omega = 2.932$ rad·s^{-1}, which ensures required translational velocity of the platform (Table 4). Hence, the torque of the electromotor on the output shaft is given by the following formulation:

$$P_v = M_h \cdot \omega = 109.22 \cdot 2.932 = 320.23 \ W. \tag{34}$$

Considering the total drivetrain efficiency η_c, the total power of the electromotor is:

$$P_m = \frac{P_v}{\eta_c} = \frac{320.23}{0.98 \cdot 0.65 \cdot 0.99} = 507.79 \ W. \tag{35}$$

The input parameters for Equations (12)–(27) are listed in Table 4, where:

f_{p1}—coefficient of friction of the upper trolley wheels;

f_{p2}—coefficient of friction of the bottom trolley wheels;

r_{p1}—radius of the pinion of the upper guidance;

r_{p2}—radius of the pinion of the bottom guidance;

e_{a1}—forward moving distance of the upper guidance wheels;

e_{b1}—forward moving distance of the bottom guidance wheels;

e_{b2}—forward moving distance of the side guiding rollers of the bottom guidance;

η_1—gearbox efficiency;

η_2—bearing efficiency;

η_3—efficiency of the gear rack and the pinion;

m_1—weight of the platform;

m_2—weight of the load;

n_m—nominal rpm under the load;

g—gravitational acceleration;

α—slope angle;

v—operational speed of the platform;

i—gear ratio;

m—module of the gear rack and the pinion;

z—number of pinion teeth;

M_{BR}—friction moment of the bearing;
R—outer radius of the annulus area of the bearing;
r—inner radius of the annulus area of the bearing;
d—reference diameter of the pinion;
ω—angular velocity of the pinion.

Table 5 contains the moments of resistance forces and the corresponding powers of the platform electromotor, which are determined for all calculated loads. The designations mean the following:

M_{a1}—friction moment of the upper trolley wheels;
M_{a2}—friction moment of an axial pin of the upper trolley;
M_{b1}—friction moment of the bottom trolley wheels;
M_{b2}—friction moment of the side guiding rollers of the bottom platform guidance;
M_S—resistance moment due to the track inclination;
P_v—electromotor power without mechanical losses;
P_m—electromotor power including mechanical losses.

Table 5. Calculated moments for individual loads of the platform and corresponding powers.

Parameter	225 kg	300 kg	375 kg
M_{a1} (N·m)	0.128	0.182	0.235
M_{a2} (N·m)	2.870	3.588	4.307
M_{b1} (N·m)	0.352	0.418	0.485
M_{b2} (N·m)	0.641	0.801	0.962
M_s (N·m)	87.013	104.232	121.451
$\sum_i M_{oi}$ (N·m)	91	109	127
P_v (W)	267	320	374
P_m (W)	423	508	593

Using the described calculation, we detected that the electromotor with the power of $P_m = 500$ W and with the torque overload coefficient of $\xi = 3.3$ is the sufficient solution for the created platform. The dependence of resistance moments and the electromotor power on the load is shown in Figure 10. It indicates values of the electromotor power needed for overcoming resistance moments resulting from the load.

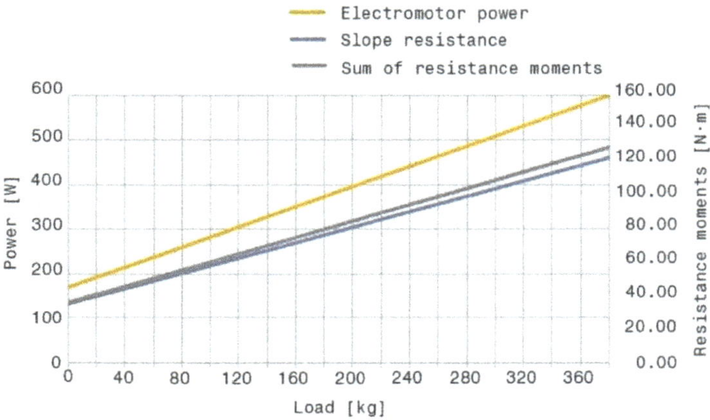

Figure 10. Dependence of resistance moments and the electromotor power on the load.

Figure 11 depicts the dependence of the electromotor power on the slope angle at constant loads, namely at 225 kg, 300 kg and 375 kg. We can see that the proposed electromotor is a suitable drivetrain for the designed platform (Figure 1) for the real slope angle of $\alpha = 30°$ (Figure 3).

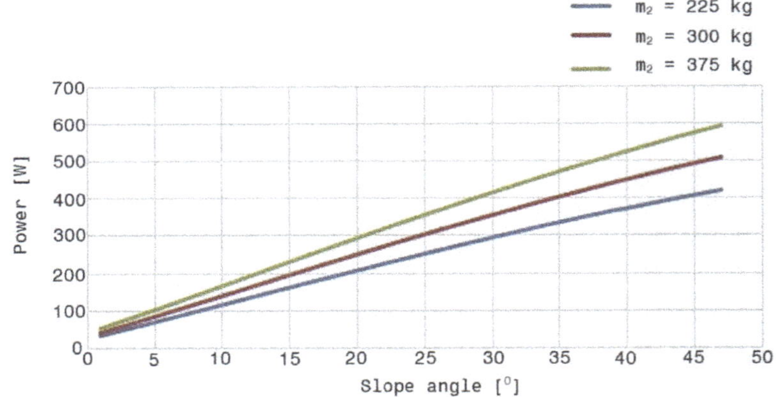

Figure 11. The dependence of the track slope angle on the electromotor power under individual platform loads m_2.

4. Determination of the Most Loaded Component of the Platform

The output shaft of the gearbox is the most loaded component. The drive-shaft of a drive-train mechanism is often the most loaded part, which is imposed not only by mechanical tension resulting from the loads, but also by dynamic loads and vibration occurring during its operation [30,31]. The calculation of the electromotor has proven that the electromotor can be used even for greater loads, but it is necessary to analyse the output shaft (Figures 8b and 12).

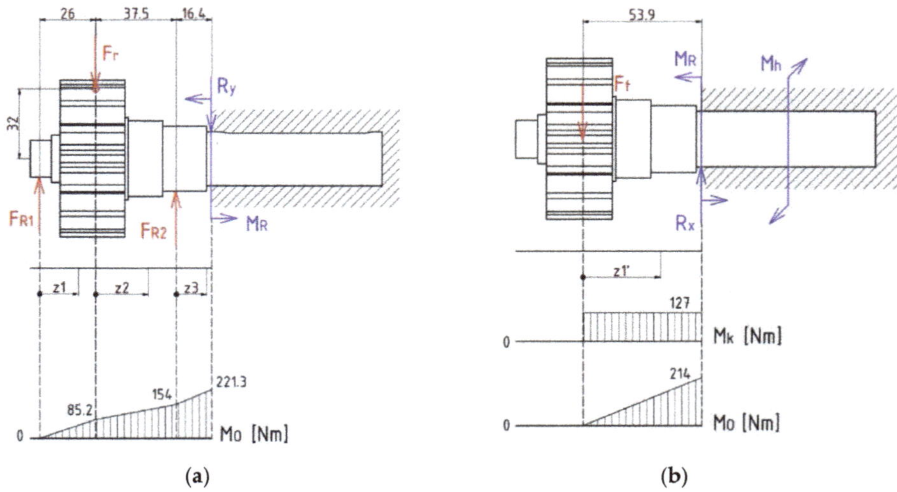

(a)   (b)

Figure 12. Scheme and distribution of the load of the output shaft: (**a**) z-y plane; (**b**) z-x plane.

Individual loading forces are given by the following formulations:

$$F_t = \frac{M_o}{\frac{d}{2}},$$

(36)

$$F_r = F_t \cdot \tan 20°, \tag{37}$$

$$\sum_i M_{iFR1} = 0 \rightarrow F_{R2} \cdot 63.5 - (R_{bn} + F_r) \cdot 26 = 0, \tag{38}$$

$$\sum_i F_{iy} = 0 \rightarrow F_{R1} + F_{R2} - R_{bn} - F_r = 0, \tag{39}$$

$$\sum_i M_{iORy} = 0 \rightarrow M_R - F_{R2} \cdot 16.4 - F_{R1} \cdot 79.9 + F_r \cdot 53.9 = 0, \tag{40}$$

$$M_{O1} = F_{R1} \cdot z_1, \tag{41}$$

$$M_{O2} = F_{R1} \cdot (26 + z_2) - F_r \cdot z_2, \tag{42}$$

$$M_{O3} = F_{R1} \cdot (63.5 + z_3) - F_r \cdot (37.5 + z_2) + F_{R2} \cdot z_3. \tag{43}$$

We can assume that the maximal value of stress will be in the considered coupling, because the maximal value of the bending moment is at the shaft constraint end (diameter is d_1 = 25 mm) and other diameters of the shaft are bigger as well as the bending moments in these locations being less. The corresponding bending modulus of section is:

$$W_O = \frac{\pi \cdot d_1^2}{32}. \tag{44}$$

The maximal stress in the z-y plane is given as the following:

$$\sigma_{O1} = \frac{M_o}{W_o}. \tag{45}$$

The individual forces depicted in Figure 12 are the following: forces F_{R1} and F_{R2} are forces in rolling bearings, which transmit loads from the platform to a bottom trolley. The end of the shaft is mounted rigidly in a gearbox, which causes reaction forces R_x, R_y and reaction torque M_R. The drive force is decomposed in the contact to a radial component of a drive force F_r and a tangential component. Forces depicted in Figure 12 always act in the radial direction regarding the shaft axis. Furthermore, Figure 12 shows the tangential component of the drive force F_t, the drive torque M_h (in the scheme, it is considered as the reaction torque due to the F_t force), the reaction force R_x and the reaction torque M_R. The torque M_h is in the equilibrium with the torque due to the force F_t on the radius of the pitch circle of the pinion. Thus, the force F_t loads the shaft by torsion and by bending. Figure 12 represents the static strength analysis in two planes, namely in the y-z plane (Figure 12a) and in the x-z plane (Figure 12b).

Substituting particular values into Equations (36)–(45) we obtain the maximal value of the bending stress of σ_{o1} = 144.3 MPa. Similar equations are also valid for Figure 12b. The analytical calculation gives the maximal value of the bending stress of σ_{o2} = 139.5 Mpa and the maximal value of the torsion stress of τ_k = 41.4 Mpa. The resulting bending stress in two perpendicular planes is determined by the following equation:

$$\sigma_O = \sqrt{\sigma_{o1}^2 + \sigma_{o2}^2}. \tag{46}$$

When we substitute values to the Equation (45), the resulting value of the bending stress is σ_o = 200.7 Mpa. The calculation of the reduced stress according to the von-Misses theory is:

$$\sigma_{RED} = \sqrt{\sigma_o^2 + 3 \cdot \tau^2} = \sqrt{(200.7)^2 + 3 \cdot (41.4)^2} = 213.3 \text{ MPa}. \tag{47}$$

Numerical calculations of the reduced stress in the analysed shaft were performed using the Inventor Nastran software package (Autodesk Inc., San Rafael, CA, USA), which is based on the finite element method [32–35]. The results are shown in Figures 13 and 14. Both analyses reached very similar results. The material of the shaft with the pinion is steel marked 42CrMo4. The yield strength of of the material is of $R_e = 650$ Mpa. It means that the reached shaft safety coefficient is $k = 2.84$. It is a sufficient value even for the maximal load of the platform of 375 kg.

(a)   (b)

Figure 13. (**a**) Definition of the boundary conditions for calculation; (**b**) the shaft mesh model.

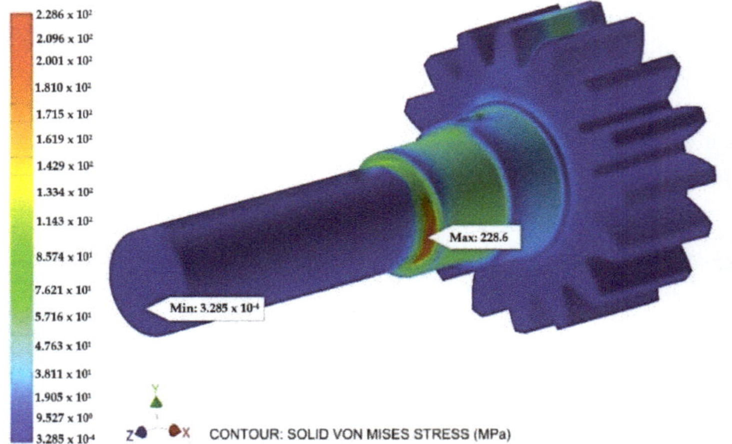

Figure 14. Distribution of the stress in the analysed shaft for the maximal considered load.

5. A Mathematical Model of a Platform Motion

This section contains the mathematical model of the platform motion during its actuation. The aim of this model is to analyse the waveform of the main acting moments in the mechanism. The mathematical model is described by an equation of motion derived by means of Lagrange's equations of the second kind, which the general form is known as the following:

$$\frac{d}{dt}\frac{\partial E_k}{\partial \dot{q}_j} - \frac{\partial E_k}{\partial q_j} + \frac{\partial E_D}{\partial \dot{q}_j} + \frac{\partial E_P}{\partial q_j} = Q_j, j = 1, 2, \ldots, n, \tag{48}$$

where E_k is the kinetic energy of a mechanical system, E_D is the Rayleigh dissipative function, E_P is the potential energy of a mechanical system, Q_j is the vector of external loads and q_j are generalised coordinates of the mechanical system.

We apply this method for our solved mechanism based on the scheme in Figure 15. We can consider the simplified mechanical model of the platform, which consists of rigid bodies. As is identified in the previous chapter, the drive shaft of the gearbox is the most loaded component. Figure 15 depicted the entire platform mechanism as well as the detail of the driving system.

Figure 15. Indication of parameters for derivation of the equation of motion of the platform motion.

We consider that the simplified mechanical system has one degree of freedom, i.e., the rotation of the drive-shaft marked as φ. Moreover, we can consider, that other bodies of the platform perform translation movement in the x-axis direction. We need to substitute into Lagrange's equation of the second kind their first time derivation, which represent velocities as the following:

- The rotational velocity of the drive-shaft is $\dot{\varphi}$;
- The translational velocity of other components is $\dot{x}$.

Other rotating components of the platform have a negligible effect on the dynamic properties of the platform in terms of the investigated problem, and therefore their moments of inertia are much smaller in comparison with the moment of inertia of the drive-shaft.

Based on the described assumptions, the mechanical system of the platform is described by means of one equation of motion. For its derivation, we need to identify the individual energies of the mechanical system (Equation (48)), which are as follows:

- The kinetic energy E_k:

$$E_k = \frac{1}{2} \cdot I_S \cdot \dot{\varphi}^2 + \frac{1}{2} \cdot m_2 \cdot \dot{x}^2, \tag{49}$$

where I_S is the moment of inertia of the drive-shaft and m_2 is total weight of the platform.

As we consider that the motion of the mechanical system of the platform will be described only by the one coordinate, namely the x coordinate, we must express the angle φ by a constraint equation. The platform is driven by the gearing of the drive-shaft without slipping. Thus, the translational velocity of the platform $\dot{x}$ equals the peripheral velocity $\dot{\varphi}$ of the drive-shaft on a pitch circle with the radius R_G. The same relation is valid between translation motion of the platform x and angular deviation φ of the drive-shaft. Then, the constraint equation is:

$$x = R_G \cdot \varphi. \tag{50}$$

The final form of the kinetic energy is:

$$E_k = \frac{1}{2} \cdot \left(I_S + m_2 \cdot R_G^2\right) \cdot \dot{\varphi}^2. \tag{51}$$

- The potential energy E_p:

$$E_P = m_2 \cdot g \cdot h, \tag{52}$$

where g is the gravitational acceleration and h is the total height which the platform must overcome during operation (Figure 1). Again, we must express the potential energy by means of the φ coordinate. When we take into account the inclination angle α, we obtain the equation:

$$h = x \cdot \sin \alpha \tag{53}$$

and, considering the Equation (50), we obtain the final form of the potential energy as the following:

$$E_P = m_2 \cdot g \cdot \varphi \cdot R_G \cdot \sin \alpha. \tag{54}$$

- The Rayleigh dissipative function E_D:

$$E_D = \frac{1}{2} \cdot b \cdot \dot{\varphi}^2, \tag{55}$$

where b is the coefficient, which includes energy losses in the mechanical system. In this case, it represents mainly friction losses in the mechanism which we can also call passive resistances.

Now, we derivate all energies in accordance with the Equation (41) and we obtain the final form of the equation of motion:

$$\left(m_2 \cdot R^2 + I_S\right) \cdot \ddot{\varphi} + b \cdot \dot{\varphi} = M_h - m_2 \cdot g \cdot R_G \cdot \sin \alpha, \tag{56}$$

where M_h is the drive torque.

Solving of the derived equation of motion of the platform mechanism is performed using the Matlab 2017a program (MathWorks Inc., Natick, MA, USA).

The driving moment is the input to the solution by means of an input file, which contains characteristics of the driving electromotor. For calculation we consider the initial conditions as the following:

- The platform starts from rest,
- For time $t = 0$ s, the initial angular velocity $\dot{\varphi} = 0$ rad $\cdot$ s^{-1}, the initial angular deviation $\varphi = 0$ rad,
- Time interval for calculation was set for 0.5 s.

Figures 16–18 show waveforms of the driving torque of the electromotor, the torque of passive resistances and the torque of slope resistance for weight m_2 of 225 kg, 300 kg and 375 kg, respectively.

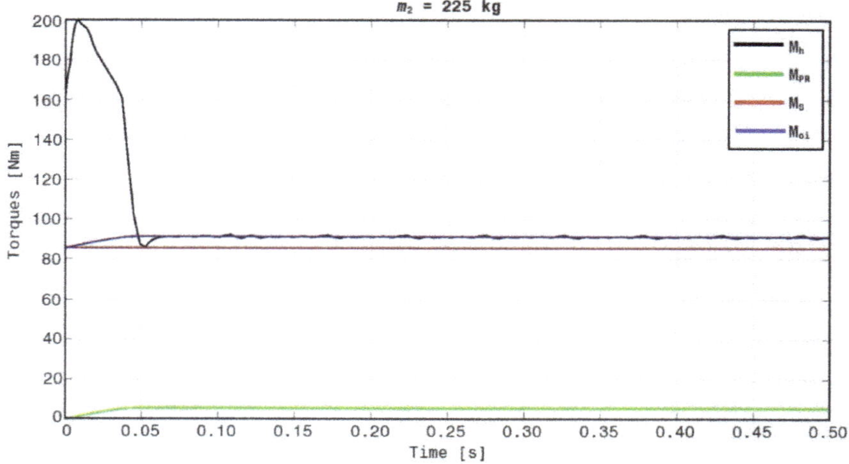

Figure 16. Waveforms of investigated torques for the weight $m_2 = 225$ kg.

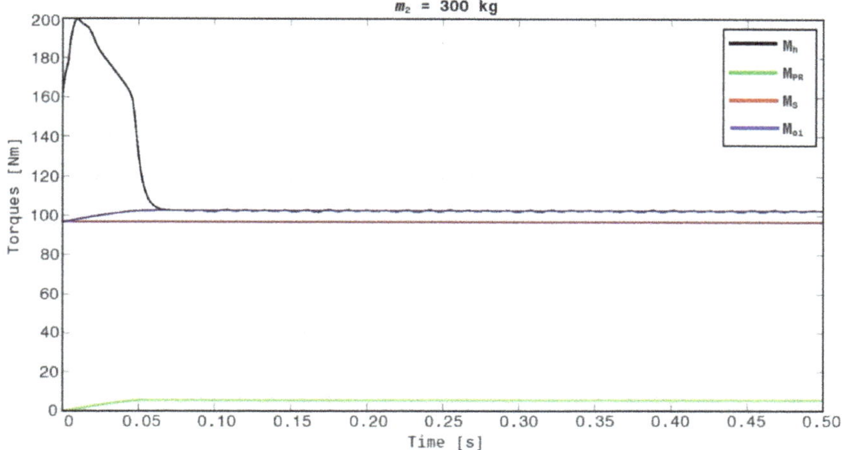

Figure 17. Waveforms of investigated torques for the weight $m_2 = 300$ kg.

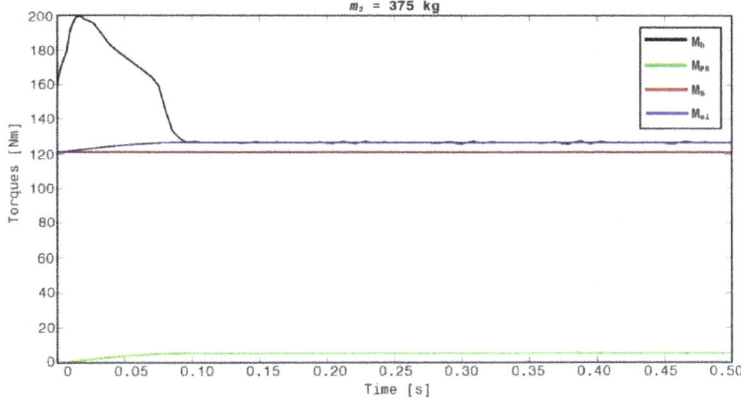

Figure 18. Waveforms of investigated torques for the weight $m_2 = 375$ kg.

These figures show that waveforms of the investigated torques correspond to values introduced in Table 5. Individual torques are marked as the following: M_h is the drive torque of the electromotor, M_{PR} is the torque of the passive resistance, M_S is the torque of slope resistance and M_{oi} is the total resistance torque, i.e., $M_{oi} = M_S + M_{PR}$. We can observe that the start of the platform motion is accompanied with increasing drive torque of the electromotor. Then, the drive torque gradually decreases, while the equilibrium of the total resistance torque and the drive torque is reached. Further waveforms depict the steady state of the platform motion.

It is obvious from the figures, that achieving equilibrium of the drive torque and the torque of total resistance depends on the weight. This weight causes directly proportional inertial effects of the mechanism. The change of the weight from 225 kg to 375 kg increases the achieving the steady state approx. twice.

Figure 19 shows acceleration of the platform for its starting. We can see that acceleration of the platform reaches high values. This phenomenon is caused due to the fact that we do not know the exact characteristics of the control system used for controlling the electromotor. Finally, it should be mentioned, that the wavy waveforms of the drive torques in Figures 15–17 and acceleration in Figure 18 are caused due to set errors in the integration method used for calculation of the mathematical model as well as due to dynamic effects of the platform motion.

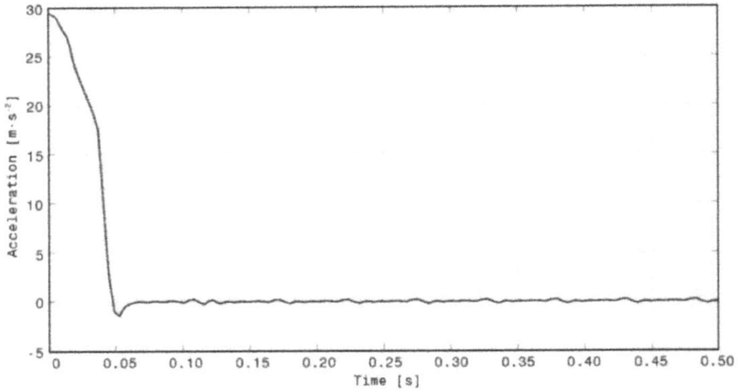

Figure 19. Waveform of acceleration of the platform mechanism for time interval of 0.0–0.5 s.

Eigenfrequencies of the Drive Shaft

Other dynamic analyses were aimed at the identification of a basic dynamic characteristic of the drive shaft, namely at finding its angular eigenfrequencies Ω_i, or eigenfrequencies f_i. The observed eigenfrequency is compared with the angular excitation frequency ω during operation. Their similar values could lead to undesirable excessive vibrations of the mechanism.

The equation of motion for free oscillation of the drive shaft is:

$$I_S \cdot \ddot{\varphi} + k_S \cdot \varphi = 0, \tag{57}$$

where I_S ($I_S = 0.462$ kg·m^2) is the moment of inertia of the shaft and k_S ($k_S = 360{,}430$ N·m·rad^{-1}) is its torsion stiffness. This form of the equation of motion (Equation (57)) and basic characteristics of the shaft are valid when we consider the drive shaft as a flexible body defined by its torsion stiffness. When we rewrite Equation (57) to the form:

$$\ddot{\varphi} + \frac{k_S}{I_S} \cdot \varphi = 0, \tag{58}$$

and substitute the known parameters, we obtain the eigenfrequency $f_S = 104.565$ Hz.

In reality, the drive shaft is a flexible body represented as a continuum mass, in which eigenmodes can be investigated by means of, e.g., the finite element method (FEM). Then, the equation of motion for free oscillation will be in the form:

$$\mathbf{I_S} \cdot \ddot{\boldsymbol{\varphi}} + \mathbf{k_S} \cdot \boldsymbol{\varphi} = \mathbf{0}, \tag{59}$$

where $\mathbf{I_S}$ is the matric of the moments of inertia, $\mathbf{k_S}$ is the stiffness matrix, $\ddot{\boldsymbol{\varphi}}$ is the vector of angular accelerations and $\boldsymbol{\varphi}$ is the vector of angular deviations. After performing the modal analysis in the FEM software, we found out, that the first six eigenvalues are as follows: $f_{S1} = 89.298$ Hz, $f_{S2} = 96.923$ Hz, $f_{S3} = 125.863$ Hz, $f_{S4} = 160.963$ Hz, $f_{S5} = 171.385$ Hz and $f_{S6} = 223.951$ Hz.

When we compare results of eigenfrequencies for the flexible body represented just by its torsion stiffness (Equation (58)) and for the flexible body modelled as a continuum (Equation (59)), we can see that results are slightly different. It is caused because the continuum better represents the real body of the analysed drive shaft. The angular excitation frequency of the drive shaft is $\omega = 2.932$ rad·s^{-1}, or $f = 0.466$ Hz. This excitation frequency is much lower than the eigenfrequencies calculated by means of Equation (58) as well as Equation (59). We can conclude that the drive shaft will not reach to the resonance due to the driving in the operation mode.

6. Analysis of the Platform Deflection

As the STN EN 81-40 standard allows the maximal deflection (w_{max}) of the platform to be $w_{max} = 20$ mm [22], it is necessary to analyse this deflection. In the device operation, such a situation can happen that the total load of the platform will be transmitted to columns either by one tube (upper or bottom) or the load will be distributed to both tubes. For both exceptional cases, dependencies of the load were made (Figures 15 and 16). These dependencies influence the deflection of the end platform point when the distance between columns is changed.

The authors introduce an example of the calculation in the case of transmitting of loads by means of the upper guiding tube.

The load, which is caused by the platform weight and the transported goods, represents only one component of force A_{yII} (the reaction force due to the own weight load of the platform and the goods), because the axis force F_{oII} is considered in the pinion gearing (this is the drive force of the pinion and it acts in the platform motion direction), i.e., it influences the strain of the bottom guiding tube. Figure 20 shows the considered loading forces. The acting force A_{xII} is depicted in the cutting-plane *A-A*.

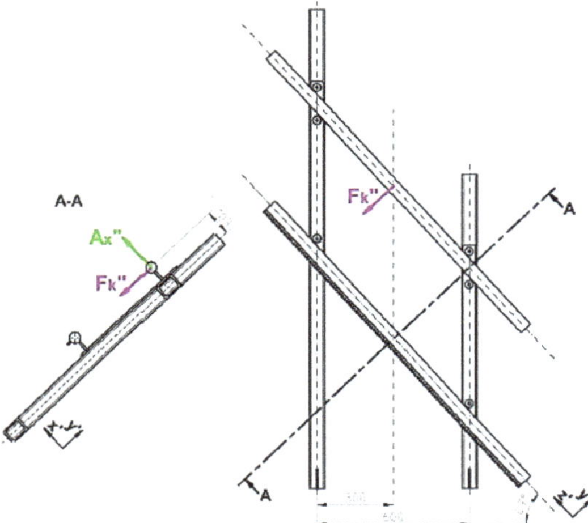

Figure 20. Waveform of acceleration of the platform mechanism for time interval of 0.0–0.5 s.

For the relation for calculation of deformation, we can use a simplified model of a beam with two supports representing columns. These supports remove three degrees of freedom. A distribution of the resulting bending moment and the lateral force regarding to the upper guiding tube is shown in Figure 21.

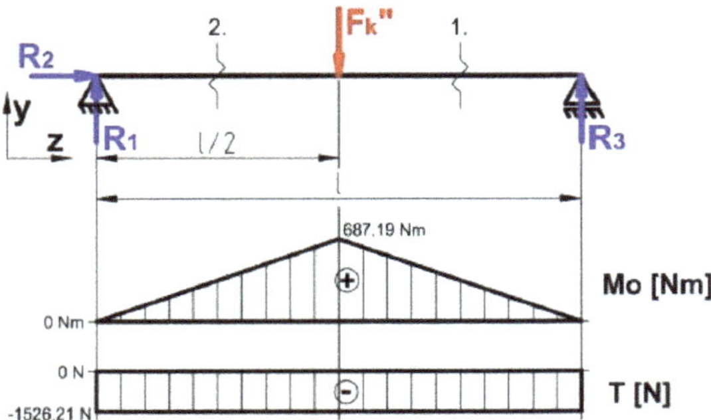

Figure 21. A simplified model of a beam and the distribution of the bending moment and the lateral force.

Forces marked as F_{kII} and A_{xII} cause deformation in two different planes. Therefore, we analyse deflections individually (we use the superposition method) and subsequently, we determine the resulting value of the deflection by means of the Pythagorean theorem as follows:

$$\sum_{i=1}^{n} F_{iy} = 0 \Rightarrow R_1 + R_3 - F_k^{II} = 0, \tag{60}$$

$$\sum_{i=1}^{n} F_{iz} = 0 \Rightarrow R_2 = 0, \tag{61}$$

$$\sum_{i=1}^{n} M_{iR_1} = 0 \Rightarrow R_3 \cdot l - F_k^{II} \cdot \frac{l}{2} = 0. \tag{62}$$

We get Equations (63)–(65) for the calculation of reaction force from Equations (60)–(62), respectively:

$$R_3 = \frac{F_k^{II}}{2}, \tag{63}$$

$$R_2 = 0, \tag{64}$$

$$R_1 = F_k^{II} - R_3 = F_k^{II} - \frac{F_k^{II}}{2} = \frac{F_k^{II}}{2}. \tag{65}$$

Calculation of deformation and lateral forces is performed applying the Schwedler's theorem, which comes from Figure 22.

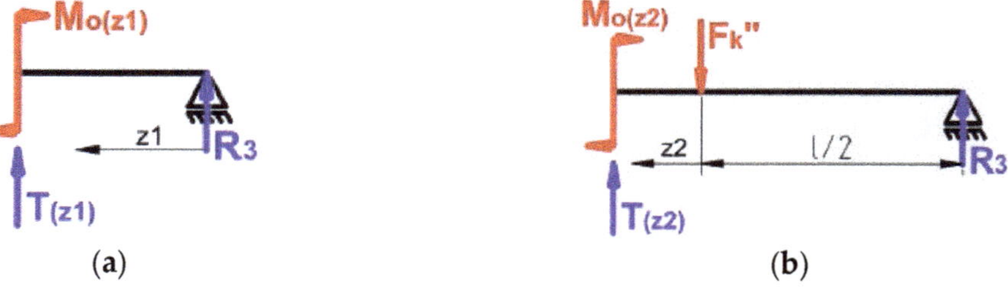

Figure 22. Application of the Schwedler's theorem: (**a**) a first cut of the upper guiding tube, (**b**) a second cut of the upper guiding tube.

Then, we proceed as follows:

- The first section:

$$z_1 \in \left\langle 0; \frac{l}{2} \right\rangle. \tag{66}$$

The bending moment $M_{O(z_1)}$ is:

$$M_{O(z_1)} = R_3 \cdot z_1 = \frac{F_k^{II} \cdot z_1}{2} \tag{67}$$

and the lateral force $T_{(z_1)}$ is calculated as:

$$T_{(z_1)} = -\frac{d\left(M_{O(z_1)}\right)}{d(z_1)} = -\frac{d\left(\frac{F_k^{II} \cdot z_1}{2}\right)}{d(z_1)} = -\frac{F_k^{II}}{2}. \tag{68}$$

- The second section:

$$z_2 \in \left\langle 0; \frac{l}{2} \right\rangle. \tag{69}$$

The bending moment $M_{O(z_2)}$ is:

$$M_{O(z_2)} = R_3 \cdot \left(\frac{l}{2} + z_2 \right) - F_k^{II} \cdot z_2 = \frac{F_k^{II} \cdot l}{4} - \frac{F_k^{II} \cdot z_2}{2} \tag{70}$$

and the lateral force $T_{(z_2)}$ is calculated as:

$$T_{(z_2)} = -\frac{d\left(M_{O(z_2)} \right)}{d(z_2)} = -\frac{d\left(\frac{F_k^{II} \cdot l}{4} - \frac{F_k^{II} \cdot z_2}{2} \right)}{d(z_2)} = -\frac{F_k^{II}}{2}. \tag{71}$$

The entire calculation is performed by means of the differential equation of the elastic curve as follows:

$$w = \int_0^{z_1} \frac{M_{O(z_1)}}{E \cdot J_x} \cdot \frac{\partial M_{O(z_1)}}{\partial F_{kII}} \cdot dz_1 + \int_0^{z_2} \frac{M_{O(z_2)}}{E \cdot J_x} \cdot \frac{\partial M_{O(z_2)}}{\partial F_{kII}} \cdot dz_2. \tag{72}$$

After integration and substituting corresponding parameters, the differential equation of the elastic curve is:

$$w = \frac{1}{E \cdot J_x} \cdot \left[\int_0^{\frac{l}{2}} \frac{F_{kII} \cdot z_1}{2} \cdot \frac{\partial \frac{F_{kII} \cdot z_1}{2}}{\partial F_{kII}} \cdot dz_1 + \int_0^{\frac{l}{2}} \left(\frac{F_{kII} \cdot l}{4} - \frac{F_{kII} \cdot z_2}{2} \right) \cdot \frac{\partial \left(\frac{F_{kII} \cdot l}{4} - \frac{F_{kII} \cdot z_2}{2} \right)}{\partial F_{kII}} \cdot dz_2 \right]. \tag{73}$$

where z_1 and z_2 are sections of the analysed track and F_{kII} is the force acting on the track in the analysed sections. The complete calculation is very extensive; therefore, it cannot be completely described.

Comparing results shown in Figures 23 and 24, we can see that both load cases are very similar even with the same values of the deflection. Therefore, we assume that the load transmission by this tube does not have the significant effect on the limit value of the platform deflection.

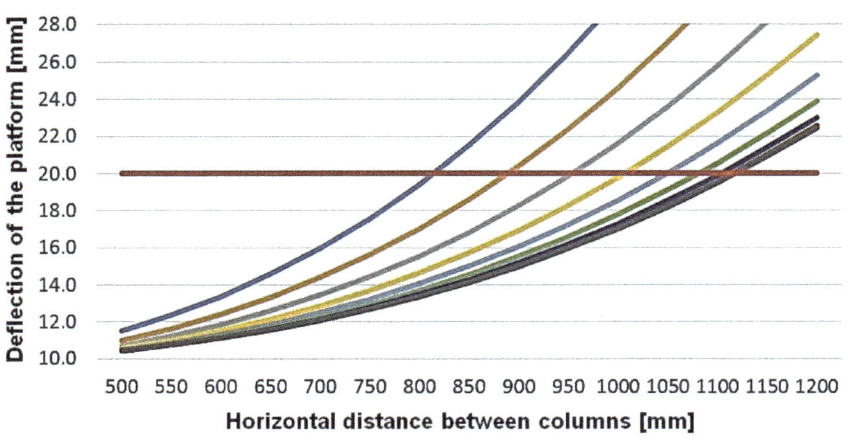

Figure 23. Dependence of the platform deflection on the horizontal distance between columns, when the load is transmitted by the upper tube. Results are shown for different track slope from $0°$ to $47°$.

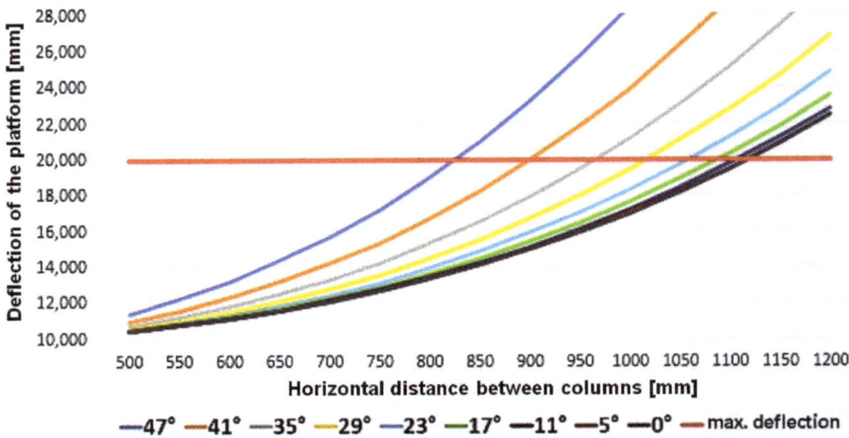

Figure 24. Dependence of the platform deflection on the horizontal distance between columns, when the load is transmitted by the bottom tube. Results are shown for different track slope from 0° to 47°.

The analysis of the load distribution state goes before the calculation of the guiding tube. The allowed overload of the platform with the weight of 375 kg is only in the location of the first column, i.e., in the location where the vector straight line of the gravitational force and the main central axis of the column are in the same plane. This plane is also perpendicular to the platform movement. Hence, it is not appropriate to analyse the tube in this location, because the tube is mounted just in this location. Suitably, the platform is imagined to move to the geometric centre of the selected computational track section, i.e., between two columns considering the load of 300 kg. In case of a greater load, the platform must not move; therefore, the situation when the platform with the load of 375 kg is in the vertical symmetry axis, cannot happen.

7. Conclusions

The article presents a partial solution to the conception of the design of a manipulator for handling with goods in particular difficult and limited operation conditions. It is considered to transport material between two destinations.

While the problem is extensive, the partial objective can be considered to have been met. The engineering design presented in this article fully complies with the requirements that will be imposed on it in operation. Analytical and numerical calculations implemented to the solution process completely support this claim.

The need of application of the manipulator in the production process of the company only demonstrates the necessity of mechanisation and automation technology and its invaluable importance in all areas.

The further solution of the device will include the design and analysis of the track and creation of a mathematical model for verifying the safety of the track in collaboration with the designed platform as well as drawing documentation, production and installation of the platform to the real operation.

Author Contributions: Conceptualization, M.B. and E.K.; methodology, J.D., M.B., M.S. and J.G.; software, J.D. and E.K.; validation, M.S., M.B. and J.G., formal analysis, J.D., M.B. and E.K.; investigation, M.B. and E.K.; J.D. and J.G.; writing—original draft preparation, J.D., M.B. and E.K.; writing—review and editing, M.S. and J.G.; visualization, M.B. and E.K.; supervision, M.S. and J.G.; funding acquisition, M.S. and J.G. All authors have read and agreed to the published version of the manuscript.

Funding: This work has been supported by grant agency grant agency KEGA project No. 001ŽU-4/2020.

Conflicts of Interest: The authors declare no conflict of interest.

References

1. Kostrzewski, M. Securing of Safety by Monitoring of Technical Parameters in Warehouse Racks, in High-Bay Warehouses and High Storage Warehouses—Literature Review of the Problem. *Logforum* **2017**, *13*, 125–134. [CrossRef]
2. Kostrzewski, M. Loads Analyzing in Pallet Racks Storage Elevation. In Proceedings of the Carpathian Logistics Congress CLC 2013, Cracow, Poland, 9–11 December 2013.
3. Fujita, M.; Domae, Y.; Noda, A.; Ricardez, G.A.G.; Nagatani, T.; Zeng, A.; Song, S.; Rodriguez, A.; Cause, A.; Chen, I.M.; et al. What Are the Important Technologies for Bin Picking? Technology Analysis of Robots in Competitions Based on a Set of Performance Metrics. *Adv. Robot.* **2019**, *34*, 560–574. [CrossRef]
4. Minkevicius, S.; Steisunas, S. About the Sojourn Time Process in Multiphase Queueing Systems. *Methodol. Comput. Appl. Probab.* **2006**, *8*, 293–302. [CrossRef]
5. Minkevicius, S.; Steisunas, S. Laws of Little in an Open Queueing Network. *Nonlinear Anal. Model. Control.* **2012**, *17*, 327–342. [CrossRef]
6. Leitner, B.; Dvorak, Z. Special Railway Crane PKP 25/20i-Dynamics Loads and a Fatigue Life Prediction of Its Load-Bearing Structure. In Proceedings of the 17th International Conference on Transport Means 2013, Kaunas, Lithuania, 24–25 October 2013.
7. Corbato, C.H.; Bharatheesha, M.; Van Egmond, J.; Ju, J.; Wisse, M. Integrating Different Levels of Automation: LESSONS from Winning the Amazon Robotics Challenge 2016. *IEEE Trans. Ind. Inform.* **2018**, *14*, 4916–4926. [CrossRef]
8. Benali, K.; Brethe, J.F.; Guerin, F.; Gorka, M. Dual Arm Robot Manipulator for Grasping Boxes of Different Dimensions in a Logistics Warehouse. In Proceedings of the 19th IEEE International Conference on Industrial Technologies ICIT 2018, Lyon, France, 19–22 February 2018.
9. Kelemen, M.; Virgala, I.; Liptak, T.; Mikova, L.; Filakovsky, F.; Bulej, V. A Novel Approach for a Inverse Kinematics Solution of a Redundant Manipulator. *Appl. Sci.* **2018**, *8*, 2229. [CrossRef]
10. Poppeova, V.; Bulej, V.; Zahoransky, R.; Uricek, J. Parallel Mechanism and Its Application in Design of Machine Tool with Numerical Control. *Robot. Theory Pract.* **2013**, *282*, 74–79. [CrossRef]
11. Osipowicz, T.; Abramek, K.F.; Barta, D.; Drozdziel, P.; Lisowski, M. Analysis of Possibilities to Improve Environmental Operating Parameters of Modern Compression-Ignition Engines. *Adv. Sci. Technol.—Res. J.* **2018**, *12*, 206–213. [CrossRef]
12. Heo, M.W.; Bae, S.J.; Jeong, J.Y.; Yoo, H.S.; Park, S.U.; Heo, H.S. Effects of an Exhaust Heat Recovery System on Performance Characteristics of a Forklift Truck. *Int. J. Automot. Technol.* **2019**, *20*, 579–588. [CrossRef]
13. Caban, J.; Drozdziel, P.; Ignaciuk, P.; Kordos, P. The Impact of Changing the Fuel Dose on Chosen Parameters of the Diesel Engine Start-Up Process. *Transp. Probl.* **2019**, *14*, 51–62. [CrossRef]
14. Puskar, M.; Jahnatek, A.; Kuric, I.; Kadarova, J.; Kopas, M.; Soltesova, M. Complex Analysis Focused on Influence of Biodiesel and Its Mixture on Regulated and Unregulated Emissions of Motor Vehicles with the Aim to Protect Air Quality and Environment. *Air Qual. Atmos. Health* **2019**, *12*, 855–864. [CrossRef]
15. Barta, D.; Mruzek, M.; Kendra, M.; Kordos, P.; Krzywonos, L. Using of Non-Conventional Fuels in Hybrid Vehicle Drives. *Adv. Sci. Technol. Res. J.* **2016**, *10*, 240–247. [CrossRef]
16. Hosseinzadeh, E.; Rokni, M.; Advani, S.G.; Prasad, A.K. Performance Simulation and Analysis of a Fuel Cell/Battery Hybrid Forklift Truck. *Int. J. Hydrogen Energy* **2013**, *38*, 4241–4249. [CrossRef]
17. Vishwakarma, R.K.; Singh, A.K.; Ahirwal, B.; Kumar, A.; Kumar, N.; Singh, J.K. Development of Dust Ignition Protected Electrically Powered Forklift Truck for Combustible Dust Environment. *Int. J. Oil Gas Coal Technol.* **2019**, *20*, 113–126. [CrossRef]
18. Lototskyy, M.V.; Tolj, I.; Parsons, A.; Smith, F.; Sita, C.; Linkov, V. Performance of Electric Forklift with Low-Temperature Polymer Exchange Membrane Fuel Cell Power Module and Metal Hydride Hydrogen Storage Extension Tank. *J. Power Sources* **2016**, *316*, 239–250. [CrossRef]
19. Nakshatharan, S.S.; Johanson, U.; Punning, A.; Aabloo, A. Modeling, Fabrication, and Characterization of Motion Platform Actuated by Carbon Polymer Soft Actuator. *Sens. Actuators A-Phys.* **2018**, *283*, 87–97. [CrossRef]
20. Rubio, J.D.; Garcia, E.; Pacheco, J. Trajectory Planning and Collisions Detector for Robotic Arms. *Neural Comput. Appl.* **2012**, *21*, 2105–2114. [CrossRef]

21. Fomin, O.; Akimova, A.; Akimova, J.; Mastepan, A. The Criteria Choice of Evaluating Effectiveness of the Process and Automatic Control Systems. In Proceedings of the 3rd International Conference on Traffic Engineering in Transportation and Logistics 2018, Budapest, Hungary, 8–10 August 2018.

22. Tadumadze, G.; Boysen, N.; Emde, S.; Weidinger, F. Integrated Truck and Workforce Scheduling to Accelerate the Unloading of Trucks. *Eur. J. Oper. Res.* **2019**, *278*, 343–362. [CrossRef]

23. Wang, J.Y.; Zhao, J.S.; Chu, F.L.; Feng, Z.J. Dynamic Stiffness of a Lift Mechanism for Forklift Truck. *Proc. Inst. Mech. Eng. Part C J. Mech. Eng. Sci.* **2013**, *227*, 387–402. [CrossRef]

24. Rubio, J.D.; Pacheco, J.; Perez-Cruz, J.H.; Torres, F. Mathematical Model with Sensor and Actuator for a Transelevator. *Neural Comput. Appl.* **2014**, *24*, 277–285. [CrossRef]

25. STN EN 81-40. *Safety Rules for the Construction and Installation of Lifts. Special Lifts for the Transport of Persons and Goods. Part. 40: Stairlifts and Inclined Lifting Platforms Intended for Persons with Impaired Mobility*; British Standards Institution: London, UK, 2020.

26. Walker, S.W. Felicity: A Matlab/C Plus Plus Toolbox for Developing Finite Element Methods and Simulation Modeling. *Saim J. Sci. Comput.* **2018**, *40*, C234–C257. [CrossRef]

27. Teixeira, M.C.M.; Marchesi, H.F.; Assuncao, E. Signal-Flow Graphs: Direct Method of Reduction and MATLAB Implementation. *IEEE Trans. Educ.* **2001**, *44*, 185–190. [CrossRef]

28. Hejma, P.; Svoboda, M.; Kampo, J.; Soukup, J. Analytic Analysis of a Cam Mechanism. In Proceedings of the 21st Polish-Slovak Scientific Conference Machine Modeling and Simulations MMS 2016, Hucisko, Poland, 6–8 September 2016.

29. SKF Plain Bearing Calculator. Available online: http://webtools3.skf.com/BearingCalc/ (accessed on 9 November 2020).

30. Wang, J.; Yao, Z.; Sun, R.; Haseen, M.F. Linear Feedback Control of Chaotic Motion in Gear Transmission Systems. In Proceedings of the 4th International Workshop on Advanced Algorithms and Control Engineering IWAACE 2020, Shenzhen, China, 21–23 February 2020.

31. Wang, Q.; Wu, Z.; Liu, L. Uncertainty and Dependence Analysis of Performance Limit State for Structural Multidimensional Fragility Evaluation. *KSCE J. Civ. Eng.* **2017**, *21*, 1386–1392. [CrossRef]

32. Korayem, M.H.; Rahimi, H.N.; Nikoobin, A.; Nazemizadeh, M. Maximum Allowable Dynamic Payload for Flexible Mobile Robotic Manipulators. *Lat. Am. Appl. Res.* **2013**, *43*, 29–35.

33. Klimchik, A.; Pashkevich, A.; Chablat, D. CAD-Based Approach for Identification of Elasto-Static Parameters of Robotic Manipulators. *Finite Elem. Anal. Des.* **2013**, *75*, 19–30. [CrossRef]

34. Lopez-Campos, J.A.; Baldonedo, J.; Suarez, S.; Segade, A.; Casarejos, E.; Fernandez, J.R. Finite Element Validation of an Energy Attenuator for the Design of a Formula Student. *Mathematics* **2020**, *8*, 416. [CrossRef]

35. Kovacikova, P.; Bezdedova, R.; Vavro, J., Jr.; Vavro, J. Comparison of Numerical Analysis of Stress-Strain States of Cast Iron with Vermicular Graphite Shape and Globular Graphite Shape. In Proceedings of the 20th International Slovak-Polish Conference on Machine Modelling and Simulations MMS 2016, Terchova, Slovakia, 7–9 September 2015.

 *applied sciences*

Article

Grasp Planning Pipeline for Robust Manipulation of 3D Deformable Objects with Industrial Robotic Hand + Arm Systems

Lazher Zaidi [1,*]**, Juan Antonio Corrales Ramon** [2]**, Laurent Sabourin** [2]**,
Belhassen Chedli Bouzgarrou** [2] **and Youcef Mezouar** [2]

[1]   LINEACT Laboratory, EA 7527, CESI ROUEN, 76800 Saint-Étienne-du-Rouvray, France
[2]   CNRS, SIGMA Clermont, Institut Pascal, Université Clermont Auvergne, F-63000 Clermont-Ferrand, France;
     juan.corrales@sigma-clermont.fr (J.A.C.R.); Laurent.Sabourin@sigma-clermont.fr (L.S.);
     chedli.bouzgarrou@sigma-clermont.fr (B.C.B.); youcef.mezouar@sigma-clermont.fr (Y.M.)
*   Correspondence: lzaidi@cesi.fr

Received: 17 October 2020; Accepted: 30 November 2020; Published: 6 December 2020

Abstract: In the grasping and manipulation of 3D deformable objects by robotic hands, the physical contact constraints between the fingers and the object have to be considered in order to validate the robustness of the task. Nevertheless, previous works rarely establish contact interaction models based on these constraints that enable the precise control of forces and deformations during the grasping process. This paper considers all steps of the grasping process of deformable objects in order to implement a complete grasp planning pipeline by computing the initial contact points (pregrasp strategy), and later, the contact forces and local deformations of the contact regions while the fingers close over the grasped object (grasp strategy). The deformable object behavior is modeled using a nonlinear isotropic mass-spring system, which is able to produce potential deformation. By combining both models (the contact interaction and the object deformation) in a simulation process, a new grasp planning method is proposed in order to guarantee the stability of the 3D grasped deformable object. Experimental grasping experiments of several 3D deformable objects with a Barrett hand (3-fingered) and a 6-DOF industrial robotic arm are executed. Not only will the final stable grasp configuration of the hand + object system be obtained, but an arm + hand approaching strategy (pregrasp) will also be computed.

Keywords: 3D deformable object; grasping; robust manipulation; robot hand

---

1. Introduction

Nowadays, manipulation has become an increasingly important standing research topic in robotics. Most of the related works in this field consider the grasping of rigid bodies as an extensively studied area, which is rich with theoretical analysis and implementations using different robotic hands ([1–8]). Robotic grasping of deformable objects has also acquired importance recently due to several potential applications in various areas, including biomedical processing, the food processing industry, service robotics, robotized surgery, etc. ([9–14]). Recent surveys covering the literature related to the robotic grasping and manipulation of deformable objects can be found in [12,15].

Nevertheless, efficient and precise (i.e., deformation/force control) robotic grasping of 3D deformable objects remains an underdeveloped area within the robotics community due to the technical and methodological difficulties of the problem. In fact, this problem involves implementing two main steps with their own subproblems: determining the initial location of the grasp points and moving the hand + arm system towards them (i.e., pregrasping strategy), and closing the fingers of the robotic hand until a robust grasp is reached while taking into account the deformation of the object (i.e., grasping strategy).

In the case of the pregrasp strategies, two main problems arise: firstly, implementing a grasp quality measure to choose the most stable grasp configuration; secondly, applying the kinematic constraints of the hand + arm system so that the chosen grasp is reachable. Several previous surveys to quantify grasp quality were developed in [1,16–19]. All of these approaches have studied stable grasps and developed various stability criteria to find optimal grasps for rigid objects. Nevertheless, the development of algorithms that can efficiently synthesize grasps in 3D deformable objects is still a challenging research problem. Wakamatsu et al. [20] analyzed the stable grasping of deformable objects based on extending the concept of force closure of rigid objects to deformable planar objects with bounded applied forces. Mira et al. [21] presented a grasp planner that can reproduce the actions of a human hand to determine the contact points. Their algorithm combines the position information of the hand with visual and tactile data in order to achieve a hand–object configuration. This method is applied to planar objects without considering applied forces by fingers and object deformation. Xu et al. [22] proposed a grasp quality metric for hollow plastic objects based on minimal work in order to combine wrench resistance and object deformation. Jorgensen et al. [23] presented a generic solution for doing pick and place operations of meat pieces with a vacuum gripper. This solution used a grasp quality metric that combines wrench resistance and object deformation. It was based on the computation of the required work to resist an external wrench and it can only be applied to planar objects, without a direct relation between applied forces and deformation. In fact, none of these works can be applied to general 3D deformable objects and they are solutions for specific types of objects.

For the second step of the grasping problem (i.e., closing of the fingers), previous grasping synthesis methods are mainly based on the generation of force closure (FC) grasps ([24–26]). This means that the robotic hand is able to maintain the manipulated object inside the palm despite any external disturbance and evaluate the contact forces including those due to friction. FC is generally used to implement the grasping of rigid objects with a low number of frictional contacts by taking into account geometric criteria of their contact surfaces. Nevertheless, traditional FC-based approaches cannot be applied to deformable objects due the complexity of their interaction with the robotic hand's fingers: the object deforms (i.e., its size and shape are changed) and the contact surfaces evolve dynamically while contact forces are applied by the fingers. New reactive methods based on the detection and avoidance of slippage have been developed in order to solve this limitation in the grasping of deformable objects. For instance, Kaboli et al. [27] computed the tangential forces required for slippage avoidance and adjusted the fingers' relative positions and grasping forces accordingly, taking into account an estimation of the weight of the grasped object. The main drawback of this approach is that the relationship between applied forces and object deformation is not considered. Zhao et al. [28] proposed a new slip detection sensor based on video processing in order to perceive multidimensional information, including absolute slip displacement, the deformation at the object surface, and force information. This method is only applied for 2D objects with transparent and reflective surface, such as pure glass. However, these types of materials are not common in deformable objects. In addition, the stress state of the gripper–object interface is not precisely analyzed when taking into account the properties of the object's material.

In order to solve all these limitations of reactive strategies, it is crucial to use a suitable contact model ([29–32]) that explains the coupling between the contact forces and the object deformations. A precise contact model can be exploited to provide a set of grasping forces and torques that can be applied to maintain equilibrium before and after the deformation [33]. Thereby, the grasp stability can be computed in real-time while the contact surfaces evolve during deformation. This evolution of the object surface is obtained by sampling mechanical deformation models [34], mainly mass-spring systems ([35,36]), finite element methods (FEM) ([37,38]), and impulse-based methods ([39]). The mass-spring system is a fast and interactive physical model with a solid mathematical foundation and well-understood dynamics. This model provides a more realistic force/deformation behavior in the case of large deformations, while maintaining the capability of real-time response. FEM provides higher accuracy, but it requires higher numerical computation, which is more appropriate for offline simulations. Therefore, this paper uses a deformation model based on nonlinear mass-spring elements distributed across a tetrahedral mesh: lumped masses are attached to the nodes and nonlinear springs represent the edges. The tracking of the mesh nodes positions is assured by solving a dynamic equation based on Newton's second law. The main interests of this deformation model are the dynamic predictions of the object's deformations and its realistic behavior, which results from the nonlinearity of the object. More details about this object deformation model can be found in previous works [40] and it will be used by the new proposed grasp synthesis algorithm.

Several previous works have taken into account the deformation of soft objects while grasping them without integrating precise contact interaction models. Berenson et al. [41] presented a vision-based deformation controller for elastic planar objects without considering the grasp properties. This method takes into account just one contact point which cannot be used with a multifingered hand for grasping deformable objects. Similarly, Nadon et al. presented a model-free algorithm in [42] based on visual information for automatically selecting the contact points between the fingers and the object's contour in order to control its shape. However, contact forces and object stability were not considered. Grasping deformable planar objects using contact analysis was proposed in [43]. It was based on an algorithm to track the contact regions during the squeezing process and determine the stick/slip mode in the contact area. Fingers' displacements were considered rather than forces, and object stability was not guaranteed. Similarly, Sanchez et al. [44] developed a manipulation pipeline that is able to estimate and control the shape of a deformable object with tactile information but the stability of the grasp is not considered. Authors in [45] proposed a new approach to lift a deformable 3D object, causing small deformations of the object by employing two rigid fingers with contact friction. This approach considered small deformations within the scope of the linear elasticity during the grasp operation and did not deal with the nonlinear relationships between deformations and forces of 3D soft objects. Jorgensen et al. [46] presented a simulation framework for using machine learning techniques to grasp deformable objects. In this framework, robot motions were parameterized in terms of grasp points, robot trajectory, and robot speed. Hu et al. [47] presented a general approach to automatically handle soft objects using a dual-arm robot. An online Gaussian process regression model was used to estimate the deformation function of the manipulated objects and low-dimension features described the object's configuration. Finally, new works (such as [48,49]) integrate deep-learning techniques, 3D vision, and tactile information in order to fold/unfold and pick-and-place clothes. Although all these works consider the deformation of the object while manipulating it, an initial stable grasp is supposed to be known and nonlinear force–deformation relations are not computed during the handling process.

None of these works combine all the previously explained elements that are required for the implementation of a general and complete grasp planning pipeline for 3D deformable objects: a pregrasp strategy for reaching the object that takes into account grasp quality metrics for 3D deformable objects, and a grasp strategy for closing the fingers that guarantees a stable configuration of the object by considering its deformation and the applied contact forces. This paper involves an extension of our previous grasping strategy in [33] in order to implement this complete grasp planning pipeline. Firstly, we adapt the stability criteria developed for robotic grasp synthesis of rigid objects to 3D deformable objects (see Section 2 for a general description of the strategy and Section 3 for the new geometric criterion of the initial grasp). We also defined a new pregrasp strategy (Sections 4 and 5) in order to approach the deformable objects in an optimal way before manipulating them by using the previous grasping metrics. This grasping strategy is based on an interaction model developed by the authors in a previous work [40]. Section 6 indicates how this finger–object contact model is used in order to compute the contact forces to be applied by the fingers for a stable grasp of the deformable object and what their precise values are in the proposed experiments. We consider the problem of grasping and manipulation of a 3D isotropic deformable objects with a 3-fingered Barrett hand. This hand is installed as the end-effector of a 6-DOF Adept Viper robotic arm S1700D. Finally, a complete grasping planning strategy combining the pregrasp and grasp synthesis strategies of the hand + arm system is developed and tested with real objects (Section 7).

2. General Description of the Grasp Planning Pipeline

As indicated in the introduction section, by using our previously developed contact model [33], we can handle highly deformable objects and give precise estimations of the contact forces generated while deforming them. Those precise estimations would guarantee the static equilibrium of the object by considering new grasps metrics (Section 3) and optimizing the pregrasp configuration of the robotic hand around the object (Sections 4 and 5).

The flowchart in Figure 1 shows all the steps that are required in order to compute and execute a robust grasp (i.e., the configuration of fingers over the surface of the object) that guarantees the stability of the object with our grasp planning pipeline.

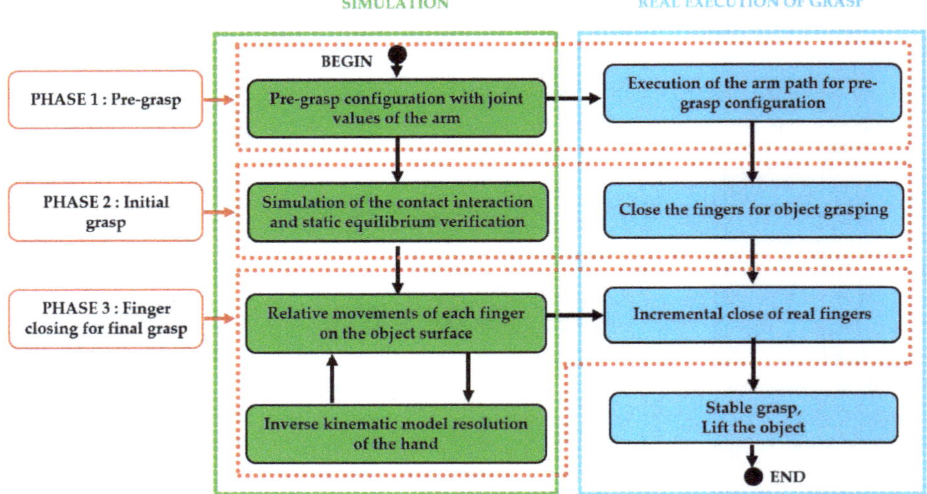

Figure 1. Flowchart of the proposed grasp planning pipeline (i.e., pregrasp and grasping strategies) for handling deformable objects with a hand + arm robot system.

In order to execute the computed grasp, the first step determines the pregrasp configuration. In fact, the complete grasp planning strategy is divided into two phases: First, determining the configuration of the robotic arm to bring the hand close to the object so that the fingers of the hand can reach the surface of the object (Section 5). The second phase is determining the appropriate initial configuration of the fingers of the hand to grasp the object based on a new geometric criterion (Sections 3 and 4). The determination of the joint parameters of the arm and the hand for executing both phases is ensured by the resolution of their inverse kinematics (IK).

Once the pregrasp step is performed, the robot moves next to the object and the fingers reach the initial grasp. No important deformation is supposed to be generated when the fingers come into contact with the object since a very small contact force will be firstly applied in order to avoid that the grasp location changes. As soon as this initial grasp is executed precisely, the third step of the algorithm is activated and an iterative closing of the fingers begins, based on the simulation of the contact interaction model (Section 6). At each iteration of the simulation, the deformations of the object are updated and the generated contact forces are computed. The computed contact forces are used to evaluate the static equilibrium of the object + fingers system. The iterative process is repeated in simulation until the static equilibrium is reached (see all these steps in simulation on the left side of the flowchart in Figure 1).

Upon completion of the simulation process, the real handling and manipulation of the object can be performed by executing the contact forces obtained from simulation (see steps on the right side of the flowchart). Firstly, the object is installed at the workspace of the robot with the same relative configuration as at the beginning of the simulation. Then, the robotic arm moves towards the pregrasp configuration (i.e., phase 1 in reality) and the hand fingers are moved towards the initial contact points by position control (i.e., phase 2 in reality). At this stage (beginning of phase 3 in reality), the fingers are closed by force control and apply progressive squeezing of the object until the contact forces (which are measured by contact sensors installed inside each finger of the Barrett hand) become equal to the contact forces computed at the simulation step. These contact forces guarantee the equilibrium of the object–hand system (as validated by the simulation). Thereby, the object can be lifted up from the table without any risk of sliding and can be robustly manipulated. In the next sections, all these steps are described in detail and are validated with real pick-and-place experiments of deformable objects by a hand + arm robot system (Section 7).

3. Synthesis of the Initial Grasp Configuration

In this section, we address the characterization of the object stability for a three-fingered grasp. This implies determining a force closure configuration based on the choice of three contact points from the set of all the points representing the outer 3D surface of the object. The following assumptions are considered in this procedure:

- The use of three fingers for the grasp operation modeled as hemispheres with radius R.
- The first contacts between the fingers and the object are point contacts.
- The outer surface of the object is represented by a set of points Ω, which are described by their position vectors $\vec{P_i}$ measured with respect to a reference frame located at the center of mass $(\vec{P_{co}})$ of the object.

In fact, a 3-fingered grasp is more reliable in terms of stability, slip avoidance, and balance of forces when it converges towards an ideal equilateral grasp [17]. Thus, each three-fingered grasp can be characterized by a value which represents its similarity to an equilateral triangle. We suggest an algorithm based on geometric criteria to find this equilateral grasp. This algorithm determines, at first, the set of all possible grasping triangles by scanning the points belonging to the contact surface Ω. Then, using the Q_1 criterion presented by Roa [17], our algorithm compares the angle values (alpha, beta, and lambda) of these triangles with $\pi/3$ (i.e., the angle that characterizes an equilateral triangle) in order to choose the configuration closest to an equilateral triangle (see Figure 2):

$$Q_1 = \frac{3}{2\pi}\left(\left|\alpha - \frac{\pi}{3}\right| + \left|\beta - \frac{\pi}{3}\right| + \left|\gamma - \frac{\pi}{3}\right|\right). \tag{1}$$

For a target angle value of $3/2\pi$, the minimum possible value of Q_1 is 0 for a perfect equilateral triangle, 2 for the most unfavorable case (the triangle degenerates into a segment), and 1 for an intermediate state.

For evaluating the angles of the triangle, we use a margin of error defined by the interval [0, 0.3]. Depending on the density of the points of the contact surface Ω, this algorithm can give several grasp configurations. Finally, a second criterion, referred to as Q_2, is used in order to choose one between them. It measures the distance between the center of the mass of the object C_O and the center of the grasping triangle C_T (Figure 3), defined as

$$Q_2 = |C_O - C_T|. \tag{2}$$

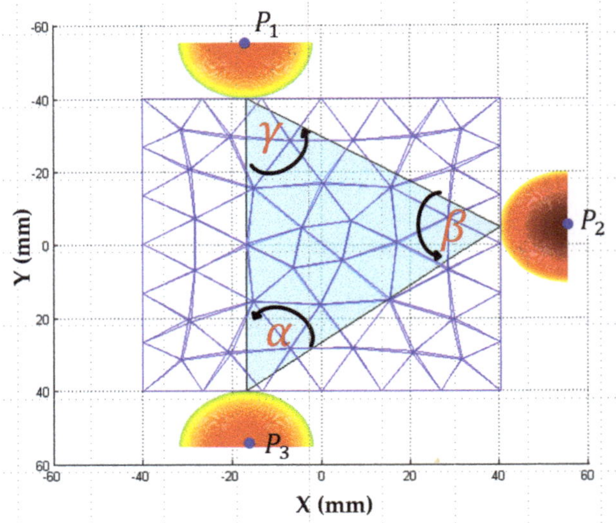

Figure 2. Force arrangement criteria.

This criterion was developed to obtain stable grasps with respect to the torques generated by gravitational and inertial forces [17]. The complete grasp synthesis algorithm based on the these two criteria Q_1 and Q_2 is represented in the diagram "Algorithm 1":

Algorithm 1 Synthesis of the 3 contact points for initial grasp configuration

---

Grasp synthesis algorithm. Input: points P_i on contact surfaces Ω with their 3D coordinates

1. Calculation of the set Γ of grasp triangles defined from points P_i.
2. Calculation of the value of the angles of the grasp triangles of the set Γ.
3. Choice of the grasp triangles closer to an equilateral triangle by applying criterion Q_1, the set of these triangles is named Ψ.
4. Calculation of the barycenter C_T of the grasp triangles of the set Ψ.
5. Calculation of the distance between C_O and the centers C_T of the triangles of the set Ψ by applying criterion Q_2.
6. Choosing the triangle that minimizes Q_2.

Output: 3 initial contact points of the grasp configuration

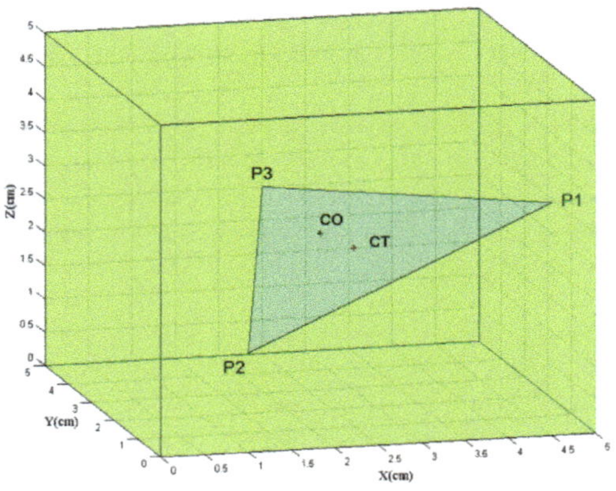

Figure 3. The distance between C_O and C_T.

4. Pregrasp Strategy for the Robotic Hand: Placement of the Fingers

In this section, we present the pregrasp strategy for reaching the three contact points of the initial grasp computed in Section 3 with a three-fingered robotic hand (e.g., the Barrett hand [50] in our experiments). In order to describe the kinematic model of the Barrett hand, we establish the joint frames of Figure 4. The movement of the fingers of the Barrett hand is controlled by defining 4 encoder values $[q_{M1} \ldots q_{M4}]$ that are connected to the joint values φ_i of the fingers (Figure 4) by the following coupling relationships:

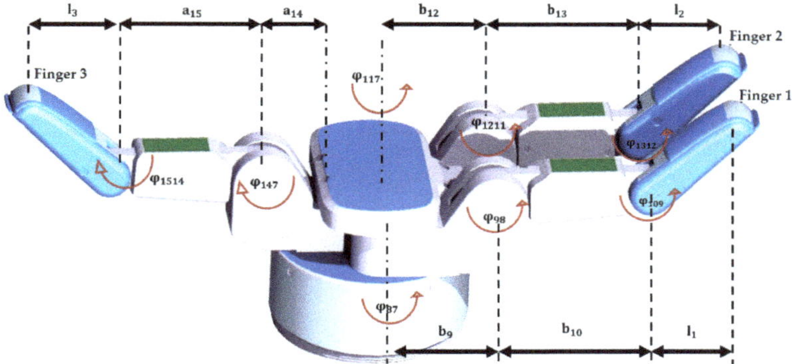

Figure 4. Barrett hand: dimensions, frames, and joints.

$$
\begin{bmatrix} \phi_{98} \\ \phi_{109} \\ \phi_{1211} \\ \phi_{1312} \\ \phi_{147} \\ \phi_{1514} \\ \phi_{87} \\ \phi_{117} \end{bmatrix} = \begin{bmatrix} 1/125 & 0 & 0 & 0 \\ 1/375 & 0 & 0 & 0 \\ 0 & 1/125 & 0 & 0 \\ 0 & 1/375 & 0 & 0 \\ 0 & 0 & 1/125 & 0 \\ 0 & 0 & 1/375 & 0 \\ 0 & 0 & 0 & 2/35 \\ 0 & 0 & 0 & -2/35 \end{bmatrix} * \begin{bmatrix} q_{M1} \\ q_{M2} \\ q_{M3} \\ q_{M4} \end{bmatrix}, \tag{3}
$$

where $q_{Mk} = \begin{bmatrix} 0 & 17,500 \end{bmatrix}$ for $k = \begin{bmatrix} 1 & 2 & 3 \end{bmatrix}$ (i.e., the range of variation of the encoders of the motors for closing/opening the three fingers), and $q_{M4} = \begin{bmatrix} 0 & 3150 \end{bmatrix}$ (i.e., the range of variation of the encoder of the motor 4 for separating fingers 1 and 2 around the palm).

The pregrasp strategy for the three-fingered robotic hand involves two steps: the first one aligns the orientation of the hand with the initial grasping triangle computed in Section 3; the second one adjusts this initial estimation of the grasp by taking into consideration the kinematic constraints of the hand. The first step consists of orientating the hand so that the TCP (Tool Center Point, defined along the axis perpendicular to the palm of the hand) coincides with the C_T grasp triangle center (as shown in Figure 5). The second step involves the resolution of the IK of the fingers in order to estimate the joint values to reach the three grasp points (P_1, P_2, and P_3) of the initial grasp triangle. If no IK solution is found, the length of the TCP line is changed and the IK resolution is recomputed. When an IK solution is found, the motor commands q_{Mk} are computed in order to attend to these joint values, by applying the inverse expression of (3). The implemented algorithm is shown in the diagram "Algorithm 2". In the context of this iterative solution search, we consider that the solution is validated for a value of joint angle difference for each of the fingers smaller than 2 degrees.

Figure 5. Intersection of the hand's Tool Center Point (TCP) with the center C_T of the grasp triangle.

Algorithm 2 Pregrasp strategy for the robotic hand

Finger placement algorithm for the initial grasp. Input: grasp triangle computed by Algorithm 1

1. Reorientation of the hand in the grasp plan and coincidence of TCP with C_T.
2. Calculation of IK and definition of finger joint values.
3. Searching for a solution for q_{M1}, q_{M2}, and q_{M3} by successive iteration over the length L in order to find a solution.
4. Reorientation of the hand around TCP and adjustment of the distance L to define a solution for q_{M4}.

Output: Definition of the values of $[q_{M1} \dots q_{M4}]$ and the 6 components (position + orientation) of the tool frame defining the hand's TCP.

---

5. Pregrasp Strategy for the Hand + Arm System: Reaching Trajectory by the Arm

After establishing the initial grasp points of the fingers (Section 3) and the corresponding pregrasp configuration of the hand around the object (Section 4), we should execute the latter by moving the hand with a robotic arm that carries it. Figure 6 shows the flowchart of the pregrasp strategy for the hand + arm system that uses the IK of the arm and the hand for achieving this initial grasp. For implementing this planning strategy, we have chosen a 6-DOF robotic arm (e.g., a Viper S1700D for our real experiments) so that the hand can attain any pose (position + orientation) in the space around the object. Initially, the object is resting on a table, modeled by a plane parallel to the XY plane of the arm base.

For performing this pregrasp strategy, we should represent the object pose (position $[x, y, z]$ in mm and orientation $[y, p, r]$ in ZYZ Euler angles) from its original *WobjO* frame (object mark) to the *WobjU* frame (user mark), known by the arm (see Figure 7). The TCP pose will also be represented firstly in the *WobjO* frame in order to easily define relative movements between the object and the hand. Later, it will be transformed into the arm frame (*WobjU*) so that final movements of the arm can be computed.

Firstly, we want to reach the relative object + hand configuration of Figure 8, which consists of putting the TCP in coincidence with C_0 (see Figure 9b). Then, we use the results from Section 3 to determine the contact points. We calculate the grasp triangle formed by these three contact points as well as its centroid C_T. Finally, we define the final pregrasp configuration (defined in Figure 5) by the translation and orientation of the hand. This step consists of matching the two TCP and C_T points and aligning the Z axis of the TCP coordinate system perpendicular to the grasp triangle (see Figure 9c) by computing the arm movements required to reach the tool frame obtained by Algorithm 2. When this final pregrasp configuration is reached by the hand, the fingers will be closed by applying the joints' angles obtained with Algorithm 2. An example of the main execution steps of this pregrasp strategy for a cube is shown in Figure 9.

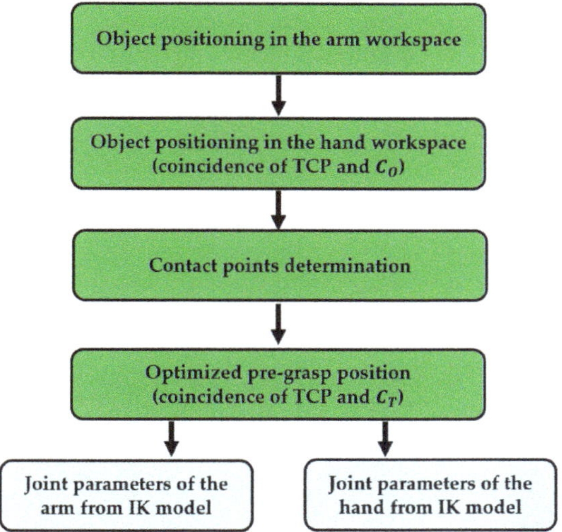

Figure 6. Flowchart for the pregrasp planning strategy of the hand + arm system. IK—Inverse Kinematics.

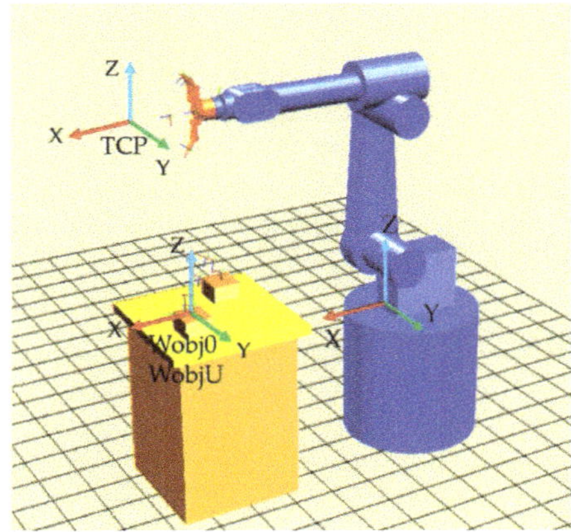

Figure 7. Main frames of the hand + arm system, simulated in Matlab.

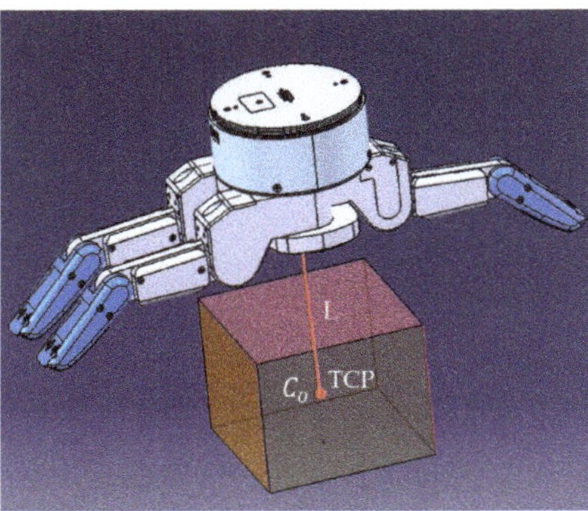

Figure 8. Intersection of TCP and C_0.

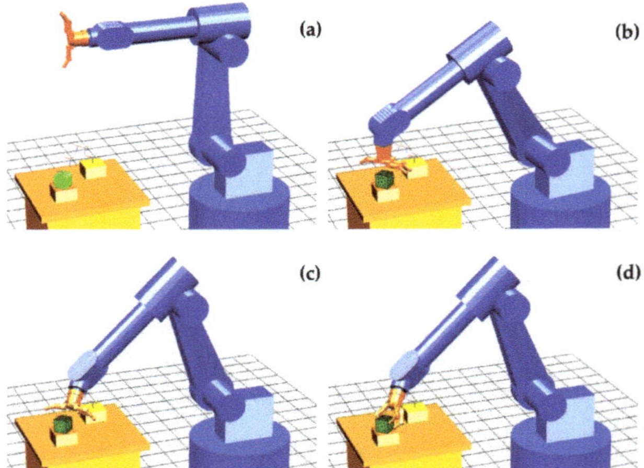

Figure 9. Main execution steps of the pregrasp strategy for the hand + arm system in order to grasp a cube: (**a**) initial robot–object configuration, (**b**) reaching trajectory (intersection of TCP and C_O), (**c**) hand orientation (intersection of TCP and C_T), (**d**) pregrasp configuration.

6. Final Stable Grasp Execution Based on Contact Interaction Modeling

In order to be able to evaluate the equilibrium with all the forces that take place in the real experiment while the fingers of the hand are closing, a model of the contact interactions between the fingertips and the object is required for precisely determining the forces–deformations relationship during grasping. This model will receive as input the initial grasp of the object without any deformation (i.e., the initial contact points of the fingertips over the object's surface or contact polygon determined at Section 3 and the fingers' configuration of the pregrasp strategy are described in Sections 4 and 5).

After this initial grasp, the fingers should close iteratively towards the center of the object until the equilibrium of all forces over the object is reached and a stable grasp is performed. With the aim of computing the required fingertip forces for this equilibrium, the simulation of this contact model is achieved in several sequential steps. Firstly, the contact between fingers and the object is detected. Secondly, contact forces are computed (by the evaluation of the relative velocity between the fingers and the object). Finally, the static equilibrium is checked for grasping stability. At each iteration of the simulation, the dynamic model updates the overall shape and contact area deformations due to the applied forces. Contact forces are individually calculated for each contact point, which gives a realistic distribution of the contact pressure in the contact zone. The contact model takes into account the normal forces and the two modes of the tangential forces due to friction: slipping and sticking modes. As the object is modeled by a set of nonlinear spring–damper pairs, the nonlinear normal force is given by

$$f_{n_j} = \begin{cases} 0 & ,\delta_j \leq 0 \\ max(0,(K\delta_j^n + C\dot{\delta}_j)) & \end{cases} \tag{4}$$

$$\mathbf{f}_{n_j} = f_{n_j}\mathbf{n}_{fj}, \tag{5}$$

where δ_j is the penetration distance measured along the normal direction to the contact surface j; K and C are the contact stiffness and damping constants, respectively. Those parameters depend on Young's modulus and the Poisson ratios. Based on the work presented in the literature [51,52], the stiffness constant is calculated by

$$K = 2\Psi\sqrt{R}, \tag{6}$$

where Ψ is a constant, calculated according to the mechanical properties of the two objects in contact, and it is given by

$$\Psi = \frac{1}{\frac{1-v_1^2}{E_1} + \frac{1-v_2^2}{E_2}}, \tag{7}$$

where E_1 and E_2 are the Young's modulus; v_1 and v_2 are the Poisson ratios of the finger and the object, respectively. The damping constant is determined by

$$C = 4\pi R\gamma, \tag{8}$$

where γ is a constant.

To characterize these mechanical parameters, compression tests are done for obtaining curves of stress evolution according to compressive deformation (see [33] for photos and detailed results of this calibration process). Thereby, the Young's modulus (4.928 MPa) and the Poisson's ratio (0.39) of the material of the objects (i.e., foam) to be grasped during the real experiments in Section 7 are identified.

Modeling of the tangential force, acting along each contact, is required in order to prevent any slippage and to ensure grasp stability. The contact model takes into account the two modes of the tangential forces due to friction: slipping and sticking modes. In this model, a parallel spring–damper is attached to the ground at one end, via a slider element, and to the fingertip at the other end. Therefore, the contact point location dynamically changes in the slipping condition. These variables correspond to the relative displacements at the contact point due, respectively, to sticking and slipping. They are dynamically reset to zero if the contact is broken. The sliding friction force can be defined in terms of the Coulomb law as follows:

$$\mathbf{f}_{slip} = \begin{cases} 0, & \|\mathbf{V}_t\| = 0 \\ -\mu f_n \frac{\mathbf{V}_t}{\|\mathbf{V}_t\|} \end{cases}, \tag{9}$$

where μ, f_n, and V_t are friction coefficient, normal force, and tangential velocity, respectively. With this model, the contact point location dynamically changes in the slipping condition. If the tangential force norm is less than the threshold of sliding, then we have a sticking mode whose tangential force is defined by

$$\mathbf{f}_{stick} = \begin{cases} 0, & v = 0 \\ -(k_t v - c_t \dot{v}) \frac{\mathbf{P}_c \mathbf{P}_s}{\|\mathbf{P}_c \mathbf{P}_s\|}, & v > 0 \end{cases}, \tag{10}$$

where v is the tangential deformation at the contact facet, P_c is the contact point position, and P_s is the contact point position at the sticking regime. The parameters k_t and c_t are respectively tangential stiffness and damping coefficients estimated by using Dopico's method [30].

As indicated at the beginning of this section, this contact model will be executed after the initial grasp configuration (obtained as output of the pregrasp strategy) is reached and the fingers begin to touch the surface of the object. In fact, a simulation of the grasping execution strategy is implemented in Matlab in order to determine the contact forces that should be applied to reach a robust grasp. This simulation will iteratively close the fingers towards the center of the object, evaluate the forces transmitted to the object by solving the contact model, update the state of the contact surface due to deformations, and verify the static equilibrium of the hand + object system. When this static equilibrium is obtained in the simulation, the contact forces applied by the simulated fingers will be used as references for the force control of the real robotic hand. An example of this grasping execution strategy will be shown in Section 7 in order to justify its application in our pipeline. Nevertheless, more experiments of this grasp execution strategy can be found in the previous work [33] by the authors, where it was first proposed.

7. Experimental Validation of the Grasp Planning Pipeline for the Hand + Arm System

The different steps of our complete grasp planning pipeline are validated with simulation and real experiments. First of all, the synthesis algorithm for the first initial grasp configuration proposed in Section 3 (i.e., Algorithm 1) is validated for three different shapes: a sphere, a cylinder, and a cube (Figure 10). This algorithm has been implemented in Matlab, by integrating a discrete 3D tetrahedral representation of the shape of these objects. The number of nodes of this 3D representation and the values of the two grasping quality metrics computed by the algorithm for the three objects are shown in Table 1. The initial grasping points obtained are shown in Figure 10 and they do not consider any deformation of the objects.

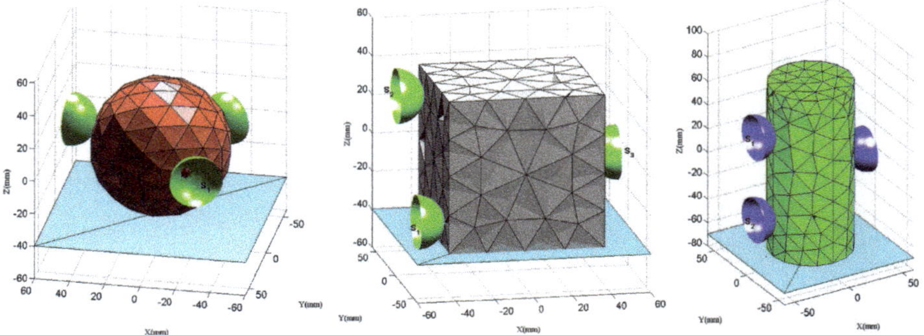

Figure 10. Configuration of the initial grasping points for a sphere, a cube, and a cylinder.

Table 1. Evaluation of the initial grasping criteria for a sphere, a cube, and a cylinder.

	Sphere (80 mm)	Cube (80*80*80mm)	Cylinder (R35 mm*L140 mm)
Number of nodes	182	162	226
Q1	0.0491	0.2302	0.1145
Q2 (mm)	4.8659	13.334	11.6667

These initial grasp configurations are used as input for the pregrasp and grasp execution steps of the grasp planning pipeline. For validating both steps, real experiments with the Adept arm and the Barrett hand grasping a foam cube (Figure 11) and a cylinder (Figure 12) are performed. Both objects represent typical shapes found in grasping applications of industrial products: a planar-faced surface (e.g., a cube) and a revolution surface (e.g., a cylinder). After the grasping, the object is lifted and carried to another position over the table in order to validate the stability of the grasp. The first step consists of generating and executing the path of the Adept arm for the pregrasp configuration in simulation, ensuring that the object is within the Barrett hand workspace (as explained for the cube in Figure 9a–c). The trajectory of the arm and the closing of the hand are then performed in the simulation (Figure 9d) until the fingers come into contact with the object.

The second step is to model the interaction between the fingers and the object for closing the fingers against the object surface in order to get the required contact forces in simulation (as explained in Section 6). The real contact forces at each finger are obtained by strain gauges installed in the phalanxes of the Barrett hand. Thereby, the Barrett hand is iteratively closed until the contact forces obtained from the simulation of the contact model are attained (as shown in Figure 13). Thus, a stable final grasp of the deformable object is obtained and the object can be lifted without any danger of falling down. Figure 13 shows the variation in the norm of the three forces applied to the cube during the simulation of the contact model. At the initial step (t_0) of the simulation, the fingers move towards the object in order to reach the initial contact points. When the fingers come into contact with the object (t_1), the contact model starts to evaluate the applied forces. This figure also shows that the contact forces are continuous and directly related to the amount of local deformation inside the contact area. The fingers continue to apply forces until equilibrium conditions are satisfied (i.e., reaching the three fingers' contact forces thresholds, identified by horizontal dashed lines in Figure 13). Once this state is reached and the three force thresholds are obtained, the fingers should keep these contact forces in order to ensure stable handling and then the simulation can stop.

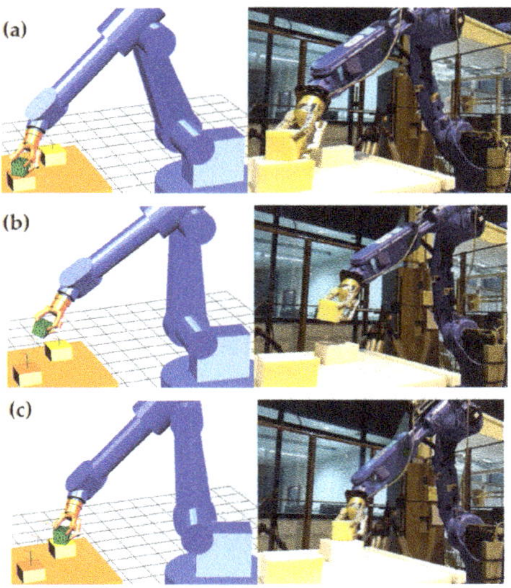

Figure 11. Pick-and-place of the cube by the Adept robot and the Barrett hand: (**a**) grasping, (**b**) lifting, and (**c**) leaving.

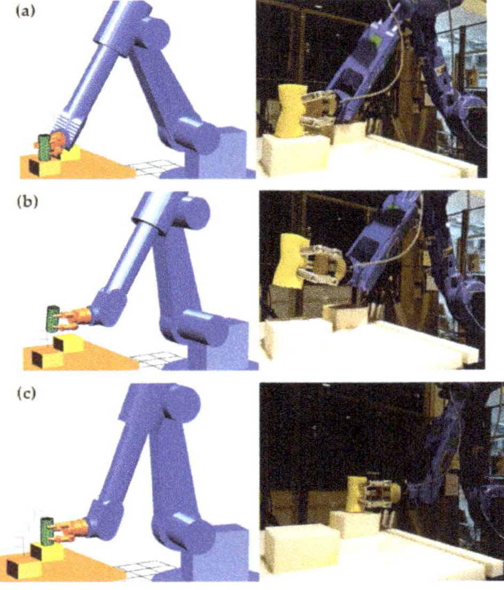

Figure 12. Pick-and-place of a cylinder by the Adept robot and the Barrett hand: (**a**) grasping, (**b**) lifting, and (**c**) leaving.

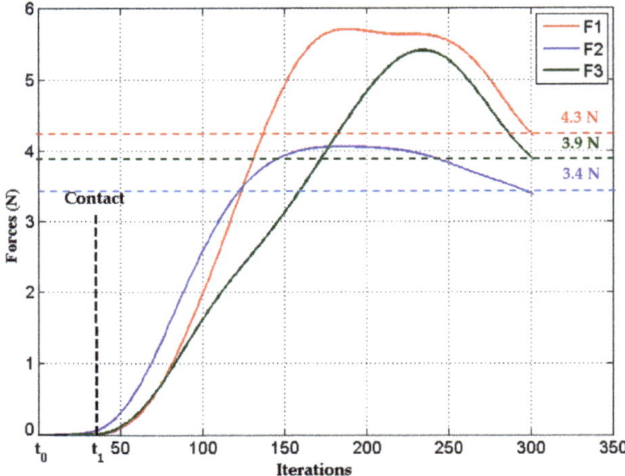

Figure 13. Grasping forces applied by the three fingers of the Barrett hand (F1, F2 and F3) in simulation.

The relation between these force thresholds and penetration of the fingers inside the object (i.e., the deformation of its surface) is obtained by the simulation of the contact interaction model while closing the fingertips over the object. The real relation force–penetration is very similar to the one computed in the simulation (see real validation experiments in [33]). Thereby, our strategy of applying the contact forces obtained from simulation in real grasping is suitable. In fact, the mean error of the contact force obtained by our model is of $0.3N$, which implies that only 15% of typical contact forces is required for grasping most common life deformable objects. In addition, the nonlinear relation force–penetration shows that simplified linear models cannot be used for grasping significantly deformable objects and justifies the necessity of the proposed model. The proposed grasp planning strategy is validated by the correct execution of two pick-and-place tasks (for a cube and a cylinder): lifting and stable transportation of both objects with the real robot without any slippage (as shown in the real photographs of Figures 11 and 12).

8. Conclusions

This work has been devoted to the complete process of deformable object grasp planning with an industrial manipulator (robotic hand + robotic arm). First of all, this requires the definition of the grasp points over the surface of the object and thus, the determination of an initial grasp configuration. Later, we presented our approach for determining the pregrasp strategy for reaching this initial grasp by taking into account the kinematic constraints of the arm and the hand. Thereby, the grasp planning has been divided into two phases: firstly, determining the configuration of the robot arm to bring the hand close to the object; secondly, determining the appropriate configuration of the fingers of the hand to grasp the object. The determination of the joint parameters of the arm and the hand is ensured by the resolution of their inverse kinematics. In the case of the hand, the implemented strategy consists first of putting the TCP (line perpendicular to the palm of the hand) at an intersection with the center of the grasp triangle (obtained in the initial grasp synthesis) and aligning it with the normal vector of this grasp triangle. The second step is to iteratively search for a solution to the articular parameters of the three fingers of the Barrett hand in order to bring them into correspondence with the grasping points previously defined by force-closure-type stability conditions. Thereby, this final grasp solution combines the initial grasp based on a general geometric criteria for force closure (i.e., it only considers the shape of the object) with the kinematic constraints of the robotic hand by applying the proposed pregrasp strategy. Nevertheless, this grasp configuration does guarantee the robustness of the grasp only for rigid objects but not for deformable objects. Thus, a force–deformation serving scheme is activated when the fingers come into contact with the deformable object to be grasped. This scheme is based on a simulation of the object–fingers interaction (previously developed by the authors in [33]), which obtains the contact force thresholds that should be applied by the fingers in order to guarantee a stable grasp while deforming the surface of the object. Finally, an iterative closing of the real fingers is performed in order to attain these contact forces with the real object.

Objects with simple geometrical shapes (sphere, cube, and cylinder) were chosen for the experiments to more efficiently simulate the interaction model described in Section 6. The proposed pregrasp/grasp pipeline is not restricted to simple shapes and can be applied to any shape of object. In fact, the mesh construction process required by the interaction model has been applied by the authors to very irregular shapes of meat pieces in a previous work [10]. This grasp planning pipeline is designed for nonlinear deformable objects to be manipulated by three-fingered robotic hands, but its architecture could be redesigned as a group of interchangeable modules in future works. These modules could include the following: other types of objects (e.g., changing the contact interaction model for rigid objects, granular media [53]), new sensing data (i.e., combining tactile and vision information for estimating the 3D shape of the object and the real contact forces in real-time [54]), multiple robotic fingers (e.g., dexterous robotic hands [55]), or even more complex manipulation tasks (e.g., in-hand manipulation [56]).

Author Contributions: Conceptualization, Methodology: all authors; Software: L.Z., J.A.C.R., L.S. and B.C.B.; Validation: L.Z. and L.S.; Writing—original draft: L.Z.; Writing—review & editing: L.Z and J.A.C.R. All authors have read and agreed to the published version of the manuscript.

Funding: This work has received funding from the European Union's Horizon 2020 research and innovation programme under grant agreement n° 869855 (Project 'SoftManBot').

Acknowledgments: L.Z. thanks the research support by the LINEACT EA 7527 laboratory at CESI Rouen.

Conflicts of Interest: The authors declare no conflict of interest

References

1. Sahbani, A.; El-Khoury, S.; Bidaud, P. An overview of 3D object grasp synthesis algorithms. *Robot. Auton. Syst.* **2012**, *60*, 326–336. [CrossRef]

2. Ramón, J.A.C.; Medina, F.T.; Perdereau, V. Finger readjustment algorithm for object manipulation based on tactile information. *Int. J. Adv. Robot. Syst.* **2013**, *10*, 9. [CrossRef]

3. Adjigble, M.; Marturi, N.; Ortenzi, V.; Rajasekaran, V.; Corke, P.; Stolkin, R. Model-free and learning-free grasping by local contact moment matching. In Proceedings of the 2018 IEEE/RSJ International Conference on Intelligent Robots and Systems (IROS), Madrid, Spain, 1–5 October 2018; pp. 2933–2940.

4. Mavrakis, N.; Stolkin, R. Estimation and exploitation of objects' inertial parameters in robotic grasping and manipulation: A survey. *Robot. Auton. Syst.* **2020**, *124*, 103374. [CrossRef]

5. Wang, C.; Zhang, X.; Zang, X.; Liu, Y.; Ding, G.; Yin, W.; Zhao, J. Feature Sensing and Robotic Grasping of Objects with Uncertain Information: A Review. *Sensors* **2020**, *20*, 3707. [CrossRef] [PubMed]

6. Deng, Z.; Jonetzko, Y.; Zhang, L.; Zhang, J. Grasping Force Control of Multi-Fingered Robotic Hands through Tactile Sensing for Object Stabilization. *Sensors* **2020**, *20*, 1050. [CrossRef]

7. Oravcová, J.; Košťál, P.; Delgado Sobrino, D.R.; Holubek, R. Clamping Fixture Design Methodology for the Proper Workpiece Insertion. In *III Central European Conference on Logistics, Applied Mechanics and Materials*; Trans Tech Publications Ltd.: Cham, Switzerland, 2013; Volume 309, pp. 20–26. [CrossRef]

8. Velíšek, K.; Košťál, P.; Zvolenský, R. Clamping Fixtures for Intelligent Cell Manufacturing. In *Intelligent Robotics and Applications*; Xiong, C., Liu, H., Huang, Y., Xiong, Y., Eds.; Springer: Berlin/Heidelberg, Germany, 2008; pp. 966–972.

9. Faure, F.; Duriez, C.; Delingette, H.; Allard, J.; Gilles, B.; Marchesseau, S.; Talbot, H.; Courtecuisse, H.; Bousquet, G.; Peterlik, I.; et al. Sofa: A multi-model framework for interactive physical simulation. In *Soft Tissue Biomechanical Modeling for Computer Assisted Surgery*; Springer: Berlin/Heidelberg, Germany, 2012; pp. 283–321.

10. Long, P.; Khalil, W.; Martinet, P. Force/vision control for robotic cutting of soft materials. In Proceedings of the 2014 IEEE/RSJ International Conference on Intelligent Robots and Systems, Chicago, IL, USA, 14–18 September 2014; pp. 4716–4721.

11. Nabil, E.; Belhassen-Chedli, B.; Grigore, G. Soft material modeling for robotic task formulation and control in the muscle separation process. *Robot. Comput.-Integr. Manuf.* **2015**, *32*, 37–53. [CrossRef]

12. Nadon, F.; Valencia, A.J.; Payeur, P. Multi-modal sensing and robotic manipulation of non-rigid objects: A survey. *Robotics* **2018**, *7*, 74. [CrossRef]

13. Pirník, R.; Hruboš, M.; Nemec, D.; Mravec, T.; Božek, P. Integration of Inertial Sensor Data into Control of the Mobile Platform. In *Advances in Intelligent Systems and Computing, Proceedings of the 2015 Federated Conference on Software Development and Object Technologies, Žilina, Slovakia, 19–20 November 2017*; Janech, J., Kostolny, J., Gratkowski, T., Eds.; Springer International Publishing: Cham, Switzerland, 2017; pp. 271–282.

14. Dodok, T.; Čuboňová, N.; Císar, M.; Kuric, I.; Zajačko, I. Utilization of Strategies to Generate and Optimize Machining Sequences in CAD/CAM. *Procedia Eng.* **2017**, *192*, 113–118. 12th international scientific conference of young scientists on sustainable, modern and safe transport, doi:10.1016/j.proeng.2017.06.020. [CrossRef]

15. Sanchez, J.; Corrales, J.A.; Bouzgarrou, B.C.; Mezouar, Y. Robotic manipulation and sensing of deformable objects in domestic and industrial applications: A survey. *Int. J. Robot. Res.* **2018**, *37*, 688–716. [CrossRef]

16. Chinellato, E.; Morales, A.; Fisher, R.B.; Del Pobil, A.P. Visual quality measures for characterizing planar robot grasps. *Syst. Man Cybern. Part C Appl. Rev. IEEE Trans.* **2005**, *35*, 30–41. [CrossRef]

17. Roa, M.A.; Suárez, R. Grasp quality measures: Review and performance. *Auton. Robot.* **2015**, *38*, 65–88. [CrossRef] [PubMed]

18. Pozzi, M.; Malvezzi, M.; Prattichizzo, D. On grasp quality measures: Grasp robustness and contact force distribution in underactuated and compliant robotic hands. *IEEE Robot. Autom. Lett.* **2016**, *2*, 329–336. [CrossRef]

19. Soler, F.; Rojas-de Silva, A.; Suárez, R. Grasp quality measures for transferring objects. In *Advances in Intelligent Systems and Computing, Proceedings of the Iberian Robotics Conference, Seville, Spain, 22–24 November 2017*; Springer: Cham, Switzerland, 2017; pp. 28–39.

20. Wakamatsu, H.; Hirai, S.; Iwata, K. Static analysis of deformable object grasping based on bounded force closure. In Proceedings of the EEE International Conference on Robotics and Automation, Minneapolis, MN, USA, 22–28 April 1996; Volume 4, pp. 3324–3329.

21. Mira, D.; Delgado, A.; Mateo, C.; Puente, S.; Candelas, F.; Torres, F. Study of dexterous robotic grasping for deformable objects manipulation. In Proceedings of the 2015 23rd Mediterranean Conference on Control and Automation (MED), Torremolinos, Spain, 16–19 June 2015; pp. 262–266.

22. Xu, J.; Danielczuk, M.; Ichnowski, J.; Mahler, J.; Steinbach, E.; Goldberg, K. Minimal Work: A Grasp Quality Metric for Deformable Hollow Objects. In Proceedings of the 2020 IEEE International Conference on Robotics and Automation (ICRA), Paris, France, 31 May–31 August 2020; pp. 1546–1552.

23. Jørgensen, T.B.; Jensen, S.H.N.; Aanæs, H.; Hansen, N.W.; Krüger, N. An adaptive robotic system for doing pick and place operations with deformable objects. *J. Intell. Robot. Syst.* **2019**, *94*, 81–100. [CrossRef]

24. Bicchi, A. On the closure properties of robotic grasping. *Int. J. Robot. Res.* **1995**, *14*, 319–334. [CrossRef]

25. Lei, Q.; Wisse, M. Object grasping by combining caging and force closure. In Proceedings of the 2016 14th International Conference on Control, Automation, Robotics and Vision (ICARCV), Phuket, Thailand, 13–15 November 2016; pp. 1–8.

26. Rakesh, V.; Sharma, U.; Murugan, S.; Venugopal, S.; Asokan, T. Optimizing force closure grasps on 3D objects using a modified genetic algorithm. *Soft Comput.* **2018**, *22*, 759–772. [CrossRef]

27. Kaboli, M.; Yao, K.; Cheng, G. Tactile-based manipulation of deformable objects with dynamic center of mass. In Proceedings of the 2016 IEEE-RAS 16th International Conference on Humanoid Robots (Humanoids), Cancun, Mexico, 15–17 November 2016; pp. 752–757.

28. Zhao, K.; Li, X.; Lu, C.; Lu, G.; Wang, Y. Video-based slip sensor for multidimensional information detecting in deformable object grasp. *Robot. Auton. Syst.* **2017**, *91*, 71–82. [CrossRef]

29. Gonthier, Y.; McPhee, J.; Lange, C.; Piedboeuf, J.C. A regularized contact model with asymmetric damping and dwell-time dependent friction. *Multibody Syst. Dyn.* **2004**, *11*, 209–233. [CrossRef]

30. Dopico, D.; Luaces, A.; Gonzalez, M.; Cuadrado, J. Dealing with multiple contacts in a human-in-the-loop application. *Multibody Syst. Dyn.* **2011**, *25*, 167–183. [CrossRef]

31. Mohammadi, M.; Baldi, T.L.; Scheggi, S.; Prattichizzo, D. Fingertip force estimation via inertial and magnetic sensors in deformable object manipulation. In Proceedings of the 2016 IEEE Haptics Symposium (HAPTICS), Philadelphia, PA, USA, 8–11 April 2016; pp. 284–289.

32. Lu, Z.; Huang, P.; Liu, Z. High-gain nonlinear observer-based impedance control for deformable object cooperative teleoperation with nonlinear contact model. *Int. J. Robust Nonlinear Control* **2020**, *30*, 1329–1350. [CrossRef]

33. Zaidi, L.; Corrales, J.A.; Bouzgarrou, B.C.; Mezouar, Y.; Sabourin, L. Model-based strategy for grasping 3D deformable objects using a multi-fingered robotic hand. *Robot. Auton. Syst.* **2017**, *95*, 196–206. [CrossRef]

34. Arriola-Rios, V.E.; Guler, P.; Ficuciello, F.; Kragic, D.; Siciliano, B.; Wyatt, J.L. Modeling of Deformable Objects for Robotic Manipulation: A Tutorial and Review. *Front. Robot. AI* **2020**, *7*, 82. [CrossRef]

35. Hammer, P.E.; Sacks, M.S.; Pedro, J.; Howe, R.D. Mass-spring model for simulation of heart valve tissue mechanical behavior. *Ann. Biomed. Eng.* **2011**, *39*, 1668–1679. [CrossRef] [PubMed]

36. Jarrousse, O. *Modified Mass-Spring System for Physically Based Deformation Modeling*; KIT Scientific Publishing: Karlsruhe, Germany, 2014.

37. Garg, S.; Dutta, A. Grasping and manipulation of deformable objects based on internal force requirements. *Int. J. Adv. Robot. Syst.* **2006**, *3*, 107–114. [CrossRef]

38. Han, L.; Hipwell, J.; Taylor, Z.; Tanner, C.; Ourselin, S.; Hawkes, D.J. Fast deformation simulation of breasts using GPU-based dynamic explicit finite element method. In *Digital Mammography*; Springer: Berlin/Heidelberg, Germany, 2010; pp. 728–735.

39. Tagawa, K.; Hirota, K.; Hirose, M. *Manipulation of Dynamically Deformable Object Using Impulse-Based Approach*; INTECH Open Access Publisher: London, UK, 2010.

40. Zaidi, L.; Bouzgarrou, B.C.; Sabourin, L.; Mezouar, Y. Modeling and analysis of 3D deformable object grasping. In Proceedings of the 2014 23rd International Conference on Robotics in Alpe-Adria-Danube Region (RAAD), Smolenice, Slovakia, 3–5 September 2014; pp. 1–8. [CrossRef]

41. Berenson, D. Manipulation of deformable objects without modeling and simulating deformation. In Proceedings of the 2013 IEEE/RSJ International Conference on Intelligent Robots and Systems, Tokyo, Japan, 3–7 November 2013; pp. 4525–4532. [CrossRef]

42. Nadon, F.; Payeur, P. Automatic Selection of Grasping Points for Shape Control of Non-Rigid Objects. In Proceedings of the 2019 IEEE International Symposium on Robotic and Sensors Environments (ROSE), Ottawa, ON, Canada, 17–18 June 2019; pp. 1–7.

43. Jia, Y.B.; Guo, F.; Lin, H. Grasping deformable planar objects: Squeeze, stick/slip analysis, and energy-based optimalities. *Int. J. Robot. Res.* **2014**, *33*, 866–897. [CrossRef]

44. Sanchez, J.; Mohy El Dine, K.; Corrales, J.A.; Bouzgarrou, B.C.; Mezouar, Y. Blind Manipulation of Deformable Objects Based on Force Sensing and Finite Element Modeling. *Front. Robot. AI* **2020**, *7*, 73. [CrossRef]

45. Lin, H.; Guo, F.; Wang, F.; Jia, Y.B. Picking up a soft 3D object by feeling the grip. *Int. J. Robot. Res.* **2015**, *34*, 1361–1384. [CrossRef]

46. Jorgensen, T.B.; Holm, P.H.S.; Petersen, H.G.; Krüger, N. Optimizing Pick and Place Operations in a Simulated Work Cell For Deformable 3D Objects. In *Lecture Notes in Computer Science, Proceedings of the International Conference on Intelligent Robotics and Applications, Portsmouth, UK, 24–27 August 2015*; Springer: Cham, Switzerland, 2015; pp. 431–444.

47. Hu, Z.; Sun, P.; Pan, J. Three-Dimensional Deformable Object Manipulation Using Fast Online Gaussian Process Regression. *IEEE Robot. Autom. Lett.* **2018**, *3*, 979–986. [CrossRef]

48. Verleysen, A.; Holvoet, T.; Proesmans, R.; Den Haese, C.; wyffels, F. Simpler Learning of Robotic Manipulation of Clothing by Utilizing DIY Smart Textile Technology. *Appl. Sci.* **2020**, *10*, 4088. [CrossRef]

49. Caporali, A.; Palli, G. Pointcloud-based Identification of Optimal Grasping Poses for Cloth-like Deformable Objects. In Proceedings of the 2020 25th IEEE International Conference on Emerging Technologies and Factory Automation (ETFA), Vienna, Austria, 8–11 September 2020; Volume 1, pp. 581–586.

50. Hasan, M.R.; Vepa, R.; Shaheed, H.; Huijberts, H. Modelling and Control of the Barrett Hand for Grasping. In Proceedings of the 2013 UKSim 15th International Conference on Computer Modelling and Simulation, Cambridge, UK, 10–12 April 2013; pp. 230–235. [CrossRef]

51. Shi, X.; Polycarpou, A.A. Measurement and modeling of normal contact stiffness and contact damping at the meso scale. *J. Vib. Acoust.* **2005**, *127*, 52–60. [CrossRef]

52. Johnson, K.L.; Johnson, K.L. *Contact Mechanics*; Cambridge University Press: Cambridge, UK, 1987.

53. Mateo, C.M.; Corrales, J.A.; Mezouar, Y. A Manipulation Control Strategy for Granular Materials Based on a Gaussian Mixture Model. In *Advances in Intelligent Systems and Computing, Proceedings of the Robot 2019: Fourth Iberian Robotics Conference, Porto, Portugal, 20–22 November 2019*; Silva, M.F., Luís Lima, J., Reis, L.P., Sanfeliu, A., Tardioli, D., Eds.; Springer International Publishing: Cham, Switzerland, 2020; pp. 171–183.
54. Sanchez, J.; Mateo, C.M.; Corrales, J.A.; Bouzgarrou, B.; Mezouar, Y. Online Shape Estimation based on Tactile Sensing and Deformation Modeling for Robot Manipulation. In Proceedings of the 2018 IEEE/RSJ International Conference on Intelligent Robots and Systems (IROS), Madrid, Spain, 1–5 October 2018; pp. 504–511. [CrossRef]
55. Aranda, M.; Corrales, J.A.; Mezouar, Y. Deformation-based shape control with a multirobot system. In Proceedings of the 2019 International Conference on Robotics and Automation (ICRA), Montreal, QC, Canada, 20–24 May 2019; pp. 2174–2180. [CrossRef]
56. Corrales Ramón, J.A.; Perdereau, V.; Torres Medina, F. Multi-fingered robotic hand planner for object reconfiguration through a rolling contact evolution model. In Proceedings of the 2013 IEEE International Conference on Robotics and Automation, Karlsruhe, Germany, 6–10 May 2013; pp. 625–630. [CrossRef]

Publisher's Note: MDPI stays neutral with regard to jurisdictional claims in published maps and institutional affiliations.

Article

Inverse Kinematics Data Adaptation to Non-Standard Modular Robotic Arm Consisting of Unique Rotational Modules

Štefan Ondočko [1], Jozef Svetlík [1,*], Michal Šašala [1], Zdenko Bobovský [2], Tomáš Stejskal [1], Jozef Dobránsky [3], Peter Demeč [1] and Lukáš Hrivniak [1]

[1] Department of Manufacturing Machinery and Robotics, Faculty of Mechanical Engineering, The Technical University of Košice, Letná 9, 04001 Košice, Slovakia; stefan.ondocko@tuke.sk (Š.O.); michal.sasala@tuke.sk (M.Š.); tomas.stejskal@tuke.sk (T.S.); peter.demec@tuke.sk (P.D.); lukas.hrivniak@tuke.sk (L.H.)

[2] Department of Robotics, Faculty of Mechanical Engineering, VSB—TU Ostrava, 17. listopadu 2172/15, 708 00 Ostrava-Poruba, Czech Republic; zdenko.bobovsky@vsb.cz

[3] Department of Automotive and Manufacturing Technologies, Faculty of Manufacturing Technologies with a seat in Prešov, Technical University of Košice, Štúrova 31, 08001 Prešov, Slovakia; jozef.dobransky@tuke.sk

* Correspondence: jozef.svetlik@tuke.sk; Tel.: +421-55-602-2195

Citation: Ondočko, t.; Svetlík, J.; Šašala, M.; Bobovský, Z.; Stejskal, T.; Dobránsky, J.; Demeč, P.; Hrivniak, L. Inverse Kinematics Data Adaptation to Non-Standard Modular Robotic Arm Consisting of Unique Rotational Modules. *Appl. Sci.* **2021**, *11*, 1203. https://doi.org/10.3390/app11031203

Academic Editor: José Luis Guzmán Sánchez

Received: 6 January 2021

Accepted: 25 January 2021

Published: 28 January 2021

Abstract: The paper describes the original robotic arm designed by our team kinematic design consisting of universal rotational modules (URM). The philosophy of modularity plays quite an important role when it comes to this mechanism since the individual modules will be the building blocks of the entire robotic arm. This is a serial kinematic chain with six degrees of freedom of unlimited rotation. It was modeled in three different environments to obtain the necessary visualizations, data, measurements, structural changes measurements and structural changes. In the environment of the CoppeliaSim Edu, it was constructed mainly to obtain the joints coordinates matching the description of a certain spatial trajectory with an option to test the software potential in future inverse task calculations. In Matlab, the model was constructed to check the mathematical equations in the area of kinematics, the model's simulations of movements, and to test the numerical calculations of the inverse kinematics. Since the equipment at hand is subject to constant development, its model can also be found in SolidWorks. Thus, the model's existence in those three environments has enabled us to compare the data and check the models' structural designs. In Matlab and SolidWorks, we worked with the data imported on joints coordinates, necessitating overcoming certain problems related to calculations of the inverse kinematics. The objective was to compare the results, especially in terms of the position kinematics in Matlab and SolidWorks, provided the initial joint coordinate vector was the same.

Keywords: Matlab; CoppeliaSim Edu; V-Rep; SolidWorks; kinematics; inverse kinematics (IK); manufacturing technology; modular robots; coordinate transformation

1. Introduction

The paper addresses data processing (in particular those of joints coordinates, which are, in the case at hand, the angles of rotation) designated for a stationary robotic arm composed of the so-called URM modules, described in detail in [1–3]. We briefly mention their main attributes, which are unlimited rotation of the module around its own axis, availability of intrinsic power and modularity units. Thanks to the modularity and the advantages it represents [4–7], individual modules can be used for building various configurations and thus to adjust the arm to a function required of it, to create machines with different mobility capabilities, several degrees of freedom. In general, modular and reconfigurable robots offer great versatility, robustness and—thanks to their series production—low costs, which is mentioned by many authors addressing this issue [8–10]. Those modular, reconfigurable robots that consist of many modules (the number of their degrees of freedom is usually much greater than 6) have the ability to reconfigure themselves into a large number of shapes. If this is the case, the robot can change its shape to meet the requirements of different tasks.

Modular robotic systems are systems composed of modules that can be disconnected and reconnected in various configurations, subject to maintaining the precision and rigidity parameters, which can thus create a new system enabling new or value-added functions [11]. The concept of the URM system has been conceived in accordance with the total productive maintenance principal implementation rules [12,13]. Individual rotation positions are the result of the inverse kinematics of the mentioned robot's arm. When it comes to kinematics of an inverse position, we look for such joint coordinate vectors of the open kinematic chain (assuming that the mechanism's size is known) as would suit the required position and the orientation of the end effector's coordinate system. Unlike the forward task, where the position vector is a function of the joint coordinate vector, the inverse task is substantially more complex because it necessitates solving a set of strongly nonlinear algebraic equations. These problems were first postulated by Paden B. [14] and Kahan W [15]. In most cases, these systems cannot be solved analytically. Thus, various types of iterative numerical methods are used, most often employing the Jacobian [16–21]. In calculating the inverse function, the inability to solve the task stems from the (configuration) workspace limitation, delimited by the configuration itself and the mechanism's physical properties. If we are located outside this workspace, there is, of course, no solution. Several solutions may exist since the end effector's defined position may be obtained in several manners. The situation can be even more varied in the case of the so-called redundant manipulators, where the joint coordinate vector's number of elements is greater than the degree of freedom in the given workspace or plane. This results in an infinite number of configurations through which the defined position and orientation can be reached. We do not deal directly with the inverse function in this paper. The respective joint coordinates are obtained from the model's CoppeliaSim Edu simulation (Coppelia Robotics AG, 8049 Zürich, Switzerland). The software calculating the inverse kinematics uses the pseudoinverse computational method, also called the Moore–Penrose method and the method of dampened least squares (DLS), also called the Levenberg–Marquardt method. More information about these and other methods can be found in [16,22–24]. Since other URM module configurations are contemplated for the future, or the redundant manipulator creation will be requested, it was mandatory to search for flexible solutions. To this end, the inverse kinematics is addressed in the CoppeliaSim Edu software. When the model is created in this environment, it enables the simulation of many types of movements and also the design of a working trajectory taking into account hurdles in the robotic arm's configuration space. Joint coordinate vector coordinates are obtained from the model created in the CoppeliaSim Edu. Thus, we have the individual joints coordinates available, which is necessary for their subsequent implementation into the models, this time in Matlab (MathWorks, 1 Apple Hill Drive, Natick, MA 01760, USA) and in SolidWorks (Dassault Systèmes, 10 rue Marcel Dassault, CS 40501, 78946 Vélizy-Villacoublay CEDEX, France). These data are subsequently compared to check the models' designs in the respective software environments. Should the designer's work show ideal precision, these results should be, especially for the position kinematics, identical. This should hold regardless of the fact that two different software environments are used.

2. Model Creation and Data Processing

Relations describing the position kinematics by means of position vectors p_i, where $i \epsilon <1;7>$ at individual open kinematic chain segments have also been derived in [25,26]. This is the case of the robotic arm's forward kinematics. The relations serve the purpose of calculating the position of a point that would enable the attachment of either an effector or another device. Thus, the position of a point $[p_{xi}, p_{yi}, p_{zi}]$ on the module will be defined by the function of the rotation positions φ_i, ϑ_i and the segment size of the structure at hand—$|r_i|$; $p_i = f(\varphi_i, \vartheta_i, r_i)$. The basis of the arm [1,2] is an autonomous, cylinder-shaped module. Such modules have a single degree of freedom, namely the rotational one. The rotation happens around the module's main axis and is not limited in the interval under consideration ranging from 0 to 360 degrees. The rotation may occur continuously in either direction of rotation without limitation. The modules and joints are contemplated to be perfectly solid bodies. The individual modules' axes of rotation cross each other at the point where they are joined by passive joints, Figure 1. The distances of section points in space are quantified by the r_i vector (a kinematic chain segment), and this vector has its own Cartesian system of $S_i\{O_i, x_i, y_i, z_i\}$. The segment's rotation around the z_i axis is defined by the rotation matrix $R_{zi}(\varphi_i)$ [3–5]. The segment's rotation around the y_i axis is defined by the rotation matrix $R_{yi}(\vartheta_i)$.

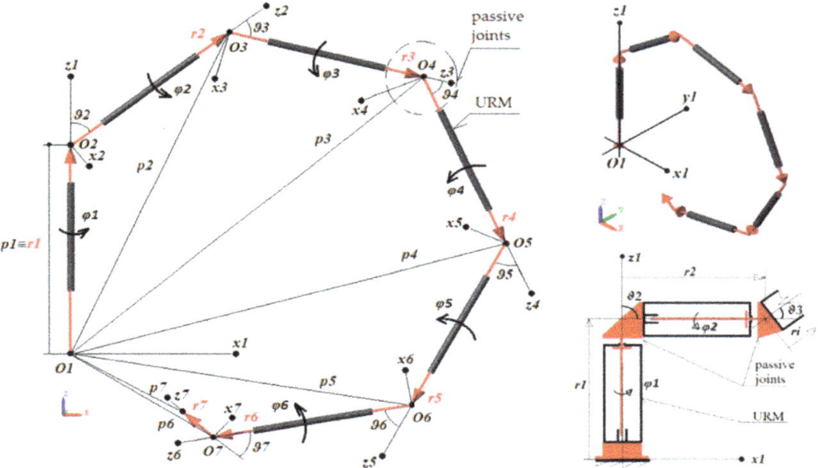

Figure 1. Robotic arm vector model composed of the universal rotational modules (URM) modules.

Thus, we can write the following for $R_{zi}(\varphi_i)$:

$$R_{zi}(\varphi_i) = \begin{bmatrix} \cos(\varphi_i) & -\sin(\varphi_i) & 0 \\ \sin(\varphi_i) & \cos(\varphi_i) & 0 \\ 0 & 0 & 1 \end{bmatrix} \tag{1}$$

Moreover, the following for $R_{yi}(\vartheta_i)$:

$$R_{yi}(\vartheta_i) = \begin{bmatrix} \cos(\vartheta_i) & 0 & \sin(\vartheta_i) \\ 0 & 1 & 0 \\ -\sin(\vartheta_i) & 0 & \cos(\vartheta_i) \end{bmatrix} \tag{2}$$

The r_1 vector is connected to the base perpendicularly to the plane with the x_1, y_1 axes. $p_1 \equiv r_1$, because the coordinate system is orthogonal. According to Figure 1, the following applies to the position vector p_1:

$$p_1 = R_{z1}(\varphi_1) \times r_1 \tag{3}$$

According to Figure 1, the following applies to the position vectors p_2, p_3, p_4, to p_i:

$$p_2 = p_1 + R_{z1}(\varphi_1) \times R_{y2}(\vartheta_2) \times R_{z2}(\varphi_2) \times r_2 \tag{4}$$

$$p_3 = p_2 + R_{z1}(\varphi_1) \times R_{y2}(\vartheta_2) \times R_{z2}(\varphi_2) \times R_{y3}(\vartheta_3) \times R_{z3}(\varphi_3) \times r_3 \tag{5}$$

$$p_4 = p_3 + R_{z1}(\varphi_1) \times R_{y2}(\vartheta_2) \times R_{z2}(\varphi_2) \times R_{y3}(\vartheta_3) \times R_{z3}(\varphi_3) \times R_{y4}(\vartheta_4) \times R_{z4}(\varphi_4) \times r_4 \tag{6}$$

A general position vector formula of this p_i structure will thus be:

$$p_{i+1} = R_{z1}(\varphi_1) \times \left[r_1 + \sum_{k=1}^{i} \left[\prod_{j=1}^{k} \left(R_{y(j+1)}(\vartheta_{j+1}) \times R_{z(j+1)}(\varphi_{j+1}) \right) \right] \times r_{k+1} \right] \tag{7}$$

The resulting orientation of the r_i vector defined by the Euler's angle can be seen in [27] according to the $R_z(\gamma)R_y(\beta)R_x(\alpha)$ option will be determined from the relation below based on the final arm rotation:

$$R_{ZYX(i+1)} = R_{z1}(\varphi_1) \times \prod_{j=1}^{i} \left(R_{y(j+1)}(\vartheta_{j+1}) \times R_{z(j+1)}(\varphi_{j+1}) \right) \tag{8}$$

Generally speaking, the final shape of the $R_{ZYX(i+1)}$ in terms of its i-segment will look as follows:

$$R_{ZYX(i+1)} = \begin{bmatrix} a_{11} & a_{12} & a_{13} \\ a_{21} & a_{22} & a_{23} \\ a_{31} & a_{32} & a_{33} \end{bmatrix} \tag{9}$$

where a_{ij} for $i = 1, 2, 3, j = 1, 2, 3$ are the matrix elements. Considering the $R_z(\gamma)R_y(\beta)R_x(\alpha)$ option, the Euler's angles will then be as follows:

$$\alpha = \arctan\left(\frac{a_{32}}{a_{33}}\right) \tag{10}$$

$$\beta = \arcsin(-a_{31}) \tag{11}$$

or also:

$$\beta = \arctan\left(\frac{-a_{31}}{\sqrt{a_{32}^2 + a_{33}^2}}\right) \tag{12}$$

$$\gamma = \arctan\left(\frac{a_{21}}{a_{11}}\right) \tag{13}$$

Thus, for example, in the case of an arm with 3 active degrees of freedom, i.e., for the r_4 vector, the final rotation according to Equation (8) will be as follows:

$$R_{ZYX4} = R_{z1}(\varphi_1) \times R_{y2}(\vartheta_2) \times R_{z2}(\varphi_2) \times R_{y3}(\vartheta_3) \times R_{z3}(\varphi_3) \times R_{y4}(\vartheta_4) \tag{14}$$

Our case involves 6 degrees of freedom.

2.1. Modeling in CoppeliaSim Edu

An open kinematic chain model was created in CoppeliaSim Edu, in Figure 2, composed of modules featuring 6 degrees of freedom.

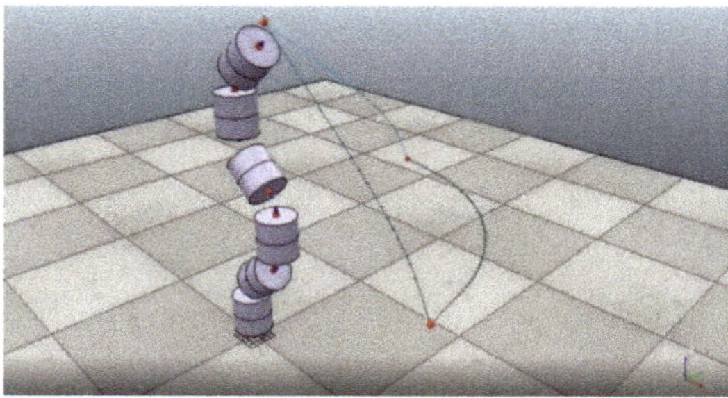

Figure 2. A model created in CoppeliaSim Edu.

The purpose of this model was especially calculating the inverse task from an arbitrary spatial trajectory the effector is to travel under 37 s. Joint coordinate vector data $\varphi = [\varphi_1, \varphi_2, \dots, \varphi_6]^T$ are obtained using a sampling period of T_{sp} each 0.005 s. The only (software-imposed) limitation is the degree of freedom itself, namely the possibility of rotation in the $<-\pi, +\pi>$ interval. This problem had to be subsequently addressed since it was causing a discontinuity in the simulation's operation as per the condition (18) and was thus disrupting the continuity of the trajectory on the models in the environments into which data were imported., i.e., Matlab, SolidWorks, but the same would have been the case in other environments, too.

2.2. Modeling in Matlab

The kinematic model in Figure 3 was created in Matlab using the Simscape/Multibody toolbox, the dimensions of which reflect the reality and are given in Table 1. The module diameter or volume need not be of interest when making a kinematic assessment, as it has no effect on kinematic parameters.

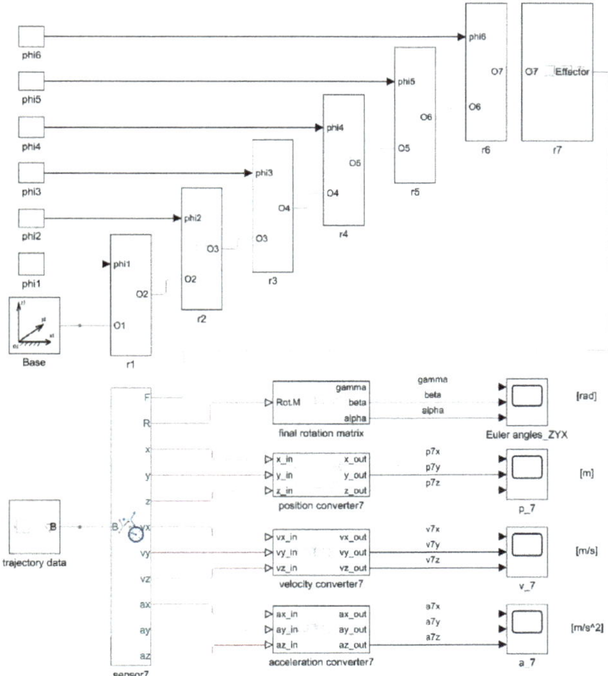

Figure 3. Model created in Matlab-Simulink using the Simscape/Multibody toolbox.

The model is composed of individual blocks that define its parts. In a nutshell, the model consists of a base identical to the "base" of the reference system and the individual blocks representing the r_i vector (a kinematic chain segment). As we can see in Figure 1, the "vector r_1" block then consists of the "URM" module and a passive joint part. The modules are connected in places where each joint represents its own Cartesian system of the module, starting in point O_i. In this experiment, the last part of the passive joint, vector r_7, was considered to be an effector. A measuring device (sensor 7) is connected to the effector r_7, checking physical quantities (position, velocity, acceleration). Such measuring devices can be mounted on other modules, too. The Euler's angles' values are also available in three consecutive rotations around the axes, often marked ZYX. The values apply to the effector and are the result of the final product of the individual spatial transformations

(8), as is usually the case with the open kinematic chain [27]. A model constructed in this way is then visualized in the Mechanics explorer window and appears in the form as seen in Figure 4.

Table 1. Dimensions table of the robotic arm of Figure 2 and also Figure 3, Figure 4, respectively.

Diameters	(mm)	Diameters	(mm)	Diameters	(mm)
r1	243.215	a1	73	b1	128
r2	212.430	a2	42.215	b2	128
r3	212.430	a3	42.215	b3	128
r4	212.430	a4	42.215	b4	128
r5	212.430	a5	42.215	b5	128
r6	212.430	a6	42.215	b6	128
r7	42.215 [1]				

[1] Effector (last part of the passive joint).

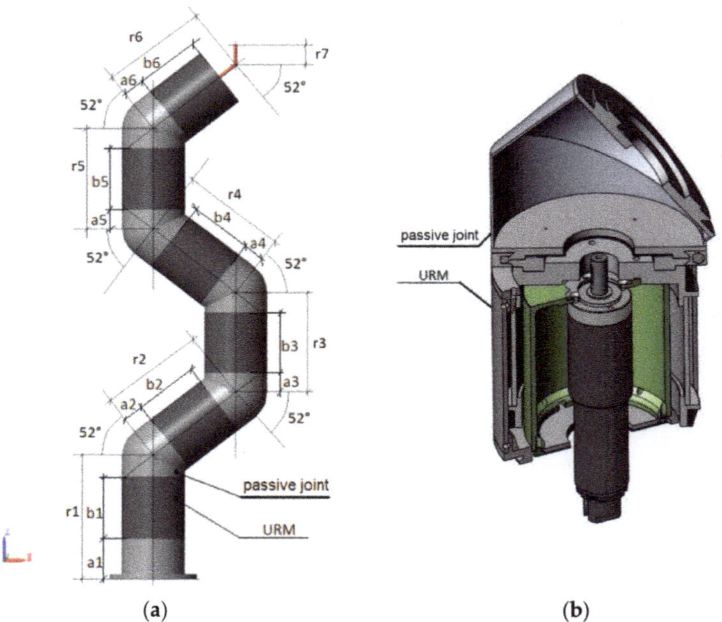

Figure 4. (**a**) Model visualization with hints of dimensions in Matlab-Simulink using the Simscape/Multibody toolbox. (**b**) Detail URM with a passive joint.

2.3. The Problem of a Sudden Change in Data Continuity and Its Solution in Matlab

The task was to use the computational core of CoppeliaSim Edu and apply it to the inverse kinematic task, to calculate the joint coordinates of $\varphi = [\varphi_1, \varphi_2, \dots, \varphi_6]^T$ for the entered trajectory. When the joint coordinates data were imported, in some cases, the above-mentioned problem with value repolarization emerged as a result of the software-imposed limitation on the joint rotation. In other words, in the case of a requirement arising from the calculation, the range of the degree of freedom upon exceeding the rotation values in the $<-\pi, +\pi>$ interval must be switched to the value with the opposite sign to remain in the defined working interval. The values above the upper limit $+\pi$ will rise again, but this time starting with the lower value of $-\pi$. Moreover, the opposite values below the lower limit $-\pi$ will fall again, this time starting from the upper value of $+\pi$. This sudden change causes deformation or even computational instability of the p_7 trajectory of the effector, see

Figure 5. With time derivations of the position, the characteristics of instantaneous velocity and accelerations are deformed even more.

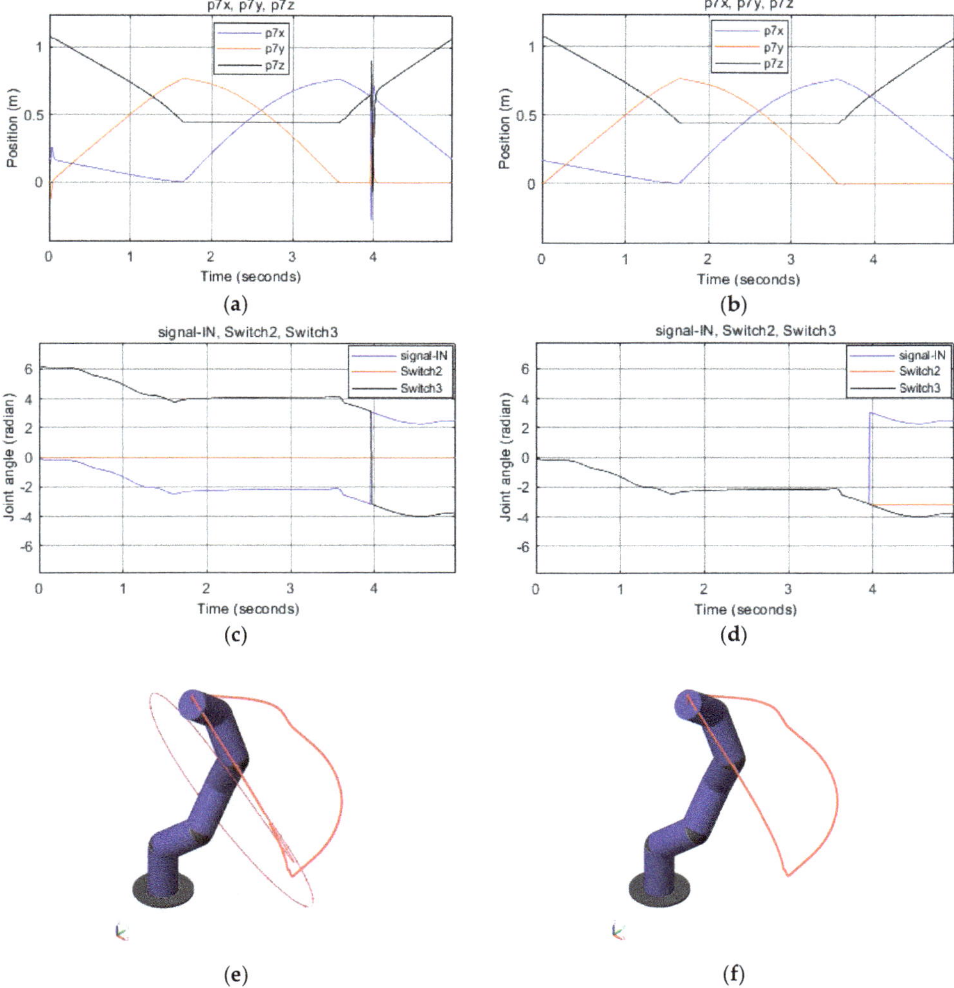

Figure 5. Illustration of the $p_7 = [p_{7x}, p_{7y}, p_{7z}]^T$ position vector (effector) depending on the joint coordinate vector $\varphi = [\varphi_1, \varphi_2, \varphi_3, \varphi_4, \varphi_5, \varphi_6]^T$ and its effect on the trajectory: (**a**) before repolarization; (**b**) after repolarization; the value of the joint coordinate vector depending on time $\varphi = [\varphi_1, \varphi_2, \varphi_3, \varphi_4, \varphi_5, \varphi_6]^T$; (**c**) before repolarization; (**d**) after repolarization; effect on the trajectory; (**e**) before repolarization; (**f**) after repolarization.

Since the precision of the trajectory traveled by any manipulator is critical in practical applications, it was necessary to somehow address this problem. Either through a labor-intensive rewriting of the imported data upon each change in the defined trajectory subject to the presence of those repolarizations in the joints or through a suitable solution. One such solution is the so-called repolarize, described below.

2.4. Composing a Repolarize

The search for a suitable and undemanding solution to this situation rested on the following conditions:

- To maintain the original trajectory, the initial values should not be skewed, i.e., if possible, they should not be averaged or approximated;
- There should be no phase shift of the initial values, as is the case with many filters.

For these reasons, we were trying to apply different approaches in the simulation, starting with a change in the initial data $\varphi = [\varphi_1, \varphi_2, \varphi_3, \varphi_4, \varphi_5, \varphi_6]^T$, which was very labor-intensive. The rate limiter has not met the expectations either [28] in switching the data flow at the moment of reaching the limits of the angles of rotation $+/-\pi$, because the characteristics of the p_7 trajectory were undulated anyway. Rate limiter is a marginalized derivation of an infinitely small difference to a difference of finite size. Or similar approaches based on a sequential omission of the undesired sample, such as in [29]. Or through averaging the values [30,31]. Neither of the approaches mentioned in this case has met our expectations. The best, if not ideal, results were finally achieved by the system working on this principle Figure 6 and explained in Figure 7. Joint coordinate data were measured in the sampling period T_{sp} every 0.005 s.

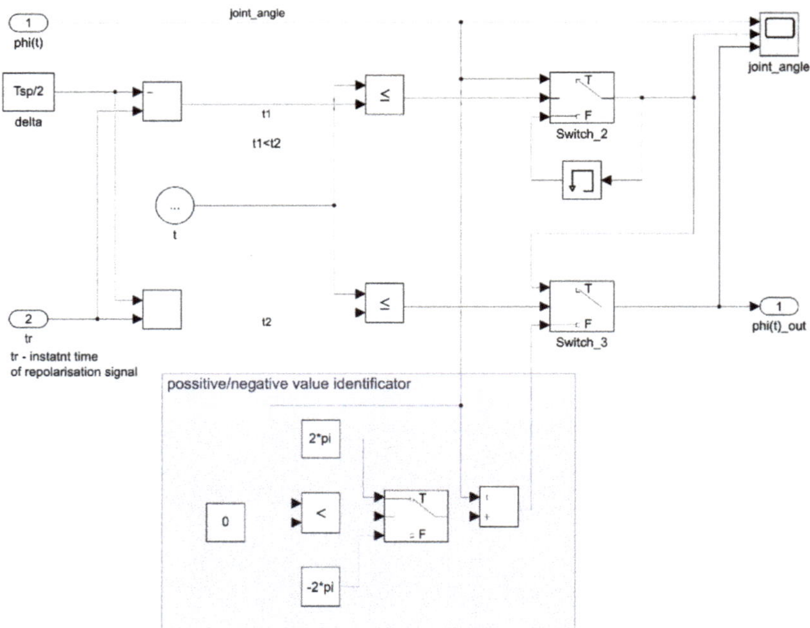

Figure 6. Constructed repolarizer.

When the data were uploaded, it was first necessary to establish the moment of repolarization t_r with the precision of $\delta = +/-0.0025$ s when the variable $\varphi(t) = 0$, i.e., cuts through the time axis. Under the assumption of reaching the upper or the lower working interval limit $<-\pi, +\pi>$ of the joint coordinate. The variable at the output from the repolarize is $\varphi_{out}(t)$. It was necessary to meet the following conditions:

$$t \leq t_1 | \varphi_{out}(t) = \varphi(t) \tag{15}$$

$$t_1 < t_2 \tag{16}$$

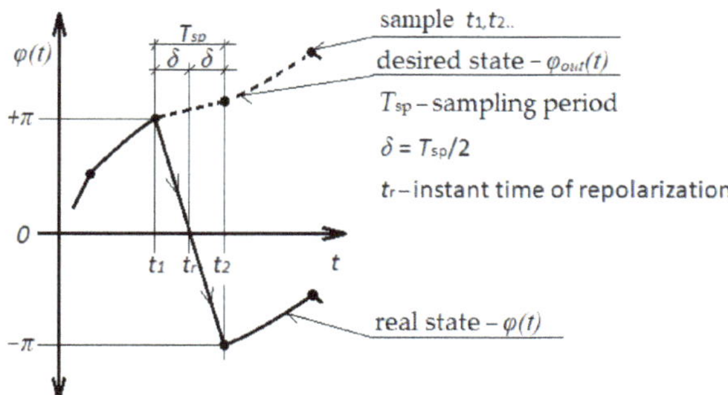

Figure 7. Plotted repolarization of rotation in the joint.

Conditions for identification of the sign and carrying out repolarization over time t_r:

$$t \geq t_2 \left| \varphi_{out}(t) = \left\{ \begin{array}{l} 2\pi - \varphi(t),\ \varphi(t) < 0 \\ \varphi(t) - 2\pi,\ \varphi(t) > 0 \end{array} \right. \right. \tag{17}$$

For the condition that can be considered discontinuation in the angle variable data $\varphi(t)$, the following applies:

$$\frac{\left| \varphi(t + T_{sp}) - \varphi(t) \right|}{T_{sp}} = \frac{2\pi}{T_{sp}} \tag{18}$$

The model was thus "excited" by the values of the joint coordinate vector $\varphi(t) = [\varphi_1(t),\ \varphi_2(t),\ \dots,\ \varphi_n(t)]^T$ obtained from the model created in CoppeliaSim Edu in the period of time T_{sp}. Under the condition of the prescribed maximum permissible deviation $[\Delta_{max\,x},\ \Delta_{max\,y},\ \Delta_{max\,z}] = +/-0.0035$ m between the real trajectory and the one calculated on the basis of inverse kinematics. The data of joint coordinates $\varphi(t)$ correspond to the effectors of the trajectory traveled. The trajectory is made of a system of fixed points $A_s\{x_{As},\ y_{As},\ z_{As}\}$, where the sample $s \epsilon <1; 7383>$ in Cartesian space Figure 8. Its real shape is shown in Figure 2.

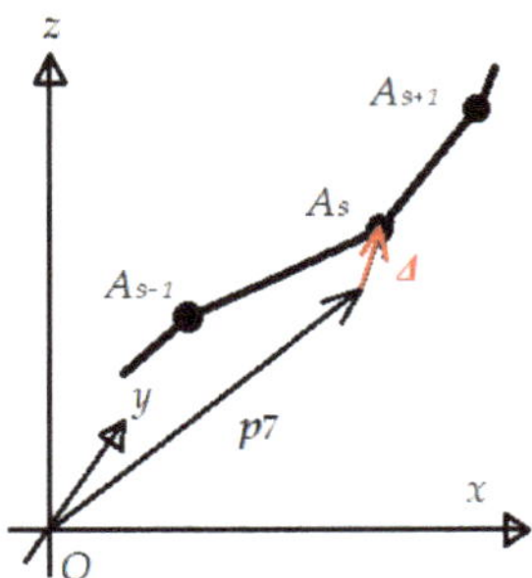

Figure 8. A vector difference Δ between the calculated position of the position vector of the p_7 effector and the position of the A_s points.

The mutual calibration of the models was done for the initial position of the robotic arm effector over time $t = 0$ s, $\varphi(0) = [0,\ \pi,\ 0,\ \pi,\ 0,\ \pi]^T$ radians see position in Figure 4, $p_7 = [0.167397,\ 0,\ 1.102644]^T$ meters. In order to check the models' correctness, the vector difference $\Delta = [\Delta_x,\ \Delta_y,\ \Delta_z]^T$ between the position of the position vector effector $p_7 = [p_{7x},\ p_{7y},\ p_{7z}]^T$ and the position of the A_s points, composing the trajectory, shown in the graph of Figure 9 (Matlab), Figure 12 (SolidWorks) over the period of time T_{sp}. This was done for each vector component. The following applies to Δ vector components, its norm $\|\Delta\|$ and the instantaneous time t of the sample in seconds:

$$\Delta_x = x_{As} - p_{7x}(\boldsymbol{\varphi}) \tag{19}$$

$$\Delta_y = y_{As} - p_{7y}(\boldsymbol{\varphi}) \tag{20}$$

$$\Delta_z = z_{As} - p_{7z}(\boldsymbol{\varphi}) \tag{21}$$

$$t = 0.005(s-1)|s \in \langle 1; 7383 \rangle \tag{22}$$

$$\|\Delta\| = \sqrt{\Delta_x{}^2 + \Delta_y{}^2 + \Delta_z{}^2} \tag{23}$$

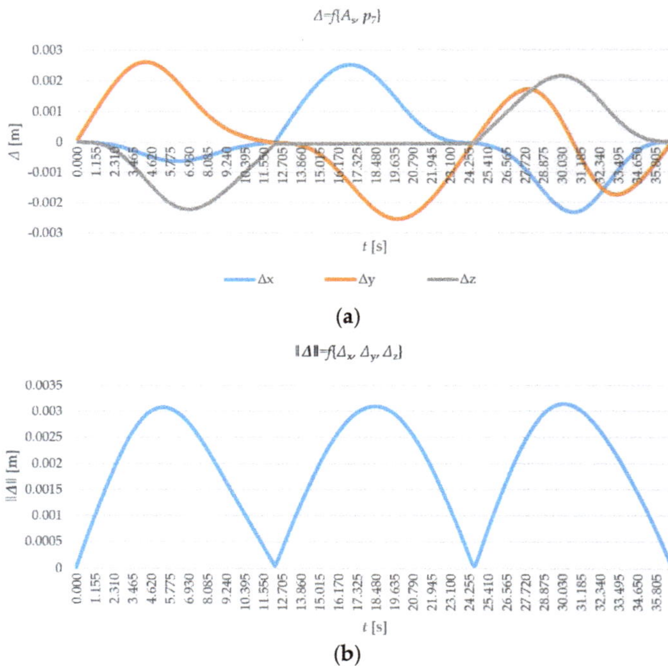

Figure 9. (a) The chart showing the difference Δ_x, Δ_y, Δ_z between the real trajectory (composed of the $A_s\{x_{As}, y_{As}, z_{As}\}$ points) and the one calculated (from the components of the position vector effector p_{7x}, p_{7y}, p_{7z}) in Matlab; (b) norm (size) of the vector $\|\Delta\|$.

As we can see, the CoppeliaSim Edu software complied with the tolerance range requirement $+/- 0.0035$ m for each vector component, for the inverse calculation of the joint coordinate data $\varphi(t)$ with respect to the trajectory consisting of a finite set of A_s points. The characteristics of physical quantities from sensor 7 in the model see Figure 3 (position, velocity, acceleration) are shown in Figure 10. We note that this is a different data set (different trajectory) than the one used in the illustration of Figure 5.

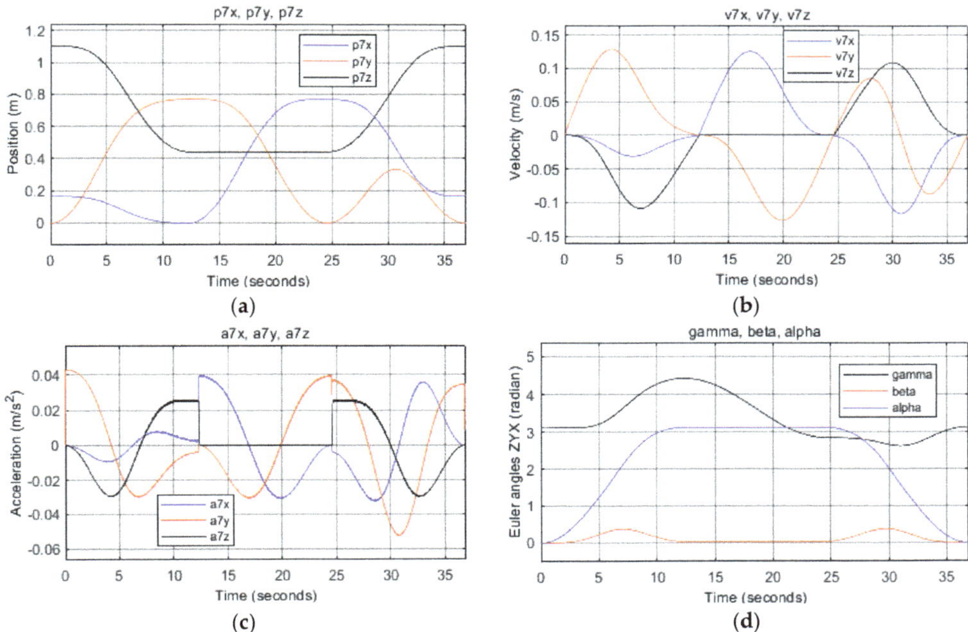

Figure 10. Effector values measured in Matlab: (**a**) position vector $p_7 = [p_{7x}, p_{7y}, p_{7z}]^T$ depending on the joint coordinate vector $\boldsymbol{\varphi}(t)$; (**b**) velocity vector $v_7 = [v_{7x}, v_{7y}, v_{7z}]^T$ depending on the joint coordinate vector $\boldsymbol{\varphi}'(t)$; (**c**) acceleration vector $a_7 = [a_{7x}, a_{7y}, a_{7z}]^T$ depending on the joint coordinate vector $\boldsymbol{\varphi}''(t)$ and $(\boldsymbol{\varphi}'(t))^2$; (**d**) Euler's angles (γ, β, α) depending on the joint coordinate vector $\boldsymbol{\varphi}(t)$.

2.5. Modeling in SolidWorks

In order to carry out structural changes, various structural compositions of the robotic arm composed of the URM modules are created in SolidWorks are shown in Figure 11. This mechanically detailed model corresponding to the reality was in the same dimensional configuration Table 1, as was used for making a comparison of kinematic indices and also the Δ_x, Δ_y, Δ_z difference between the real and the calculated trajectory.

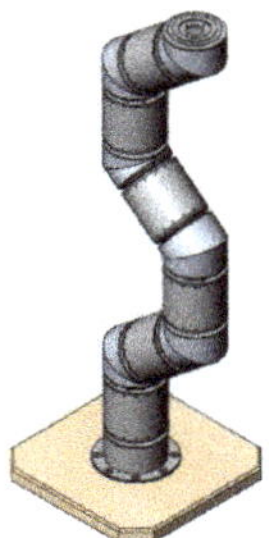

Figure 11. Model created in SolidWorks.

In this case, too, the graph of Figure 12 has confirmed that the requirement for a tolerance range of $+/-0.0035$ m for each vector component was adhered to by the inverse calculation of the joint coordinate data $\varphi(t)$ with respect to the trajectory composed of a finite set of the A_s points.

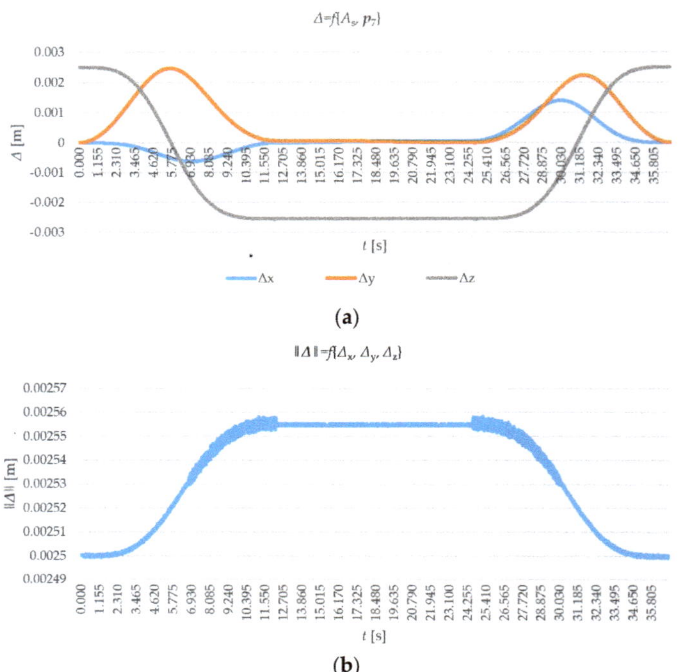

Figure 12. (**a**) A graph showing the Δ_x, Δ_y, Δ_z difference between the real trajectory (composed of $A_s\{x_{As}, y_{As}, z_{As}\}$ points) and the one calculated (from the components of the position vector effector p_{7x}, p_{7y}, p_{7z}) in SolidWorks; (**b**) norm (size) of the vector $\|\Delta\|$.

Here, however, certain changes in the shape of the course can be observed, which is logical, as we are dealing with two different software environments and the model structure itself as well. The norm, i.e., the size of the vector of deviation from the original trajectory $\|\Delta\|$ already has a permanent value of 0.0025 m here, and no significant minima can be seen in the three places where the trajectory "breaks" see Figure 2. The characteristics of physical quantities corresponding to the effector in the model see Figure 11 (position, velocity) are illustrated in Figure 13.

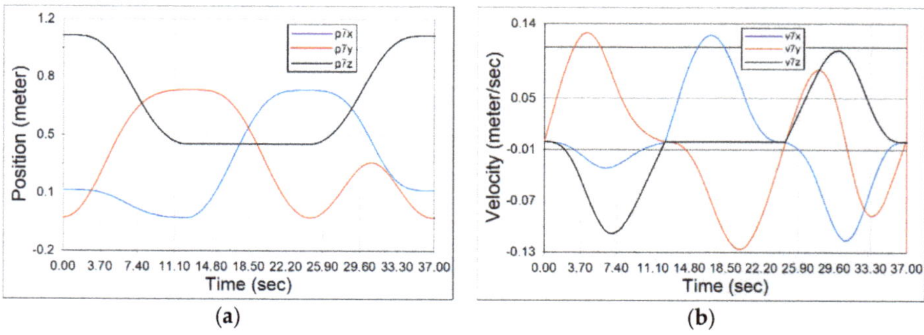

Figure 13. Effector values generated in SolidWorks: (**a**) position vector $p_7 = [p_{7x}, p_{7y}, p_{7z}]^T$ depending on the joints coordinate vector $\varphi(t)$; (**b**) velocity vector $v_7 = [v_{7x}, v_{7y}, v_{7z}]^T$ depending on the joints coordinate vector $\varphi'(t)$.

As we can see, the characteristics of position and velocity alike Figure 13a,b, corresponding to the effector in the model of Figure 11, are similar to the characteristics shown in Figure 10a,b.

3. Results

The robotic arm model was created in CoppeliaSim Edu in order to obtain the joints coordinates, corresponding to a description of the spatial trajectory consisting of the fixed points $A_s\{x_{As}, y_{As}, z_{As}\}$ with a permissible deviation of $+/-0.0035$ m from the original trajectory.

Results obtained in Matlab are shown in Figures 9 and 10. Figure 9 shows the difference Δ_x, Δ_y, Δ_z between the real trajectory (consisting of the $A_s\{x_{As}, y_{As}, z_{As}\}$ points) and the one calculated (from the components of the position vector effector p_{7x}, p_{7y}, p_{7z}). On the other hand, Figure 10 shows the values measured on the effector, namely:

(a)   The position determined by the position vector $p_7 = [p_{7x}, p_{7y}, p_{7z}]^T$ depending on the joint coordinate vector $\varphi(t)$;

(b)   The speed determined by the velocity vector $v_7 = [v_{7x}, v_{7y}, v_{7z}]^T$ depending on the first derivation (time-bound) of the joint coordinate vector $\varphi'(t)$;

(c)   The acceleration determined by the acceleration vector $a_7 = [a_{7x}, a_{7y}, a_{7z}]^T$ depending on the second derivation (time-bound) of the joint coordinate vector $\varphi''(t)$ and $(\varphi'(t))^2$;

(d)   The orientation determined by the Euler's angles according to the $[\gamma, \beta, \alpha]$ option, depending on the joint coordinate vector $\varphi(t)$.

Figure 5 shows the importance of using the repolarize. Otherwise, the joint coordinate data must be edited manually prior to their import, as was the case with SolidWorks.

The results obtained in SolidWorks are shown in Figures 12 and 13. Figure 12 shows the Δ_x, Δ_y, Δ_z difference between the real trajectory (consisting of the $A_s\{x_{As}, y_{As}, z_{As}\}$ points) and the one calculated (from the position vector effector components of p_{7x}, p_{7y}, p_{7z}). On the other hand, Figure 13 shows the values generated on the effector, namely:

(a)   The position determined by the position vector $p_7 = [p_{7x}, p_{7y}, p_{7z}]^T$ depending on the joint coordinate vector $\varphi(t)$;

(b)   The speed determined by the velocity vector $v_7 = [v_{7x}, v_{7y}, v_{7z}]^T$ depending on the first derivation (time-bound) joint coordinate vector $\varphi'(t)$.

4. Discussion

Based on the above results, we can conclude that the currently available software environments offer not only relatively powerful visualization tools but also those enabling various kinds of physical computations. The quality of results depends not only on the sophistication of the software and its preferable use but also on correct parameter specification, on the defined conditions and on the application of the software to its maximum potential. The aim of our work was especially to enable the comparison of data from the position kinematics in Matlab and in SolidWorks, subject to the same initial joint coordinate vector $\varphi(t)$. This has enabled us to check the designed models and to detect eventual structural discrepancies such as counter-rotation of the URM module or an undesirable shift of structural elements.

Matlab made it possible to choose the manner of processing the imported data, to align the sampling step with the CopeliaSim imported data step and thus ensure that a particular input datum actually corresponded to the given output value over the same time. It was possible to additionally construct a system to accommodate some changes in the input data Figure 6 and thus to make the work involved in input data changes more effective. The difference between the real and the calculated trajectory was quantified using the Δ vector. Figure 9b shows that the vector size $\|\Delta\|$ reaches its maxima at the maximum effector velocities.

In SolidWorks, the difference between the real and the calculated trajectory Δ was already different in nature; see Figure 12. Here, the $\|\Delta\|$ vector size has a permanent value and changes only slightly; no significant minima are observed here in the places where the trajectory "breaks", as shown in Figure 2.

Author Contributions: Draft concept and formal analysis, J.S., M.Š., Š.O.; model design and data processing in CoppeliaSimEdu, Z.B.; model design and data processing in Matlab, Š.O.; model design and data processing in SolidWorks, M.Š. and L.H.; supervision, methodology and project administration, J.S., J.D., P.D., T.S. All authors have read and agreed to the published version of the manuscript.

Funding: This work was supported by the Slovak Research and Development Agency under Contract no. APVV-18-0413. This paper was prepared with the support of the grant project KEGA 025TUKE-4/2019.

Conflicts of Interest: The authors declare no conflict of interest. The funders had no role in the design of the study; in the collection, analyses, or interpretation of data; in the writing of the manuscript, or in the decision to publish the results.

Abbreviations

List of the most important quantities:

p_i	Position vector between the reference coordinate system $S_1\{O_1, x_1, y_1, z_1\}$ and the coordinate system $S_i\{O_i, x_i, y_i, z_i\}$
r_i	Vector quantifying a kinematic chain segment
φ_i	Angle of rotation around the z_i axis of the $S_i\{O_i, x_i, y_i, z_i\}$ system, joint coordinate
φ	Joint coordinate vector
ϑ_i	Angle of rotation around the y_i axis of the $S_i\{O_i, x_i, y_i, z_i\}$ system
R_{yi}	Rotation matrix for the transformation of the rotational movement around the y_i axis
R_{zi}	Rotation matrix for the transformation of the rotational movement around the z_i axis
$R_{ZYX(i+1)}$	Rotation matrix for calculating Euler's angles α, β, γ
T	Sampling period
δ	A half period of the sampling period
t_r	Instant time of repolarization
t	Time
Δ	Vector difference between the p_7 position vector effector's position and the position of the $A_s\{x_{As}, y_{As}, z_{As}\}$ points that make up the trajectory
v_i	Instantaneous velocity vector to the $\{O_i, x_i, y_i, z_i\}$ system
a_i	Instantaneous acceleration vector to the $\{O_i, x_i, y_i, z_i\}$ system

References

1. Svetlík, J.; Štofa, M.; Pituk, M. Prototype development of a unique serial kinematic structure of modular configuration. *MM Sci. J.* **2016**, 994–998. [CrossRef]
2. Svetlík, J. Contribution to Construction of Manufacturing Engineering on Flexible Architecture Basis. Habilitation Thesis, Technical University of Košice, Košice, Slovakia, 14 June 2012.
3. Štofa, M. Experimentálny vývoj rotačných modulov pre stavbu sériových kinematických štruktúr vo výrobnej technike. Ph.D. Thesis, Technical University of Košice, Košice, Slovakia, 30 January 2020.
4. Gershenson, J.K.; Prasad, G.J. Modularity in product design for manufacturability. *Int. J. Agile Manuf.* **1997**, *3*, 99–110.
5. Benderbal, H.H.; Dahane, M.; Benyoucef, L. Modularity assessment in reconfigurable manufacturing system (RMS) design: An Archived Multi-Objective Simulated Annealing-based approach. *Int. J. Adv. Manuf. Technol.* **2018**, *94*, 729–749. [CrossRef]
6. Pérez, R.; Aca, J.; Valverde, A. A modularity framework for concurrent design of reconfigurable machine tools. In Proceedings of the International Conference on Cooperative Design, Visualization and Engineering, Heidelberg, Germany, 19–22 September 2004; Springer: Berlin, Germany, 2004.
7. Svetlík, J. Modularity of Production Systems. In *Machine Tools*; Šooš, Ľ., Marek, J., Eds.; IntechOpen: London, UK, 2020; pp. 1–22.
8. Yim, M.; Duff, D.G.; Roufas, K.D. PolyBot: A modular reconfigurable robot. In Proceedings of the ICRA '00, IEEE International Conference on Robotics and Automation, San Francisco, CA, USA, 24–28 April 2000; Volume 1, pp. 514–520.
9. Yim, M.; Roufas, K.; Duff, D.; Zhang, Y.; Eldershaw, C.; Homans, S. Modular reconfigurable robots in space applications. *Auton. Robot.* **2003**, *14*, 225–237. [CrossRef]
10. Acaccia, G.; Bruzzone, L.; Razzoli, R. A modular robotic system for industrial applications. *Assem. Autom.* **2008**, *28*, 151–162. [CrossRef]
11. Xu, W.; Han, L.; Wang, X.; Yuan, H. A wireless reconfigurable modular manipulator and its control system. *Mechatron.* **2021**, *73*, 102470. [CrossRef]

12. Pacaiova, H.; Isiarikova, G. Base Principles and Practices for Implementation of Total Productive Maintenance in Automotive Industry. *Qual. Innov. Prosper.* **2019**, *23*, 45–59. [CrossRef]
13. Nikitin, Y.; Bozek, P.; Peterka, J. Logical–linguistic model of diagnostics of electric drives with sensors support. *Sensors* **2020**, *20*, 4429. [CrossRef] [PubMed]
14. Paden, B. Kinematics and Control Robot Manipulators. Ph.D. Thesis, University of California, Berkeley, CA, USA, 1986.
15. Kahan, W. *Lectures on Computational Aspects of Geometry*; Department of Electrical Engineering and Computer Sciences, University of California: Berkeley, CA, USA, 1983; Unpublished.
16. Buss, S.R. Introduction to inverse kinematics with jacobian transpose, pseudoinverse and damped least squares methods. *IEEE J. Robot. Autom.* **2004**, *16*, 1–19.
17. Dulęba, I.; Opałka, M. A comparison of Jacobian-based methods of inverse kinematics for serial robot manipulators. *Int. J. Appl. Math. Comput. Sci.* **2013**, *23*, 373–382. [CrossRef]
18. Meredith, M.; Maddock, S. *Real-Time Inverse Kinematics: The Return of the Jacobian*; Technical Report No. CS-04-06; Department of Computer Science, University of Sheffield: Sheffield, UK, 2004.
19. Šoch, M.; Lórencz, R. Solving inverse kinematics—A new approach to the extended Jacobian technique. *Acta Polytech.* **2005**, *45*, 21–26.
20. Virgala, I.; Gmiterko, A.; Surovec, R.; Vacková, M.; Prada, E.; Kenderová, M. Manipulator end-effector position control. *Procedia Eng.* **2012**, *48*, 684–692. [CrossRef]
21. Virgala, I.; Tomáš, L.; Miková, Ľ. Snake robot locomotion patterns for straight and curved pipe. *J. Mech. Eng.* **2018**, *68*, 91–104. [CrossRef]
22. Nilsson, R. Inverse Kinematics. Master's Thesis, Luleå University of Technology, Luleå, Sweden, 2009.
23. Wampler, C.W. Manipulator inverse kinematic solutions based on vector formulations and damped least-squares methods. *IEEE Trans. Syst. Man, Cybern.* **1986**, *16*, 93–101. [CrossRef]
24. Lozhkin, A.; Bozek, P.; Maiorov, K. The method of high accuracy calculation of robot trajectory for the complex curves. *Manag. Syst. Prod. Eng.* **2020**, *28*, 247–252. [CrossRef]
25. Ondočko, Š.; Stejskal, T.; Svetlík, J.; Hrivniak, L.; Šašala, L. Direct kinematics of modular system. In *Automatizácia a riadenie v teórii a praxi 2020, Proceedings of the 14. ročník konferencie odborníkov z univerzít, vysokých škôl a praxe, Stará Lesná, Slovensko, 5–7 February 2020*; Šeminský, J., Mižáková, J., Šimšík, J., Balog, M., Eds.; Technická Univerzita v Košiciach: Košice, Slovakia, 2020.
26. Ondočko, Š.; Stejskal, T.; Svetlík, J.; Hrivniak, L.; Šašala, L.; Žilinský, A. Position forward kinematics of 6-DOF robotic arm. *Acta Mech. Slovaca.* (under review).
27. Murray, R.; Li, Z.; Shankar, S. *A Mathematical Introduction to Robotic Manipulation*, 1st ed.; CRC Press: Boca Raton, FL, USA, 2017; p. 519. ISBN 9781315136370.
28. Rate Limiter. Available online: https://www.mathworks.com/help/simulink/slref/ratelimiter.html?s_tid=srchtitle (accessed on 29 October 2020).
29. How to Remove Unwanted Short Time Signal Simulink. Available online: https://stackoverflow.com/questions/52384093/how-to-remove-unwanted-short-time-signal-simulink/52409614 (accessed on 29 October 2020).
30. Filtering and Smoothing Data. Available online: https://www.mathworks.com/help/curvefit/smoothing-data.html (accessed on 29 October 2020).
31. Signal Smoothing. Available online: https://www.mathworks.com/help/signal/ug/signal-smoothing.html (accessed on 29 October 2020).

Article

Industrial Robot Positioning Performance Measured on Inclined and Parallel Planes by Double Ballbar

Ivan Kuric [1], Vladimír Tlach [1,*], Milan Sága [2], Miroslav Císar [1] and Ivan Zajačko [1]

[1] Department of Automation and Production Systems, Faculty of Mechanical Engineering, University of Zilina, 010 26 Zilina, Slovakia; ivan.kuric@fstroj.uniza.sk (I.K.); miroslav.cisar@fstroj.uniza.sk (M.C.); ivan.zajacko@fstroj.uniza.sk (I.Z.)

[2] Department of Applied Mechanics, Faculty of Mechanical Engineering, University of Zilina, 010 26 Zilina, Slovakia; milan.saga@fstroj.uniza.sk

* Correspondence: vladimir.tlach@fstroj.uniza.sk

Abstract: Renishaw Ballbar QC20–W is primarily intended for diagnostics of CNC machine tools, but it is also used in connection with industrial robots. In the case of standard measurement, when the measuring plane is parallel to the robot base, not all robot joints move. The purpose of the experiments of the present article was to verify the hypothesis of the motion of all the robot joints when the desired circular path is placed on an inclined plane. In the first part of the conducted experiments is established hypothesis is confirmed, through positional analysis on a simulation model of the robot. They are then carried out practical measurements being evaluated the influence of individual robot joints to deform the circular path, shown as a polar graph. As a result, it is found that in the case of the robot used, changing the configuration of the robot arm has the greatest effect on changing the shape of the polar graph.

Keywords: industrial robot; renishaw Ballbar QC20–W; positional analysis; measurement

Citation: Kuric, I.; Tlach, V.; Sága, M.; Císar, M.; Zajačko, I. Industrial Robot Positioning Performance Measured on Inclined and Parallel Planes by Double Ballbar. *Appl. Sci.* **2021**, *11*, 1777. https://doi.org/10.3390/app11041777

Academic Editor: Subhas Mukhopadhyay

Received: 29 January 2021
Accepted: 15 February 2021
Published: 17 February 2021

Publisher's Note: MDPI stays neutral with regard to jurisdictional claims in published maps and institutional affiliations.

1. Introduction

The constant increase in automation in industrial areas is also related to the increasing share of industrial robotics applications. According to a press release issued by IFR (International Federation of Robotics), almost two million new robots are expected to be installed in manufacturing plants worldwide between 2020 and 2021 [1]. One of the first steps in creating a robotic application is a selection of a suitable robot according to certain parameters [2]. Key features include workspace size, load capacity, and also the number of controlled axes. Depending on the specific task that will be automated by the industrial robot, the so-called performance criteria are also important.

Performance criteria allow a more detailed look at the robot features. One of the best-known performance criteria of industrial robots is the pose repeatability, which expresses the ability of the robot TCP (Tool Center Point) to repeatedly return to a defined position in the same direction [3,4]. Pose repeatability is also the performance criterion most frequently mentioned in technical data sheets by industrial robot manufacturers [5]. In addition to the performance criterion above, there are many others, all of which are defined in the ISO 9283 standard. This standard defines the individual performance criteria, the method of their calculation, together with the recommended conditions for their measurement [6].

ISO 9283 is further elaborated on in the respective technical report ISO/TR 13309, which aims to provide an overview of metrological methods and measuring equipment for measuring performance criteria [7]. Numerous scientific papers also deal with the issues related to measurement of the performance criteria of industrial robots. In most cases, these measurements are performed using indicators [8–11], laser trackers [12–14], vision systems [15–17], or laser interferometers [18–20] primarily intended for measuring the properties of CNC machine tools. These measuring devices satisfy the classifications and methods described in the above technical report to varying degrees.

Furthermore, in professional publications, a double ballbar type of a device is experimentally used to measure performance parameters of industrial robots [19,21–23], which, same as the laser interferometers, are primarily intended for evaluating the condition of CNC machine tools [24]. At the Department of Automation and Production Systems, several papers have also been published related to the possibilities of using the given measuring equipment in connection with industrial robots [25–28]. The principle of measurement with the Renishaw Ballbar is based on movement along a circle or an arc with a fixed point of rotation, while a real distance of moving part from the center of the rotation is compared to the ideal one and then the deviations are recorded. Actual measurement utilizes a precise linear encoder [29]. The use of this device in connection with industrial robots is not specified in the standard (ISO 9283) nor in the technical report (ISO/TS 13309) nor in the standard that describes intended usage of such device (ISO 230–4).

The necessity for knowledge of real values of performance criteria is important not only in the creation of a new robotic cell, in connection with the selection of a suitable industrial robot, but also in robotic applications where the industrial robot has been performing a given task already for some time. These are mainly applications that are more demanding on performance criteria, their stability, or predictability of their development thru time. Such tasks are, for example, precision assembly, a dimensional inspection that implements 3D scanners or measuring probes, etc. [25,30]. In such cases, the measurement of performance criteria is necessary to monitor the robot's technical condition or to troubleshoot existing problems. Early detection of changes in the robot's technical condition may prevent a negative impact of these changes on the work process of the robotic application.

The presented paper deals with the measurement of an industrial robot using a double ballbar type of device-Renishaw Ballbar QC20–W. Specifically, its implementation for measurements with a circular motion path located on an inclined plane, resulting in the motion of all six joints of the robot performing a given TCP movement. The measurement conducted in this way provides a significant amount of information about the robot's current technical state, which can be used in connection monitoring or in the troubleshooting process.

2. Motivation

As indicated in the introduction, several measurements with the Renishaw Ballbar QC20–W involving industrial robots were carried out in the past at the Department of Automation and Production Systems, and the results have been published [25–28]. In most cases, these were basic measurements in the XY plane, parallel to the base plane of the WCS (World Coordinate System) and, at the same time, parallel to the robot base. The measurements were done on Fanuc robots, where the WCS is represented by a fixed coordinate system defined by the manufacturer. The same orientation of the measuring plane in the robot's workspace has been chosen by other authors in their papers too [19,21,22]. The said basic type of measurement requires completing two circular paths in the XY plane in a clockwise direction (CW) and two circular paths in the opposite direction (CCW) [31,32]. Figure 1 shows the course of the motion of the individual robot joints (J1 through J6), during the TCP movement in a circular path in the clockwise direction. The graph shows data that was created in the Creo Parametric 5.0.4.0 program using movement simulation of the model of the Fanuc LR Mate 200iC robot, with the simulation conditions given in Table 1.

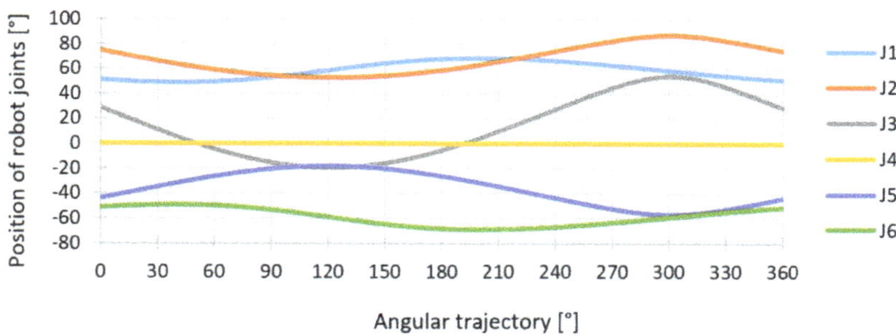

Figure 1. The course of the motion of the individual robot joints during the TCP movement in a circular path in the XY plane.

Table 1. Conditions for creating the circle in the XY plane.

Type of Industrial Robot	Fanuc LR Mate 200iC
The radius of circular paths	100 mm
Centre of circle relative to WCS (X; Y; Z; W; P; R)	300; 500; −253.5; 0; 0; 0
Coordinates of the TCP (X; Y; Z)	0; 0; 74.65

The graph in Figure 1 shows that when the measuring is done in the XY plane, the J4 joint is static for the entire duration of the TCP moving along the circle, therefore, it can be said that the results do not take into account the possible negative effect of the J4 joint motion or that the given results correspond to only one specific position from the range of the J4 joint motion. Regardless of the above statement, the measurement in the XY plane is informative and can be used to evaluate the robot's technical condition. For example, in one of our papers [25], a measurement in the XY plane was used to evaluate the effect of calibration on the accuracy of the resulting movement of the robot's TCP.

In the case of CNC machine tools, to evaluate the effect of movement of all three linear axes (X, Y, and Z), the so-called volumetric analysis, consisting of measurements in three mutually perpendicular planes XY, ZX, and ZY, is used. During the measurement in the XY plane (see Figure 2a) two standard clockwise and two counterclockwise circles are made. In the ZX and ZY planes (see Figure 2b,c) two semicircular paths in the range of 220° are measured, again one in the clockwise and the other in the counterclockwise direction [33,34].

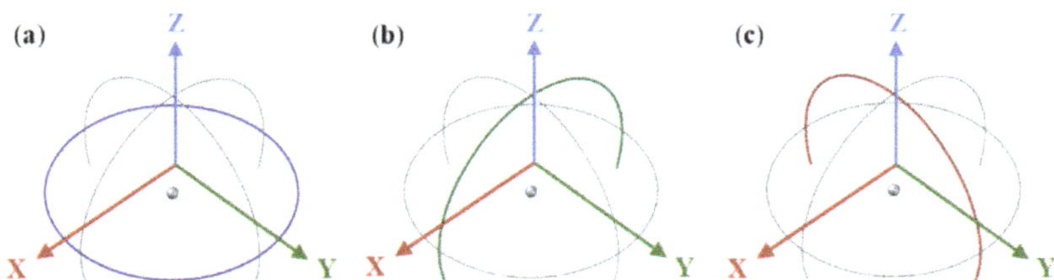

Figure 2. Circular paths in individual planes for volumetric analysis. (**a**) plane XY; (**b**) plane ZX; (**c**) plane ZY.

To verify the possibilities of the volumetric analysis application on industrial robots, a series of measurements were performed in the past on two industrial robots Fanuc LR Mate 200iD and LR Mate 200iD/7L. The circular path and circular arcs with a radius of 100 mm, necessary for measurement purposes, were created through standard programming commands for Fanuc robots. Further information on the measurement conditions is given in Table 2.

Table 2. Measurement conditions for volumetric analysis.

Type of Industrial Robot	Fanuc LR Mate 200iD (E–117096)	Fanuc LR Mate 200iD/7L (E–109256)
The radius of circular paths	100 mm	100 mm
Centre of circle relative to WCS (X; Y; Z; W; P; R)	300; 500; −253.5; 0; 0; 0	300; 500; −253.5; 0; 0; 0
Coordinates of TCP (X; Y; Z)	0; 0; 74.65	0; 0; 74.65
The payload on the end of the robot's arm	438 g	438 g
Speed of TCP	55.33 mm/s	55.33 mm/s
Number of repetitions	30	30
Running-in period before measurement	1 h	1 h

E–117096 and E–109256 represent the serial numbers of the industrial robots used.

The measurements were evaluated in the Ballbar 20 software, with Figures 3 and 4 showing polar graphs of one of thirty repeated measurements for each plane and for both Fanuc robots. In addition to a more detailed analysis of the measurement results, two observations are important for the purposes of this paper.

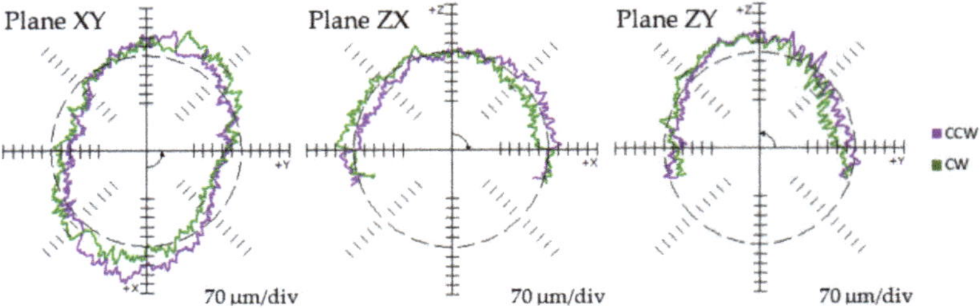

Figure 3. Polar graphs of the volumetric analysis evaluation—LR Mate 200iD.

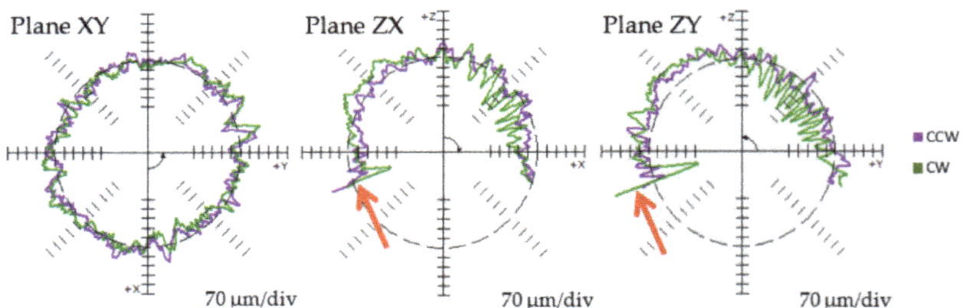

Figure 4. Polar graphs of the volumetric analysis evaluation—LR Mate 200iD/7L.

The first is the "solid line", which is apparent on the polar graphs for the ZX and ZY planes when measured on the LR Mate 200iD/7L robot (Figure 4). This phenomenon occurs due to the continuation of value recording even during the exit at the end of the measurement. As part of the measurement with the Renishaw Ballbar QC20–W, the data recording process is started and ended by entering and exiting the measuring radius. In the case of measuring a circle or a circular arc with a radius of 100 mm, the measurement is started by infeed, a linear movement from the radius of 101 mm to a value of 100 mm, followed by a movement along the circular path itself. On the contrary, after completing a circular path in one direction, the measurement is ended by a feed out, a linear movement to a radius of 101 mm. If the data were also recorded during the feed out from the measuring radius, then this fact is reflected in the already mentioned "continuous line", observed on the polar graph. This phenomenon can be prevented by extending lengths of the infeed and the feed out to more than 1 mm, or by adjusting the speed during these movements. However, in the case of the measurement conducted on the LR Mate 200iD/7L robot, neither of the measures mentioned was effective, and the said phenomenon occurred in all of the thirty repeated measurements. The results of such measurement are skewed and in order to be used, the data need to be additionally filtered. However, the need for additional filtering results in the necessity for a longer time and more complex processing and evaluation of the measurement. Furthermore, it should be noted that the source of the phenomenon described could also have been the linear sensor used in the measurement. However, in order to refute this claim, further measurements would have to be conducted.

The second of the mentioned observations in the implementation of the volumetric analysis is related to the motion of the industrial robot's individual joints. Figure 5 shows the course of the motion of all six joints of the LR Mate 200iD robot during the TCP movement along a circular path in the ZX plane, within the volumetric analysis. As the graph makes apparent, the J4 joint is static during the entire movement. The situation is the same in the case of the ZY plane. It follows from the above that even in the case of applying the volumetric analysis on an industrial robot, the motion of the J4 joint is not achieved during the measurement.

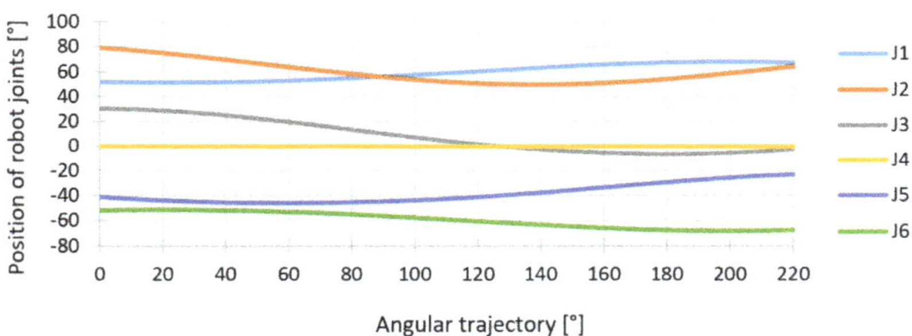

Figure 5. The course of the motion of the individual robot joints during the TCP movement in a circular path in the ZX plane.

In all the presented methods of measurement, whether using a circle or arched paths in the case of the volumetric analysis, the robot's TCP has always moved only in one of some perpendicular planes of the cartesian coordinate system. At the same time, the "Z" axes of the tool coordinate system and world coordinate system were parallel during movement in each of the measuring planes. Therefore, it can be assumed that if both mentioned axes formed a certain angle, the J4 joint would be non-zero and would participate in the given circular motion. One of the possible solutions to achieve the J4 joint motion is to place the desired circle at a certain angle in the robot's workspace. I.e., placing the circle on an inclined plane. With the measurement path designed in this way, it is necessary to verify the correctness of the statement in connection with the J4 joint motion and, at the same time, the possibilities or the ability to conduct measurements with Renishaw Ballbar QC20–W. In the present paper, this hypothesis has been verified, with a positional analysis of the robot's movement in Section 4 by the Creo Parametric 5.0.4.0 system, and inSection 5 by a real measurement done on an industrial robot Fanuc LR Mate 200iC, located in the laboratory of our department.

3. Experiment Preparation

The LR Mate 200iC robot was chosen to conduct the measurement on, with the selection of the position and orientation of the desired circular path in the robot's workspace being the first step. For this purpose, information from the ISO 9283 standard was applied, where the recommended procedure for measuring performance criteria is based on the so-called ISO cube. It is an imaginary cube, its vertices marked C1 to C8, and it is placed in the robot's workspace, with its edges parallel to the robot's basis or its world coordinate system. Based on the industrial robot type, a plane containing points is placed in the said cube, with the points defining the individual paths used for measuring the performance criteria. In the case of robots with six controlled axes, the inclined plane determined by the vertices C1–C2–C7–C8 is selected (see Figure 6) [6]. For the purpose of the experiment, a circle required for conducting the measurement with the Renishaw Ballbar QC20–W is be placed on this plane.

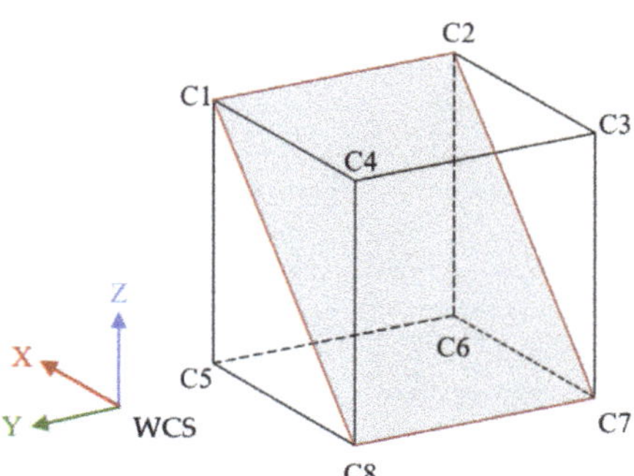

Figure 6. ISO cube with a measuring plane, for robots with six controlled axes.

As for the circle diameter, a basic nominal radius of 100 mm, which had been used in the past experiments, was chosen so that it would be easier to compare the measurement results. Furthermore, it was necessary to determine the position of the ISO cube or the center of the circle in the Fanuc LR Mate 200iC robot's workspace. The base consisted of the coordinates of the center of the circle, used in the present volumetric analysis (see Table 2). However, this position was adjusted so that when the circle was tilted below the angle of −45°, the measuring apparatus did not collide with the workbench on which the Fanuc robot was placed. At the same time, the possibilities of the robot arm's reach in the movement along a defined path were also taken into account. As a result, the coordinates of the center of the circle {X = 200; Y = 400; Z = −200} were defined by the Fanuc Roboguide HandlingPRO 8 software.

Two possible configurations of the robot arm correspond to the defined position and orientation of the circle, as well as to the circular path required for the measurement. Specifically, this is a configuration related to the J5 joint position, which is referred to as the N–NOFLIP (joint J5 < 0°) and, as an F–FLIP (joint J5 > 0°) position in Fanuc robots [35]. Both configurations are shown in Figure 7. The measurement was done for each of the two possible configurations of the robot arm.

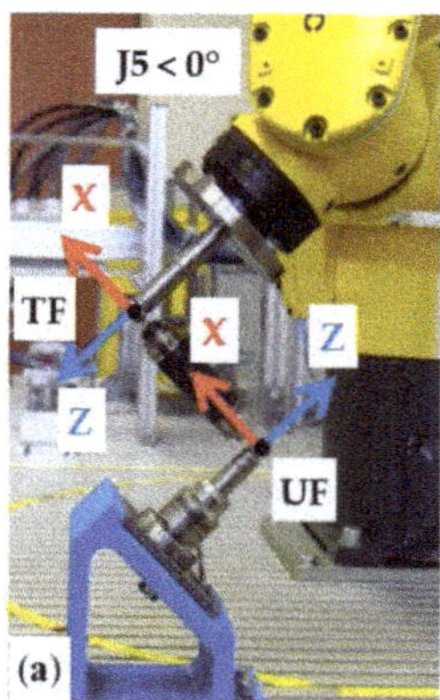

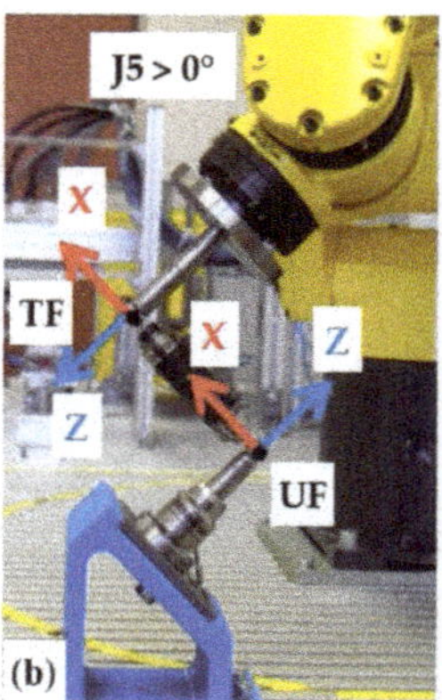

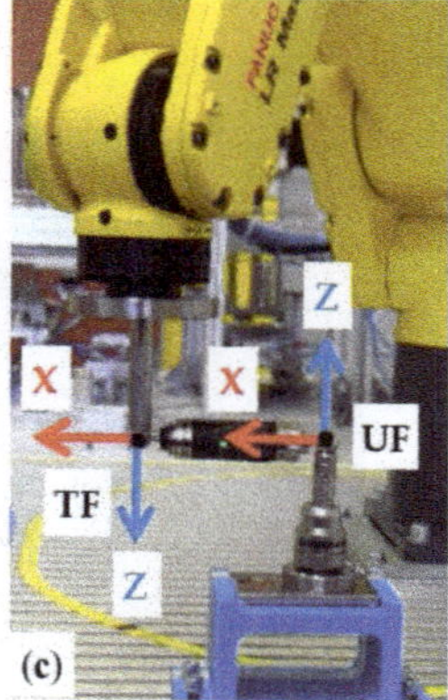

Figure 7. Jigs for measuring in two selected measuring plane. (a) NOFLIP configuration; (b) FLIP configuration; (c) NOFLIP configuration.

To carry out practical measurements, two jigs were designed and manufactured through 3D printing. These allowed the measuring apparatus to be clamped to the robot's workbench while maintaining the defined coordinates of the center of the circle. The first of the jigs enabled a measurement at an angle of −45° (Figure 7a,b). The second jig was used to conduct a standard measurement in the XY plane parallel to the benchtop (Figure 7c) while maintaining the same coordinates of the center of the circle as was the case with the inclined plane.

The robot's control program was created in such a way that the center of the circle of a given radius was at the beginning of the User Coordinate System (UF) defined concerning the robot's WCS. Through a change in the position or the orientation of the user coordinate system, the program thus created allows a change in the position and orientation of the emergent circle. Thus, the same program is actually used for the desired circle on an inclined plane at an angle of −45° and for the circle in a plane parallel to the benchtop. The programs created in this way are to some extent universal, as they can be used for any position and orientation in the robot's workspace and can also be used on another Fanuc robot. The relative position and orientation of the user (UF) and tool coordinate system (TF) used in the conducted experiments are shown in Figure 7.

4. Positional Analysis in the Creo Parametric System 5.0.4.0

To verify the influence of the selected orientation of the circle on the motion of the robot's individual joints, a positional analysis was created in the Creo Parametric system 5.0.4.0. For the purpose thereof, a simulation model of the LR Mate 200iC robot was created with the kinematic constraints applied. Zero positions of the individual robot joints and their range of motion were defined, corresponding to those of the real robot. The position of the TCP and of the individual coordinate systems was also defined, as was the case with the experiments on a real robot. The simulation conditions are given in Table 3.

Table 3. Simulation conditions for positional analysis.

Type of Industrial Robot	Fanuc LR Mate 200iC
The radius of circular paths	100 mm
Centre of circle relative to WCS (X; Y; Z; W; P; R)	200; 400; −200; 0; −45; 0
Coordinates of the TCP (X; Y; Z)	0; 0; 74.65

The center of the required circular path was located at the zero of the local coordinate system, while its position relative to the WCS robot is shown in Table 3. The circular path itself was created through an inverse kinematics task, by defining the so-called servomotors in the individual axes of motion of the Cartesian coordinate system. The cosine function was used for the direction of the X and Y axes, with its general shape in the Creo Parametric system being the following (1):

$$q = A \cdot \cos\left(360\frac{x}{T} + B\right) + C \tag{1}$$

where the coefficients A = amplitude [mm], B = phase, C = offset [mm] and T [s] represent a period [36]. For the X, Y–axis direction, the required radius of the circle of 100 mm, and the clockwise movement, the given function has the following form (2), (3):

$$q = 100 \cdot \cos\left(360\frac{x}{360}\right) \tag{2}$$

$$q = 100 \cdot \cos\left(360\frac{x}{360} + 90\right) \tag{3}$$

Other axes are defined as constants. Movement in direction of the Z–axis = 0 mm, rotations around the Y and Z axes are also zero (P = 0°; R = 0°). The rotation around the X–axis has a defined value W = −180°, which ensures that the Z–axis of the tool coordinate system has the correct orientation and is perpendicular to the measuring plane or the plane in which the required circle is located.

The results of the positional analysis of the course of motion of individual robot joints are shown in the form of a graph in Figure 8. From the individual trend lines, one can see that all joints of the robot, including the J4 joint, participated in the defined motion. The above graph applies to the clockwise movement of the TCP and to the "NOFLIP" robot arm configuration. For the second configuration ("FLIP"), the graph is similar, with all the joints of the robot participating in the movement. The absolute value of the range of motion of each of the robot's joints is shown in the graph in Figure 9. Because the TCP moved along the same path in both configurations, the displayed range value for both configurations is identical.

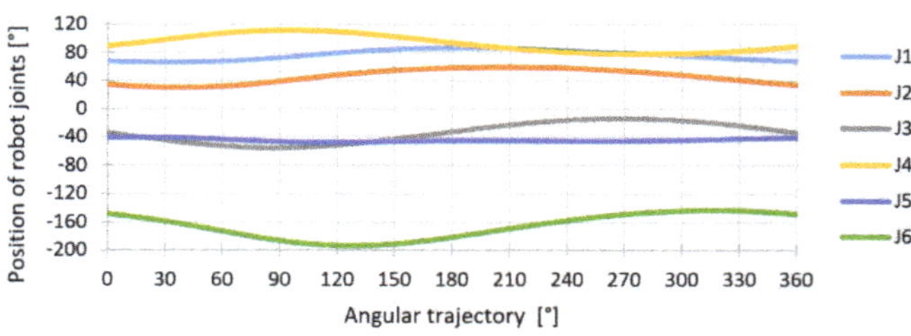

Figure 8. The course of the motion of the individual robot joints during the TCP movement in a circular path in the inclined plane.

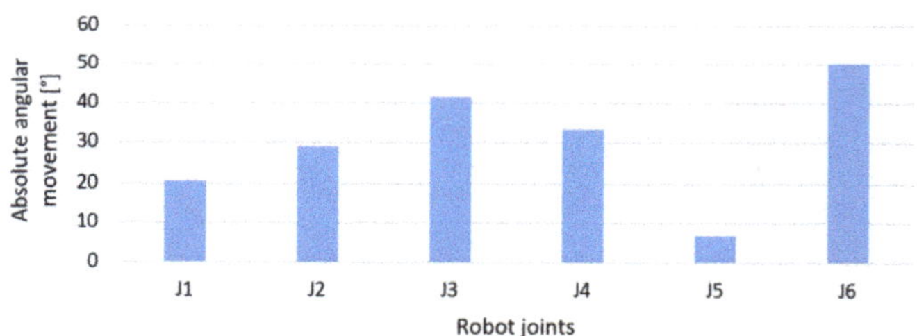

Figure 9. Absolute angular movement of robot joints during the movement along a circular path on the inclined plane.

The present positional analysis confirmed the hypothesis of the robot's individual joints' motion in connection with the location of the required circle on an inclined plane. The next step was to verify the possibilities of practical implementation of such measurement with the Renishaw Ballbar.

5. Measurement with Renishaw Ballbar QC20–W

All the practical measurements presented here were carried on an industrial robot Fanuc LR mate 200iC, located in the laboratory of our department. This robot was examined mainly due to its availability. Basic measurement with the Renishaw Ballbar QC20–W and a circle with a radius of 100 mm placed on an inclined plane (see Figure 6) consisted of two series of measurements with ten repetitions. One series for each of the robot's arm configurations. More detailed measurement conditions are given in Table 4.

The robot control program was created by standard programming commands, such as through connecting two semicircles. For the TCP speed during the measurement, the value of 55.33 mm/s was selected and entered into the robot control program. Such speed corresponds to the federate of 3320 mm/min in the Ballbar 20 software. This speed value is based on experiments conducted in the past [26]. Specifically, it was a comparison of two methods of creating a circular path. The first of the methods was based on the already mentioned connection of two semicircles. For the second method, the circle was created as a polygon, representing the most accurate approximation of the desired circle. The speed of 55.33 mm/s was then chosen as optimal for the purposes of the said comparison. Subsequently, further experiments with the Renishaw Ballbar were conducted with this speed value, in addition to the volumetric analysis mentioned in the introduction of this paper. At this speed value, the measurement process itself (circle radius 100 mm) with the Renishaw Ballbar takes approximately 53 s. This is a relatively short measurement time, also suitable for the purposes of industrial robot troubleshooting and monitoring.

Table 4. Measurement conditions.

Type of Industrial Robot	Fanuc LR Mate 200iC (E–31806)
The radius of circular paths	100 mm
Centre of circle relative to WCS (X; Y; Z; W; P; R)	300; 500; −253.5; 0; −45; 0
Coordinates of the TCP (X; Y; Z)	0; 0; 74.65
The payload on the end of the robot's arm	438 g
Speed of TCP	55.33 mm/s
Number of repetitions	10 × for configuration N, F, and F′
Running-in period before measurement	1 h

Standard Ballbar 20 software was used to record the measurements, with graphical results in the form of polar graphs shown in Figure 10. By comparing the polar graphs for both robot arm configurations, it is possible to see a significant difference in their shape. The significant difference is also shown in the average value of circularity in Table 5, where the circularity represents the difference between the largest and the smallest radii recorded by the Ballbar during the measurement excluding infeed and feed out. This value is related to the accuracy of the machine in the sense that the higher the circularity value, the worse the accuracy of the machine [37].

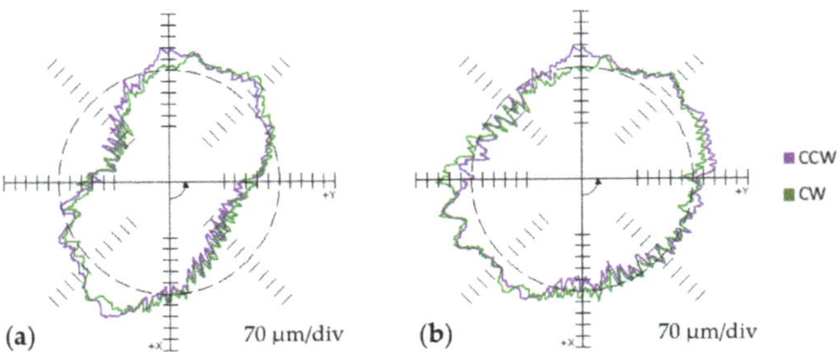

Figure 10. Polar graphs from measuring a circle on the inclined plane. (**a**) NOFLIP configuration; (**b**) FLIP configuration.

Table 5. Average circularity value for both robot arm configurations.

	Configuration NOFLIP	Configuration FLIP
Circularity [μm]	652.2	409.8

The polar graphs mainly show the squareness error and the scaling error. Both of these defects are characterized by the oval shape of the profile or its "peanut" shape. In the case of a squareness error, the axis of such deformation is inclined by 45° or 135°. In the case of CNC machine tools, this error is caused by the fact that the two axes performing the movement are not perpendicular [38]. The second of these errors, the scaling error, is caused by the different magnitude of the movement increment between the axes and is manifested by the said deformation along the axis of movement [38]. Unlike a CNC machine tool, with industrial robots, the circular motion during the measurement is the result of the simultaneous motion of several or all of the robot's joints. For this reason, these errors can be attributed mainly to the incorrect conversion of Cartesian coordinates, related to the direct kinematic task, and the issue of calibration. Specifically, this is the first level calibration, related to the relationship between the actual position of the robot arm and the information from the encoders in its individual joints [39].

For a more detailed analysis of the measurements, based on the robot's simulation model, a graph was created in Figure 11. This graph shows the range of motion of the robot's individual joints during measurements with both configurations of the robot arm. The graphical representation shows that the movements of the joints J1, J2, and J3 are identical in both configurations. On the contrary, the changes occur only in the orientation of the robot's subsystem, i.e., that of the joints J4, J5, and J6. Based on this analysis, it can be concluded that the change in the shape of the polar graphs (Figure 10) is caused by one of the joints of the robot's orientation subsystem, or a combination thereof.

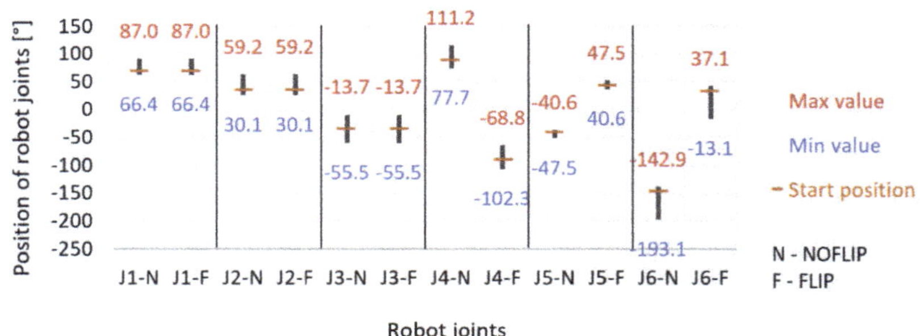

Figure 11. Comparison of the movement of robot joints for both used robot arm configurations.

A more accurate identification of the specific robot joint or joints that adversely affect the measurement results can be addressed in several ways. This involves, for example, limiting the motion of one of the robot's examined joints, or modifying the motion of the joint so that its possible adverse effect is apparent. The graph in Figure 11 shows that the motion of the J6 joint is shifted by 180° relative to each other in both configurations. At the same time, the coordinates of the TCP orientation are constant during the entire circular motion and have the following values (W = 180°; P = 0°; R = 0°). Because the TCP is in the axis of rotation of the J6 joint, the coordinate "R" (rotation around the Z–axis) can influence the way the joint J6 moves.

This theory was first verified in the robot's simulation model. For an arm configuration with J5 > 0° (FLIP), a TCP position was defined with the coordinate R adjusted to 180°. The graph in Figure 12 shows that by applying the said change, the same motion of the J6 joint was achieved as in the NOFLIP configuration. At the same time, there was no change in the J4 and J5 joints.

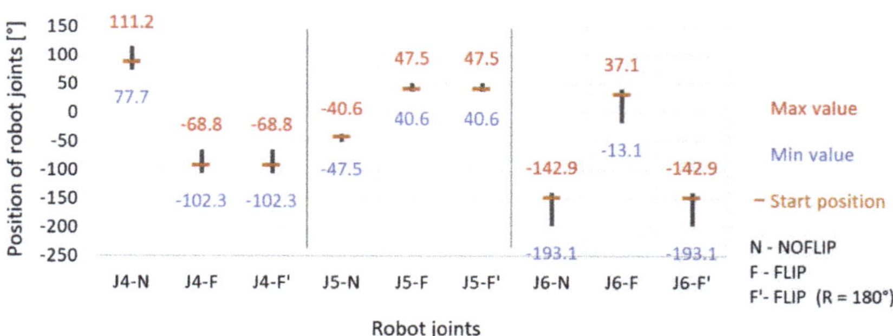

Figure 12. Impact of TCP coordinate modification on the movement of robot joints.

Subsequently, practical measurements were done, with a change applied to individual points of the control program with the adjustment of the coordinate R = 180°. The conditions of this measurement were identical to the conditions from Table 4, with the number of repeated measurements being ten. The resulting polar graph is shown in Figure 13c. In comparison to previous measurements (Figure 13a,b), it is apparent that the change in the motion of the J6 joint does not affect the shape of the polar graph. This fact is also confirmed by the average value of circularity given in Table 6, while the difference of 20.4 μm can be considered negligible. Based on the presented results, it can be argued that the change in the shape of the polar graph is caused by the joint J4 or J5, or a combination thereof.

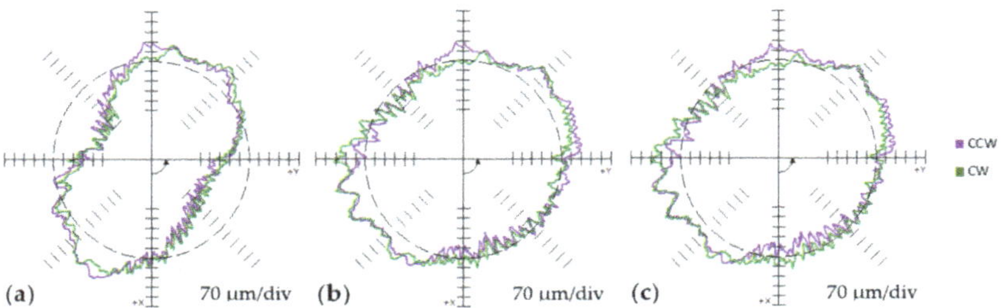

Figure 13. Polar graphs from measuring a circle on the inclined plane. (**a**) NOFLIP configuration; (**b**) FLIP configuration; (**c**) FLIP′ configuration (coordinate R = 180°).

Table 6. Average circularity value for measuring a circle on the inclined plane.

	Configuration NOFLIP	Configuration FLIP	Configuration FLIP′
Circularity [μm]	652.2	409.8	389.3

5.1. Measurement in a Plane Parallel to the Robot's Base

The series of measurements was further supplemented by the measurement of a circular path when the XY plane formed an angle of 0° with the benchtop. The second of the already mentioned jigs was used for this measurement (Figure 7c), which ensured that the central position of the circle was identical to its central position in the previous measurements. Compared to measurements on an inclined plane, the ranges of motions of the individual robot joints are different, with the J4 joint not participating in the movement and its rotation has a constant value throughout the measurement. The absolute values of the range of motion of the individual robot joints measured on an inclined plane (at an angle of −45°) and a plane parallel to the base of the robot (angle 0°) are shown in the graph in Figure 14.

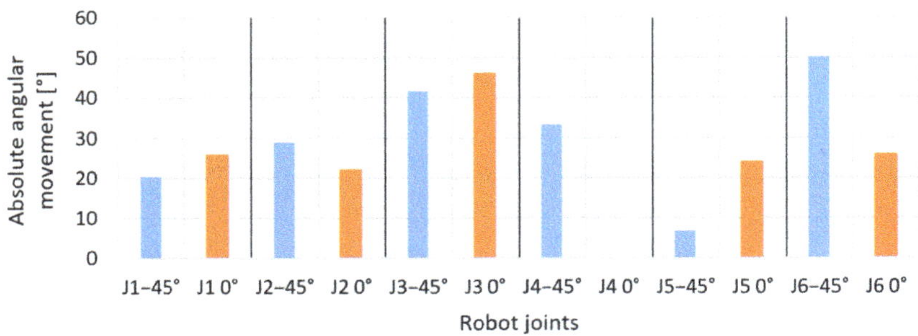

Figure 14. Absolute angular movement of robot joints during the movement along a circular path in both measuring planes.

There are also two possible configurations of the robot arm for the selected position and orientation of the circle. For the configuration with the joint position J5 < 0° (NOFLIP), the J4 joint has a constant value of 0° throughout the path traveled. The circular path can also be completed with the second configuration, i.e., J5 > 0° (FLIP), but in this case, the J4 joint acquires the limit values of its range of rotation, which are +180° or −180°. Such position of the robot arm, with limit values of one of the joints, is unsuitable due to the restriction on further movements, but it can meet the purpose of the measurement. At the same time, to ensure that the range of motion of the J6 joint was the same for all three mentioned positions of the robot arm in the measurement process, the "W" coordinate of individual points forming the circle was again adjusted to 180° in the FLIP configuration. The ranges of motions of the J4, J5, and J6 joints, when moving along the circular path, are shown in the graph in Figure 15. Again, as in the previous cases, positional analysis in the Creo Parametric system 5.0.4.0 was used to chart this graph. For the sake of simplicity, the graph does not include the range of motions of the J1, J2, and J3 joints, which are identical for all positions of the robot arm used.

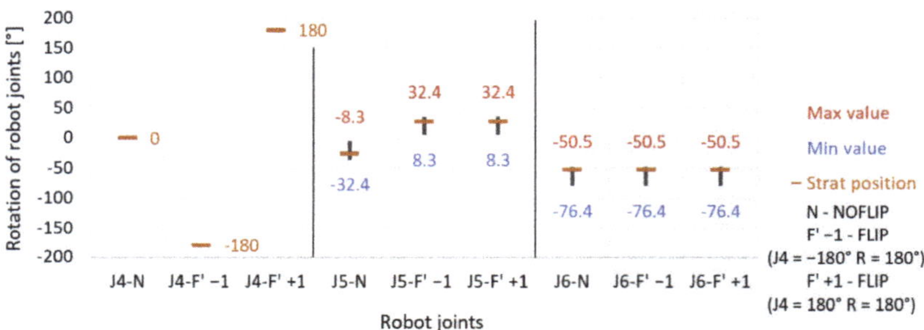

Figure 15. Position of robot joints J4, J5, and J6 when measured in the plane parallel to the base of the robot.

Measurements with the Renishaw Ballbar QC20–W were conducted under the same conditions as in Table 4, except for the orientation of the center of the circular path, which is zero (W = 0°; P = 0°; R = 0°) in the present case. The individual polar graphs are shown in Figure 16 and the average circularity values from ten repeated measurements for each robot arm position used are shown in Table 7.

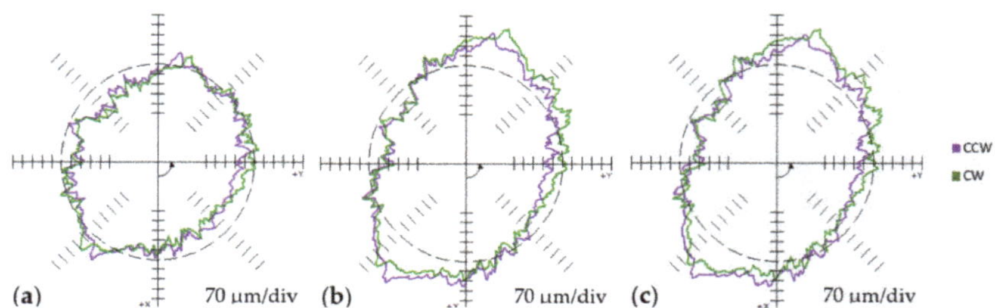

Figure 16. Polar graphs from measuring a circle on the plane parallel to the base of the robot. (a) NOFLIP configuration; (b) FLIP′ −1 configuration (J4 = −180°, R = 180°); (c) FLIP′ +1 configuration (J4 = 180°, R = 180°).

Table 7. Average circularity value for measuring a circle on the plane parallel to the base of the robot.

	Configuration NOFLIP	Configuration FLIP′ −1	Configuration FLIP″ +1
Circularity [µm]	385.2	508.3	505.6

When comparing the resulting polar graphs, the first thing to be observed is the similarity between the measurements with the FLIP configuration of the arm. The similarity is also indicated by the value of circularity shown in Table 7, the resulting difference is negligible. During both measurements in the FLIP configuration, the motions of all joints were identical except for the J4 joint, which did not take part in the movement along the circular path but acquired a different limit value in both cases. In the case of Figure 16b, the limit value of the J4 joint was $-180°$ and vice versa, when measured as shown in Figure 16c, this limit value was $+180°$. Based on these results, it can be said that the static value of the J4 joint rotation does not affect the measurement results. In terms of measurement errors, both graphs reveal the squareness error and the scaling error.

A fundamental influence on the shape of the polar graph and, at the same time, the value of circularity is exerted by a change in the configuration of the robot arm, i.e., the motion of the J5 joint, which can be seen when comparing the polar graph of Figure 16a with the previous two graphs of Figure 16b,c, respectively. At the same time, in this polar graph of the NOFLIP configuration (Figure 16a), it is possible to observe a smaller manifestation of the squareness error than in other measurements. In terms of the shape of the circular path being affected, the relevant changes can be attributed to the motion of the J5 joint.

5.2. Results Summary

As part of measurements with the Renishaw Ballbar QC20–W, the possibility of measuring a circular path placed on an inclined plane has been confirmed as viable. The conducted experiments consisted of two series of measurements. The first of them was done with the circle radius of 100 mm, placed on an inclined plane at an angle $P = -45°$ concerning the robot's WCS. Two possible arm configurations—NOFLIP and FLIP, exist for the selected position and orientation of the measuring plane. The results of the measurements and their representation in polar graphs show a significant difference in the shape of the circular path between the two configurations. This difference has also been confirmed by the average circularity values in both measurements. Based on the motion analysis of the individual robot joints in connection with polar graphs, the motion of the J4, J5, and J6 joints was identified as a possible source of these errors. Subsequent measurements ruled out the effect of the J6 joint. The mentioned measurement was based on the modification of the robot program so that in two measurements in the FLIP configuration, the motion was different only in joints J4 and J5. The motion of the other joints during the measurement was identical.

The second series of measurements took place with the same radius of the circle -100 mm, but in a plane parallel to the base of the robot or parallel to the workbench. These measurements aimed to further examine the influence of the J4 and J5 joints on the measurement results. With the selected orientation of the measuring plane, the position of the joint J4 is static during the entire travel along the circular path. Depending on the selected configuration, the respective values are $0°$, $-180°$, or $180°$. By analyzing the range of motion of individual robot joints and the polar graphs obtained from the measurement, the greatest influence on the deformation of the circular path can be attributed to the selected configuration of the robot arm and especially the motion of the J5 joint.

Although both series of measurements took place in two different orientations of the measuring XY plane, the center of the circular path was the same for both orientations. Partial similarity when comparing the polar graphs obtained in the measurement on an inclined plane and the FLIP configuration, with the polar graph in the second series of NOFLIP measurements and configurations, relates to this fact. This similarity is also indicated by the value of circularity, which takes on values of 409.8 μm, or 389.3 μm at an inclined plane, and a value of 385.2 μm when measured in a plane parallel to the robot base. At the same time, when comparing the ranges of motion of the J5 joint, it is possible to rule out their mutual similarity of motion in the above-mentioned configurations and orientations of the circular path. Based on this, it can be argued that the change in the shape of the polar graph is influenced not only by the J5 joint but also by the motion of the joint J4. When measuring was done in a plane parallel to the robot base, the mentioned influence of the J4 joint was corrected, as the J4 joint acquired only static values.

6. Conclusions

Renishaw Ballbar QC20–W is a tool designed and used specifically for troubleshooting CNC machine tools with a Cartesian kinematic system. When used in connection with industrial robots, it is necessary to proceed from the fact that, unlike CNC machine tools, the circular motion required for measurement is created by the simultaneous motion of several or all of the robot joints. Differences in the kinematics of CNC machine tools and industrial robots renders Ballbar software incapable of identifying and evaluating errors of industrial robots. Therefore it is necessary to approach measurements and evaluation differently. In the case of standard measurement of industrial robots, when the measuring plane is parallel to the robot base, not all robot joints move. The same applies to the so-called volumetric analysis, in which the circle in the XY plane is supplemented by two more circular arcs in the mutually perpendicular planes ZX and ZY. The purpose of the experiments of the present paper was to verify the hypothesis of the motion of all the robot joints when the desired circular path is placed on an inclined plane. The impetus for this work was the measurements previously done with the Renishaw Ballbar at the Department of Automation and Production Systems.

The individual experiments described in the paper can be divided into two parts, preceded by setting the conditions for and preparing the experiment. The key step was the choice of the orientation of the measuring plane in the workspace of the selected Fanuc LR Mate 200iC robot. Based on the recommendations of the ISO 9283 standard, which deals, among other things, with the conditions for measuring the operating characteristics of industrial robots, an inclined plane at an angle of $-45°$ (rotation around the Y–axis) was chosen in the robot's WCS. The starting points for determining other conditions have already been mentioned in experiments conducted in the past.

In the first part of the experiments, a positional analysis was performed with a simulation model of the LR Mate 200iC robot in the environment of the Creo Parametric software 5.0.4.0. The result is the confirmation of the hypothesis that when placing a circle on the selected inclined plane, all its joints, including the J4 joint, would participate in the circular movement of the TCP robot. Commonly, when the measuring plane is parallel to the robot base, the J4 joint is static throughout the movement. The results of the positional analysis were subsequently used to evaluate the results of measurements. The practical measurements with the Renishaw Ballbar QC20–W are described in the second part of this paper. The analysis of the measurement results was performed using polar graphs and circularity values, which represent one of the possible outputs of the Ballbar 20 software. Two series of measurements were done. In the first series, a circle with a radius of 100 mm was placed on an inclined plane, under the before-mentioned angle. The second series represented measurements in a plane parallel to the robot base. In addition to the measurement conditions, the common feature of all measurements was also the identical position of the center of the circle in the robot's workspace. Only the orientation of the measuring plane has changed. The result is the finding that in the selected position and orientation of the measuring plane, the change in the shape of the polar graph was locally caused mainly by a change in the configuration of the robot arm, which is related to the motion of the J4 and J5 joints. At the same time, the similarity of the results was observed when comparing the two series of measurements, which can be attributed to the just mentioned unchanged position of the center of the circular path.

The measurement on an inclined plane provides mainly a better view of industrial robot properties. The standard measurement of industrial robot properties that utilize the Renishaw Ballbar system is conducted on the plane parallel to its base plane. We used such a measurement method in previous works [25,26,28]. However, the parallel orientation of the measuring plane results in the uneven involvement of individual motors on the overall motion of the TCP. In particular, the joint J4 is static throughout the whole path both for measuring in the parallel plane and for volumetric measurement too. The pro-posed method places the measuring path on an inclined plane, which results in movement that involves all of the robot joints, including J4. The results of such measurements offer better information about the technical condition of the robot including the possible impact of the J4 joint. The presented measurement method can be used in process of industrial robot condition monitoring based on measurements repeated in regular intervals. The changes can be evaluated by changes in the shape of the polar chart. It is important to include all of the possible unwanted effects that can affect the performance of the industrial robot.

Possible Further Work

The LR Mate 200iC robot used had been calibrated several times in the past, with the application of the so-called Zero position master method. The accuracy of this method is based only on the visual adjustment of the individual joints of the robot to the zero position, using reference marks. Therefore, this calibration method is considered inaccurate, that fact is also pointed out in the Fanuc robot's user manual [40]. The manual emphasizes that the method should only be used in exceptional cases. Such a state of affairs was addressed in the past by refining the calibration, using original calibration data. In connection with the identified influence of the motion of the J4 and J5 joints on the final shape of the polar graph and the overall measurement results, it is possible to look for the cause probably in the still persistent inaccurate calibration of the given robot. Based on this, there is room for further work and repetition of the measurements presented, subject to a more detailed specification of the LR Mate 200iC robot calibration. There is also room for comparison with a robot of a clearly more accurate calibration than that of the robot used in the experiments described in our paper.

It is possible to use Renishaw Ballbar to evaluate individual properties of industrial robots defined in ISO 9283 standard. The double ballbar measuring technique is based on periodical measuring of deviation of the real radius from programmed one during the movement in a circular path. It is possible to utilize such a measurement technique for the evaluation of so-called path accuracy (AT according to ISO 9283) and path repeatability (RT according to ISO 9283). Path accuracy characterizes the ability of a robot to move its mechanical interface along the command path in the same direction n times [6]. Path repeatability expresses the closeness of the agreement between the attained paths for the same command path repeated n times [6]. Measured values are important for the calculation of both mentioned properties. A good example is a measurement on the inclined plane using NOFLIP configuration, measurement conditions are included in Table 4 and the polar graph is shown in Figure 10a. In this case, the average radius of ten repeated measurements (clockwise) was between 99.594 mm and 100.201 mm. In order to reliably determine the values of RT and AT, it is necessary to analyze measurement records more and to define the correct methodology of evaluation. The measurements collected with Ballbar 20 software can be too distorted for this purpose, as the software subtracts estimated eccentricity represented by the first harmonics of the measured profile processed by fast Fourier transformation. For evaluation and calculation of the performance parameters of the industrial robot, it would be necessary to get values declared by its manufacturer as they usually declare only unidirectional path repeatability (RP according to ISO 9283). Another possible way to verify results would be to compare it to measurement using a laser interferometer or a laser tracker.

Author Contributions: Conceptualization, Methodology: all authors; Writing—original draft: V.T. and I.Z.; Writing—review and editing: M.C., I.K. and M.S. All authors have read and agreed to the published version of the manuscript.

Funding: This work was supported by the Slovak Research and Development Agency under the contract No. APVV–16–0283.

Institutional Review Board Statement: Not applicable.

Informed Consent Statement: Not applicable.

Data Availability Statement: The data presented in this study are available on request from the corresponding author.

Conflicts of Interest: The authors declare no conflict of interest.

References

1. Heer, C. Top Trends Robotics 2020 (Press Release). Available online: https://ifr.org/ifr-press-releases/news/top-trends-robotics-2020 (accessed on 22 November 2020).
2. Akatov, N.; Mingaleva, Z.; Klackova, I.; Galieva, G.; Shaidurova, N. Expert Technology for Risk Management in the Implementation of QRM in a High-Tech Industrial Enterprise. *Manag. Syst. Prod. Eng.* **2019**, *27*, 250–254. [CrossRef]
3. Hu, M.; Wang, H.; Pan, X.; Tian, Y. Optimal Synthesis of Pose Repeatability for Collaborative Robots Based on the ISO 9283 Standard. *Ind. Robot Int. J. Robot. Res. Appl.* **2019**, *46*, 812–818. [CrossRef]
4. Bulej, V.; Stanček, J.; Kuric, I.; Zajacko, I. The Space Distribution and Transfer of Positioning Errors from Actuators to the TCP Point of Parallel Mechanism. In Proceedings of the MATEC Web of Conferences, Sklene Teplice, Slovakia, 5–8 September 2017; Volume 157.
5. Slamani, M.; Nubiola, A.; Bonev, I.A. Modeling and Assessment of the Backlash Error of an Industrial Robot. *Robotica* **2012**, *30*, 1167–1175. [CrossRef]
6. *ISO 9283:1998 Manipulating Industrial Robots—Performance Criteria and Related Test Methods*; International Organization for Standardization: Geneva, Switzerland, 1998.
7. *ISO/TR 13309:1995 Manipulating Industrial Robots—Informative Guide on Test Equipment and Metrology Methods of Operation for Robot Performance Evaluation in Accordance with ISO 9283*; International Organization for Standardization: Geneva, Switzerland, 1995.
8. Dandash, D.; Brethe, J.-F.; Vasselin, E.; Lefebvre, D. Micrometre Scale Performances of Industrial Robot Manipulators. *Int. J. Adv. Robot. Syst.* **2012**, *9*. [CrossRef]
9. Brethé, J.-F.; Lefebvre, D. Risk Ellipsoids and Granularity Ratio for Industrial Robots. *Int. J. Fact. Autom. Robot. Soft Comput.* **2007**, *2*, 93–101.
10. Şirinterlikçi, A.; Murat, T.; Bird, A.; Harris, A.; Kweder, K. Repeatability and Accuracy of an Industrial Robot: Laboratory Experience for a Design of Experiments Course. *Technol. Interface J.* **2009**, *9*, 1–10.
11. Pastor, M.; Zivcak, J.; Puskar, M.; Lengvarsky, P.; Klackova, I. Application of Advanced Measuring Methods for Identification of Stresses and Deformations of Automotive Structures. *Appl. Sci.* **2020**, *10*, 7510. [CrossRef]
12. Jiang, Y.; Yu, L.; Jia, H.; Zhao, H.; Xia, H. Absolute Positioning Accuracy Improvement in an Industrial Robot. *Sensors* **2020**, *20*, 4354. [CrossRef] [PubMed]
13. Icli, C.; Stepanenko, O.; Bonev, I. New Method and Portable Measurement Device for the Calibration of Industrial Robots. *Sensors* **2020**, *20*, 5919. [CrossRef]
14. Nubiola, A.; Bonev, I.A. Absolute Calibration of an ABB IRB 1600 Robot Using a Laser Tracker. *Robot. Comput. Integr. Manuf.* **2013**, *29*, 236–245. [CrossRef]
15. Jozwik, J.; Ostrowski, D.; Jarosz, P.; Mika, D. Industrial Robot Repeatability Testing with High Speed Camera Phantom V2511. *Adv. Sci. Technol. Res. J.* **2016**, *10*, 86–93. [CrossRef]
16. Abderrahim, M.; Khamis, A.; Garrido, S.; Moreno, L. Accuracy and calibration issues of industrial manipulators. In *Industrial Robotics: Programming, Simulation and Application*; I-Tech: Madrid, Spain, 2004; pp. 131–146.
17. Goetz, C.; Tuttas, S.; Hoegner, L.; Eder, K.; Stilla, U. Accuracy Evaluation for a Precise Indoor Multi-Camera Pose Estimation System. In Proceedings of the PIA11: Photogrammetric Image Analysis, Munich, Germany, 5–7 October 2011; Stilla, U., Rottensteiner, F., Mayer, H., Jutzi, B., Butenuth, M., Eds.; Copernicus Gesellschaft Mbh: Gottingen, Germany, 2011; Volume 38–3, pp. 97–102.
18. Jozwik, J.; Jacniacka, E.; Ostrowski, D. Estimation of Uncertainty of Laser Interferometer Measurement in Industrial Robot Accuracy Tests. In Proceedings of the II International Conference of Computational Methods in Engineering Science (CMES'17), Lublin, Poland, 23–25 November 2017; Borys, M., Czyz, Z., Falkowicz, K., Kujawska, J., Kulisz, M., Szala, M., Eds.; EDP Sciences: Les Ulis, France, 2017; Volume 15.
19. Slamani, M.; Joubair, A.; Bonev, I.A. A Comparative Evaluation of Three Industrial Robots Using Three Reference Measuring Techniques. *Ind. Robot. Int. J. Robot. Res. Appl.* **2015**, *42*, 572–585. [CrossRef]
20. Slamani, M.; Bonev, I.A. Characterization and Experimental Evaluation of Gear Transmission Errors in an Industrial Robot. *Ind. Robot. Int. J. Robot. Res. Appl.* **2013**, *40*, 441–449. [CrossRef]
21. Jozwik, J.; Kuric, I.; Ostrowski, D.; Dziedzic, K. Industrial Robot Accuracy Testing with QC20-W Ballbar Diagnostic System. *Manuf. Technol.* **2016**, *16*, 519–524. [CrossRef]
22. Slamani, M.; Nubiola, A.; Bonev, I. Assessment of the Positioning Performance of an Industrial Robot. *Ind. Robot. Int. J. Robot. Res. Appl.* **2012**, *39*, 57–68. [CrossRef]
23. Nubiola, A.; Slamani, M.; Bonev, I.A. A New Method for Measuring a Large Set of Poses with a Single Telescoping Ballbar. *Precis. Eng.* **2013**, *37*, 451–460. [CrossRef]

24. Cubonova, N.; Dodok, T.; Sagova, Z. Optimisation of the Machining Process Using Genetic Algorithm. *Sci. J. Silesian Univ. Technol. Ser. Transp.* **2019**, *104*, 15–25. [CrossRef]
25. Kuric, I.; Tlach, V.; Cisar, M.; Sagova, Z.; ZajaCko, I. Examination of Industrial Robot Performance Parameters Utilizing Machine Tool Diagnostic Methods. *Int. J. Adv. Robot. Syst.* **2020**, *17*. [CrossRef]
26. Tlach, V.; Ságová, Z.; Kuric, I. Circular and Quasi-Circular Paths for the Industrial Robots Measuring with the Renishaw Ballbar QC20-W. In Proceedings of the MATEC Web of Conferences, Rydzyna, Poland, 4–7 September 2018; Volume 254, p. 5007.
27. Kuric, I.; Císar, M.; Tlach, V.; Zajačko, I.; Gál, T.; Więcek, D. *Technical Diagnostics at the Department of Automation and Production Systems*; Springer: Cham, Switzerland, 2019; Volume 835, ISBN 9783319974897.
28. Tlach, V.; Císar, M.; Kuric, I.; Zajačko, I. Determination of the Industrial Robot Positioning Performance. In Proceedings of the MATEC Web of Conferences, Cluj Napoca, Romania, 12–13 October 2017; Balc, N., Ed.; EDP Sciences: Les Ulis, France, 2017; Volume 137, p. 1004.
29. Renishaw PLC. *QC20-W Wireless Ballbar System Description and Specifications*; Renishaw PLC: Wotton-under-Edge, UK, 2013.
30. Dodok, T.; Čuboňová, N. Application of strategy manager tools for optimized NC programming. In *Lecture Notes in Mechanical Engineering*; Springer International Publishing AG: Cham, Switzerland, 2019; pp. 322–334.
31. *ISO 230-4:2005 Test Code for Machine Tools—Part 4: Circular Tests for Numerically Controlled Machine Tools*; International Organization for Standardization: Geneva, Switzerland, 2005.
32. Cunningham, J.; Xu, Q.; Ford, D.G. A Case Study on Practical Error Correction Practices Utilized on a CNC Machine Tool. In Proceedings of the 18th National Conference on Manufacturing Research, Leeds, UK, 10–12 September 2002; pp. 361–366.
33. Yang, S.-H.; Lee, H.-H.; Lee, K.-I. Identification of Inherent Position-Independent Geometric Errors for Three-Axis Machine Tools Using a Double Ballbar with an Extension Fixture. *Int. J. Adv. Manuf. Technol.* **2019**, *102*, 2967–2976. [CrossRef]
34. Poppeova, V.; Bulej, V.; Zahoransky, R.; Uricek, J. Parallel Mechanism and Its Application in Design of Machine Tool with Numerical Control. In Proceedings of the Robotics in Theory and Practice, Strbske Pleso, Slovakia, 14–16 November 2013; Pachnikova, L., Hanjuk, M., Eds.; Trans Tech Publications Ltd.: Durnten-Zurich, Switzerland, 2013; Volume 282, pp. 74–79.
35. *FANUC Robotics SYSTEM R-30iA HandlingTool Setup and Operations Manual. MAROC77HT01101E REV B*; FANUC America Corporation: Michigan, MI, USA, 2015.
36. PTC To Define a Motor. Available online: http://support.ptc.com/help/creo/creo_pma/usascii/index.html#page/simulate%2Fdes_anim%2Fservomotors%2Fto_create_servo.html%23wwconnect_header (accessed on 2 December 2020).
37. Kwon, Y.; Ertekin, Y.M.; Tseng, B. In-Process and Post-Process Quantification of Machining Accuracy in Circular CNC Milling. *Mach. Sci. Technol.* **2005**, *9*, 27–38. [CrossRef]
38. Kuric, I.; Císar, M. Machine Tool Errors and Its Simulation on Experimental Device. *Acad. J. Manuf. Eng.* **2015**, *13*, 17–21.
39. Karan, B.; Vukobratovic, M. Calibration and Accuracy of Manipulation Robot Models—An Overview. *Mech. Mach. Theory* **1994**, *29*, 479–500. [CrossRef]
40. *FANUC Robot LR Mate 200iC. Maintenance Manual. B-82585EN/04*; FANUC America Corporation: Michigan, MI, USA, 2008.

applied
sciences

MDPI

Article

A New Foot Trajectory Planning Method for Legged Robots and Its Application in Hexapod Robots

Haichuang Xia [1], Xiaoping Zhang [1,*] and Hong Zhang [2,3]

1   School of Electrical and Control Engineering, North China University of Technology, Beijing 100144, China; 18153010606@mail.ncut.edu.cn
2   School of Automation, Xi'an University of Posts and Telecommunications, Xi'an 710121, China; zhmlsa@xupt.edu.cn
3   School of Electrical and Engineering, The University of New Mexico, Albuquerque, NM 87131, USA
*   Correspondence: zhangxp@ncut.edu.cn

Abstract: Compared with wheeled and tracked robots, legged robots have better movement ability and are more suitable for the exploration of unknown environments. In order to further improve the adaptability of legged robots to complex terrains such as slopes, obstacle environments, and so on, this paper makes a new design of the legged robot's foot sensing structure that can successfully provide accurate feedback of the landing information. Based on this information, a new foot trajectory planning method named three-element trajectory determination method is proposed. For each leg in one movement period, the three elements are the start point in the support phase, the end point in the support phase, and the joint angle changes in the transfer phase where the first two elements are used to control the height, distance, and direction of the movement, and the third element is used make decisions during the lifting process of the leg. For the support phase, the trajectory is described in Cartesian space, and a spline of linear function with parabolic blends is used. For the transfer phase, the trajectory is described in joint-space, and the joint angle function is designed as the superposition of the joint angle reverse-chronological function and the interpolation function which is obtained based on joint angle changes. As an important legged robot, a hexapod robot that we designed by ourselves with triangle gait is chosen to test the proposed foot trajectory planning method. Experiments show that, while the foot's landing information can be read and based on the three-element trajectory planning method, the hexapod robot can achieve stable movement even in very complex scenes. Although the experiments are performed on a hexapod robot, our method is applicable to all forms of legged robots.

Keywords: trajectory planning method; hexapod robots; legged robots; triangle gait

Citation: Xia, H.; Zhang, X.; Zhang, H. A New Foot Trajectory Planning Method for Legged Robots and Its Application in Hexapod Robots. *Appl. Sci.* **2021**, *11*, 9217. https://doi.org/10.3390/app11199217

Academic Editor: Tibor Krenicky

Received: 18 August 2021
Accepted: 29 September 2021
Published: 3 October 2021

Publisher's Note: MDPI stays neutral with regard to jurisdictional claims in published maps and institutional affiliations.

1. Introduction

With the development of AI technology, mobile robots are appearing more and more frequently in public. According to the motion mode, mobile robots can be divided into wheeled robots [1,2], tracked robots [3], and legged robots [4,5]. In the face of complex environments as well as rugged terrains, legged robots have incomparable flexibility and applicability compared with the other two kinds of robots.

The design idea of legged robots originated from bionics. Inspired by mammals, insects, amphibians, etc., legged robots try to imitate the structure as well as movement mode of different legs [6]. Up to this point, there have been many different kinds of bionic legged robots that were designed and studied. With the support of the U.S. Department of Defense and by imitating a dog, Boston Dynamics developed the quadruped robot BigDog, which possessed strong obstacle crossing ability and could overcome rugged terrain [7]. Inspired by insects, Case Western Reserve University designed a bionic insect robot that could jump, walk, turn, and avoid obstacles in a certain space [8]. In China, also from the perspective of bionics and combining motion control analysis, different kinds of legged robots were designed by Shandong University [9,10], Harbin University of Science and Technology [11], and Shanghai Jiaotong University [12].

Common legged robots include biped robots, quadruped robots, hexapod robots, and octopod robots. For biped robots , the movement balance is difficult to control [13,14]. Even very simple walking similar to those in humans will become a big challenge for biped robots . For quadruped robots, they are able to walk steadily, but have strong dependencies on each leg [15]. When one of the four legs breaks for some reason, the quadruped robot cannot continue working anymore. In contrast, hexapod robots can still adjust in the case that one leg breaks through an adaptive fault tolerant gait [16]. Up to this point, there is very little work about octopod robots in terms of complex control [17]. All in all, hexapod robots have natural advantages in movement stability and are studied a lot.

In terms of the hexapod robot's stable movement in different terrains, most research realized this by switching gait. Bai et al. presented a novel CPG (center pattern generator)-based gait generation for a curved-leg hexapod robot and enabled the robot to achieve smooth and continuous mutual gait transitions [18]. Ouyang presented an adaptive locomotion control approach for a hexapod robot by using a 3D two-layer artificial center pattern generator (CPG) network [19]. These methods all obtained promising results in some aspects, but they could not confront any complex scene because of the limited gait patterns. Our current work, from another point of view, finds a new method to help the hexapod robot improve its movement stability for all kinds of terrains, which is achieved by adaptively planning the foot's trajectory in order to stabilize the robot's attitude. With this method, the gait can remain the same.

To implement the above theory on a physical robot, the robot should have the ability to sense its landing information firstly. In [20], Zha et al. developed a new kind of free gait controller and applied it to a large-scale hexapod robot with heavy load, where sensory feedback signals of the foot position were employed in both the free gait planner and the gait regulator. In [21], Faigl and Čížek presented a minimalistic approach for a hexapod robot's adaptive locomotion control, which enabled traversing rough terrains with a small and affordable hexapod walking robot, and servomotor position feedback was also used to reliably detect the ground contact point. Up to this point, the problem of how to obtain landing information has not been sufficiently considered as well as studied in most legged robots. In [21], this information was obtained by comparing the joint error and the error threshold. In some other works, based on the robot's dynamic model, joint torque feedback was selected to judge whether the feet had touched the ground [22]. However, the sensors used for such joint torque feedback are always with large volume and high in terms of price, which limits their utility in small robots. To solve these lingering issues, we create a new design of the legged robot's foot by adding a short-stroke inching button in order to obtain accurate feedback of the foot's landing information. By conducting experiments on our designed robot in real environments, this foot sensing structure is proved to be easy to use, with low cost and strong sensitivity.

The rest of the paper is structured as follows. Section II introduces the hexapod robot we designed in detail, including its mechanical structure and electrical structure, especially the new design of the robot's foot sensing structure. To elucidate our method, the robot's mathematic model is firstly given in section III. All details of the foot trajectory planning method including simulation verification are in Section IV. Section V provides the method's application in a triangle gait. Section VI shows the experiment results. Conclusions are finally put forward in Section VII.

2. The Hexapod Robot

The robot we designed is as shown in Figure 1, where (a) is the hexapod robot prototype, and (b) is its mechanical schematic. For the sake of description, the six legs are numbered as *Leg*1 to *Leg*6 in a counter-clockwise fashion.

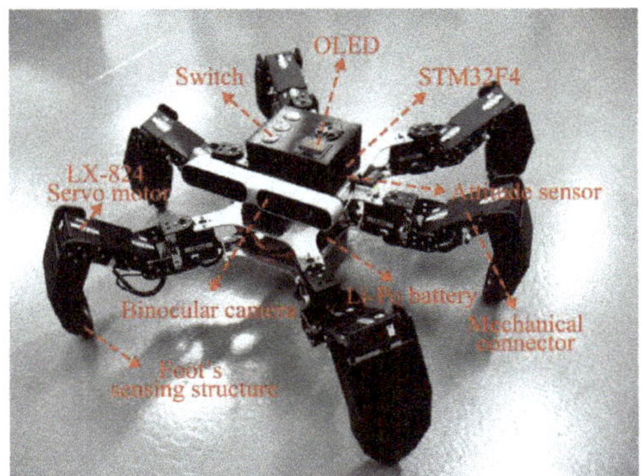

(a) The hexapod robot prototype

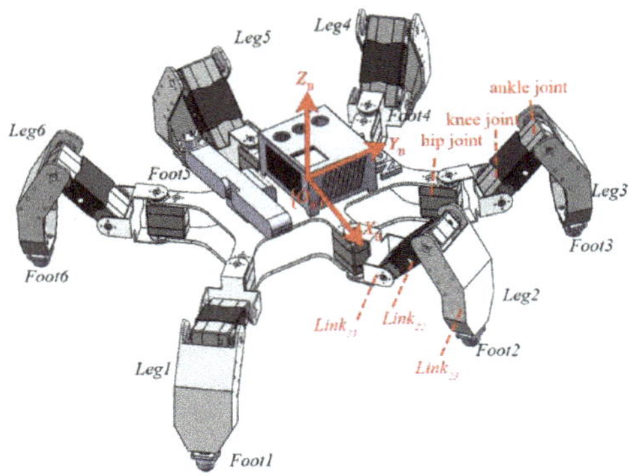

(b) The mechanical schematic of the hexapod robot

Figure 1. The hexapod robot.

2.1. The Hexapod Robot's Mechanical Structure

From the perspective of mechanical structure, the hexapod robot is mainly composed of a body, legs, and feet. Normally, to better guarantee the stability and controllability, the hexapod robot is centrally symmetric.

2.1.1. The Body

For our hexapod robot, its body is composed of two identical plates with the length and width ratio of 2:1. These two plates are placed up and down and are fixedly connected by the hip joints of six legs. Located between the two plates is the battery. The STM32F4 microcontroller, binocular camera and attitude sensor are placed on the top of the upper plate.

2.1.2. The Leg

As in Figure 1b, for each leg, there are three joints from the direction of body to foot tip, which are the hip joint, the knee joint, and the ankle joint. Different joints are connected by links named as $Link_{i1}$ (between the hip joint and the knee joint), $Link_{i2}$ (between the knee joint and the ankle joint), and $Link_{i3}$ (between the ankle joint and the foot tip) for the *ith* leg. For the hexapod robot, the total of 18 degrees of freedom is utilized.

2.1.3. The Foot

To help the robot to sense its foot landing information, this paper creates a new design of the legged robot's foot sensing structure as in Figure 2. A short-stroke inching button is used here. Once one foot touches the ground, the reaction force from the ground will press the short-stroke inching button, and then landing information will be transferred to the controller. When the foot is lifted off the ground, the button is reset and prepares for the next landing. A hemispherical foot tip is designed to ensure that the short-stroke inching button can be pressed regardless of the direction of the foot's fall.

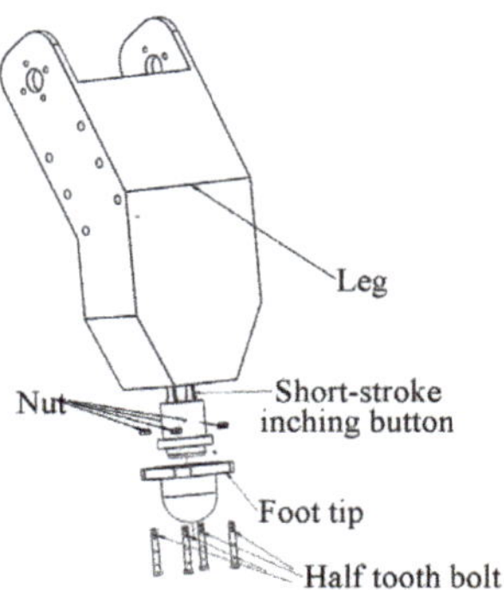

Figure 2. The hexapod robot's foot sensing structure.

2.2. The Hexapod Robot's Electrical Structure

The hexapod robot's electrical structure is shown in Figure 3.

The sensory unit includes five parts as follows.

(1) Binocular camera for object capture and tracking.

(2) Attitude sensor MPU-9250 for measuring the robot's posture. MPU-9250 is a System in Package (SiP) and contains a 3-axis gyroscope, a 3-axis accelerometer, and AK8963, a 3-axis digital compass. As accelerometers and magnetometers have high frequency noise while the gyroscope has low frequency noise, by utilizing their complementary characteristics in frequency and fusing the low-pass filtered accelerometer and magnetometer data with high-pass filtered gyroscope data, the attitude feedback information with high precision can be obtained.

(3) WIFI module for remote data transmission.

(4) Remote control receiver for instructions receiving.

(5) Foot sensing structure for landing information.

The control unit is STM32, and it outputs the commands to the motion unit, more specifically, the 18 servo motors through serial ports. For each leg, its three servo motors cooperate with each other in a certain time sequence and finally achieve the foot to reach the specified point in space.

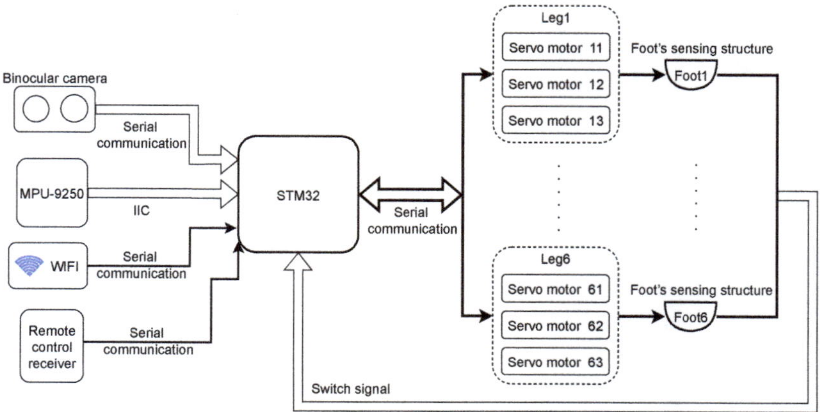

Figure 3. The hexapod robot's electrical structure.

The hexapod robot's technical specifications are provided in Table 1.

Table 1. Technical specifications of the hexapod robot prototype.

Structure	Parameters		
		Value	Unit
Servo motor	Number	18	\
	Weight	57.0	g
Power supply	Voltage	7.4	DC(V)
Body	Length	254.1	mm
	Width	179.3	mm
	Height	3.0	mm
Leg $Link_{i1}$	Length	45.5	mm
	Weight	18.8	g
Leg $Link_{i2}$	Length	75.0	mm
	Weight	14.7	g
Leg $Link_{i3}$	Lenght	109.6	mm
	Weight	41.1	g
Foot	Radius	9.5	mm
	Lenght	19.5	mm
	Weight	6.3	g

3. Hexapod Robot Kinematics Modeling

The robot's kinematic model can help us quantitatively analyze the robot's velocity, acceleration, attitude, and so on from a mathematical point of view. To establish the kinematic model of the hexapod robot, for each leg i, three coordinates are defined at the hip joint ($\{O_{i1}\}$), the knee joint ($\{O_{i2}\}$), and the ankle joint ($\{O_{i3}\}$), respectively, as depicted in Figure 4. The reference coordinate is $\{O_{i0}\}$. For each coordinate $\{O_{ij}\}(i = 1, \cdots, 6; j = 1, \cdots, 3)$, the Z_{ij} axis coincides with the axis of the jth joint. The X_{ij} axis is the common perpendicular between the axes of the jth joint and the $(j+1)$th joint. The Y_{ij} axis is then determined according to the right-hand rule. In this paper, the kinematic model is based on the DH (Denavit–Hartenberg) method [23], and the parameters are provided in Table 2. Since the six legs are the same, we do not distinguish the leg number i anymore, and only the joint number j is considered in the description that follows.

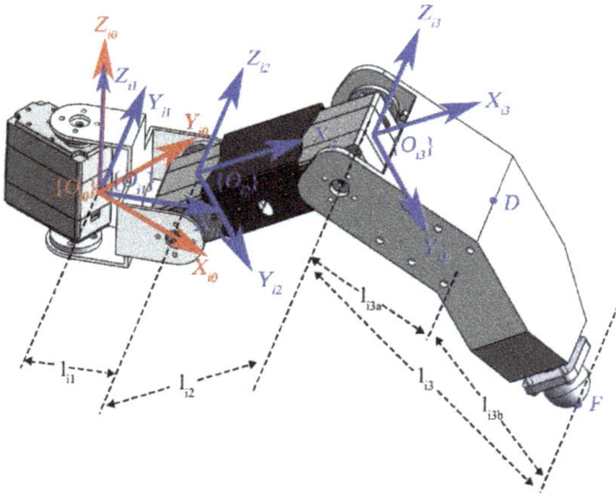

Figure 4. Coordinates at different joints for each leg.

Table 2. Denavit-Hartenberg parameters.

Joint j	α_{j-1}	a_{j-1}	d_j	θ_j
1	0	0	0	θ_1
2	$-\pi/2$	l_{i1}	0	θ_2
3	0	l_{i2}	0	θ_3

The relative translations and rotations between the $(j-1)$th and the jth joint coordinates are computed by the transformation matrix (1).

$$
{}^{j-1}_{j}T = \begin{bmatrix} \cos\theta_j & -\sin\theta_j & 0 & a_{j-1} \\ \sin\theta_j \cos\alpha_{j-1} & \cos\theta_j \cos\alpha_{j-1} & -\sin\alpha_{j-1} & -\sin\alpha_{j-1} d_j \\ \sin\theta_j \sin\alpha_{j-1} & \cos\theta_j \sin\alpha_{j-1} & \cos\alpha_{j-1} & \cos\alpha_{j-1} d_j \\ 0 & 0 & 0 & 1 \end{bmatrix}.
\tag{1}
$$

3.1. The Forward Kinematics

Based on (1), the transformation matrix from the coordinate $\{O_{i3}\}$ to $\{O_{i0}\}$ is the following.

$$
\begin{aligned}
{}^{0}_{3}T &= {}^{0}_{1}T {}^{1}_{2}T {}^{2}_{3}T \\
&= \begin{bmatrix} \cos\theta_1 & -\sin\theta_1 & 0 & 0 \\ \sin\theta_1 & \cos\theta_1 & 0 & 0 \\ 0 & 0 & 1 & 0 \\ 0 & 0 & 0 & 1 \end{bmatrix} \begin{bmatrix} \cos\theta_2 & -\sin\theta_2 & 0 & l_{i1} \\ 0 & 0 & 1 & 0 \\ -\sin\theta_2 & -\cos\theta_2 & 0 & 0 \\ 0 & 0 & 0 & 1 \end{bmatrix} \begin{bmatrix} \cos\theta_3 & -\sin\theta_3 & 0 & l_{i2} \\ \sin\theta_3 & \cos\theta_3 & 0 & 0 \\ 0 & 0 & 1 & 0 \\ 0 & 0 & 0 & 1 \end{bmatrix} \\
&= \begin{bmatrix} \cos(\theta_2+\theta_3)\cos\theta_1 & -\sin(\theta_2+\theta_3)\cos\theta_1 & -\sin\theta_1 & \cos\theta_1(l_{i1}+l_{i2}\cos\theta_2) \\ \cos(\theta_2+\theta_3)\sin\theta_1 & -\sin(\theta_2+\theta_3)\sin\theta_1 & \cos\theta_1 & \sin\theta_1(l_{i1}+l_{i2}\cos\theta_2) \\ -\sin(\theta_2+\theta_3) & -\cos(\theta_2+\theta_3) & 0 & -l_{i2}\sin\theta_2 \\ 0 & 0 & 0 & 1 \end{bmatrix}.
\end{aligned}
\tag{2}
$$

The coordinate of the foot tip with respect to the coordinate $\{O_{i3}\}$ is the following.

$$
{}^{3}P_F = [\; x_{\{O_{i3}\}} \quad y_{\{O_{i3}\}} \quad z_{\{O_{i3}\}} \quad 1 \;]^T.
\tag{3}
$$

Then, the coordinate of the foot tip with respect to the coordinate $\{O_{i0}\}$ is described in Equation (4).

$$
\begin{aligned}
{}^{0}P_F &= {}^{0}_{3}T\, {}^{3}P_F \\
&= \begin{bmatrix} \cos\theta_1(l_{i1}+l_{i2}\cos\theta_2) - z_{\{O_{i3}\}}\sin\theta_1 + x_{\{O_{i3}\}}\cos(\theta_2+\theta_3)\cos\theta_1 - y_{\{O_{i3}\}}\sin(\theta_2+\theta_3)\cos\theta_1 \\ z_{\{O_{i3}\}}\cos\theta_1 + \sin\theta_1(l_{i1}+l_{i2}\cos\theta_2) + x_{\{O_{i3}\}}\cos(\theta_2+\theta_3)\sin\theta_1 - y_{\{O_{i3}\}}\sin(\theta_2+\theta_3)\sin\theta_1 \\ -y_{\{O_{i3}\}}\cos(\theta_2+\theta_3) - x_{\{O_{i3}\}}\sin(\theta_2+\theta_3) - l_{i2}\sin\theta_2 \\ 1 \end{bmatrix}.
\end{aligned}
\tag{4}
$$

3.2. The Inverse Kinematics

A geometric method is used in this work for the inverse kinematics analysis. It is intuitive and clear, and the solution can be simultaneously unique by restricted conditions. The leg's diagram for the solution to its inverse kinematics is shown in Figure 5. The angle of CDF is 150°, resulting from the robot leg's fixed mechanical structure design, as shown in Figures 1 and 4. Based on this diagram and by combing geometric knowledge, we have the following.

$$
d = \sqrt{(\sqrt{x^2+y^2}-l_{i1})^2 + z^2},
\tag{5}
$$

$$
l_{i3} = \sqrt{l_{i3a}{}^2 + l_{i3b}{}^2 - 2l_{i3a}l_{i3b}\cos\frac{5\pi}{6}},
\tag{6}
$$

$$\cos \beta = \frac{d^2 + l_{i2}^2 - l_{i3}^2}{2l_{i2}d}, \tag{7}$$

$$\cos \gamma = \frac{l_{i2}^2 + l_{i3}^2 - d^2}{2l_{i2}l_{i3}}, \tag{8}$$

$$\cos \sigma = \frac{l_{i3}^2 + l_{i3a}^2 - l_{i3b}^2}{2l_{i3}l_{i3a}}, \tag{9}$$

$$\sin \alpha = \frac{-z}{d}. \tag{10}$$

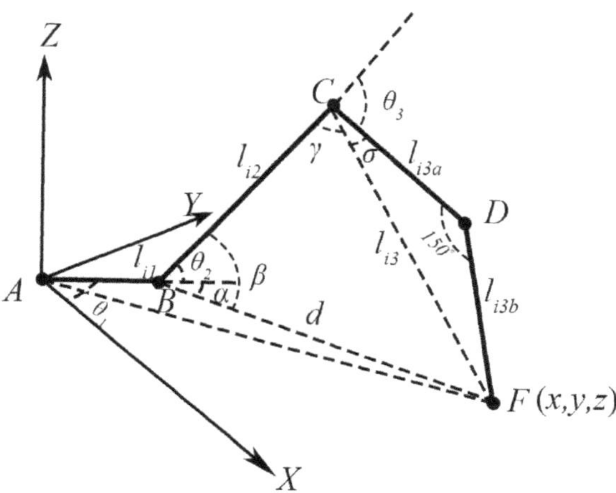

Figure 5. The leg's diagram for the solution of inverse kinematics.

Thus, the inverse kinematics can be easily obtained as follows.

$$\begin{cases} \theta_1 = \arctan \frac{y}{x}, \\ \theta_2 = \beta - \alpha = \arccos \frac{d^2 + l_{i2}^2 - l_{i3}^2}{2l_{i2}d} + \arcsin \frac{z}{d}, \\ \theta_3 = \pi - \gamma - \sigma \\ \quad = \pi - \arccos \frac{l_{i2}^2 + l_{i3}^2 - d^2}{2l_{i2}l_{i3}} - \arccos \frac{l_{i3}^2 + l_{i3a}^2 - l_{i3b}^2}{2l_{i3}l_{i3a}}. \end{cases} \tag{11}$$

4. The Three-Element Trajectory Determination Method

For a hexapod robot, planning its foot trajectory is very important, and it will affect the robot's flexibility and stability. While moving, the robot's leg may be in the support phase in order to drive the robot towards a certain direction or off the ground and in the transfer phase. The gait is then realized by switching between the support phase and transfer phase in different sequences. For each leg, its trajectory planning is worth studying in order to enrich the movement of legged robots.

In this paper, we design a three-element trajectory determination method to help the robot plan its foot's movement. The three elements used are the start point in support phase $P_0 = (x_0, y_0, z_0)$, the end point in support phase $P_1 = (x_1, y_1, z_1)$, and the joint angle changes in transfer phase $\Delta\theta = [\Delta\theta_1, \Delta\theta_2, \Delta\theta_3]$. P_0 and P_1 are used to control the height, distance, and the direction of the movement, while $\Delta\theta$ is used to make decisions on the lifting motion of the leg. Applying this method to legged robots, this paper transfers the six variables of P_0 and P_1 to another three variables that are related to the movement of legged robots, which are height, direction, and distance. Such a transfer is based on some restrictions such as symmetry, equal altitude, vertical landing, etc. $\Delta\theta$ is set according to the feedback information from the foot. The trajectory planning methods in the support phase and transfer phase are different, and they are elaborated separately.

4.1. Trajectory Planning in Support Phase

By using MATLAB, Figure 6 simulates the trajectory planning process in the support phase, where J_1, J_2, and J_3 represent the hip joint, the knee joint, and the ankle joint, respectively, while J_5 represents the foot tip. J_4 is fixed in $Link_{i3}$. In Figure 6a, P_0 and P_1 are the start point and the end point of the foot, while O_0 and O_1 are the start point and the end point of the hip joint in Figure 6b.

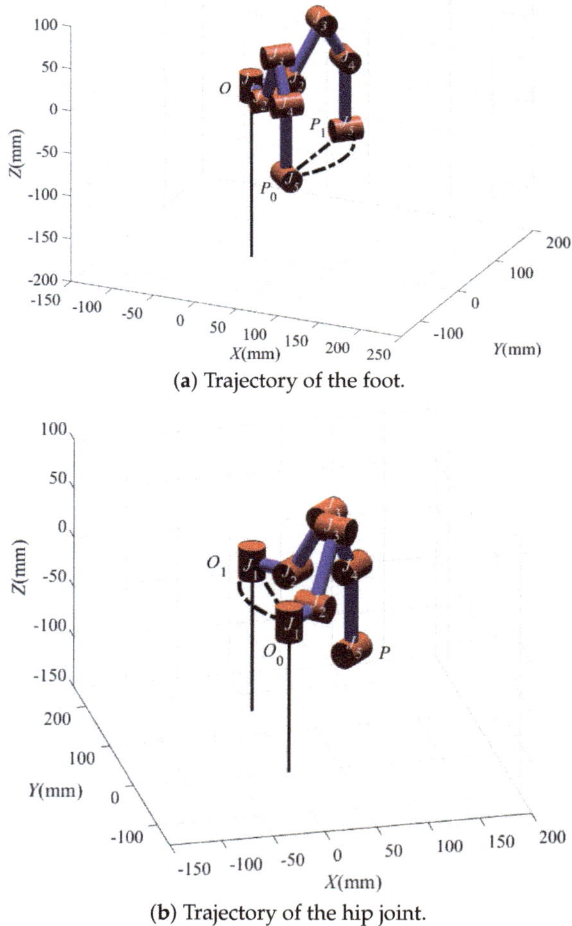

(**a**) Trajectory of the foot.

(**b**) Trajectory of the hip joint.

Figure 6. Trajectory planning in support phase.

Traditionally, during the hexapod robot's movement, the angles of the knee joint J_2 and the ankle joint J_3 are always fixed and not considered. Then, based on the forward kinematics model, the foot will rotate around the hip joint J_1, and its trajectory is $\overparen{P_0 P_1}$, as shown in Figure 6a. When the foot is on the ground, the body needs to move forward with the foot J_5 as the fulcrum, and the joint J_1 will then move along the trajectory $\overparen{O_0 O_1}$, as shown in Figure 6b. As the relative position of the hip joint to the robot's body center remains unchanged, the center of the robot will also move according to $\overparen{O_0 O_1}$, which means the robot will shake. Therefore, to ensure accuracy in direction as well as stability while moving, the hexapod robot's foot trajectory in the support phase should better be a straight line $\overline{P_0P_1}$, as shown in Figure 6a. Then, J_1 will move along $\overline{O_0O_1}$, as shown in Figure 6b. Here, the robot's moving direction keeps constant to ensure that the robot moves in a straight line.

According to the inverse kinematics model, the foot's trajectory in straight line is planned in Cartesian space. To ensure the robot's speed, the trajectory's continuity, and the servo motor's angular velocity, a spline of linear function with parabolic blends is used in (12).

$$F(t) = \begin{cases} a_f t^2 + b_f t + c_f, & 0 \leq t \leq T_b; \\ k_f t + d_f, & T_b < t \leq T_f - T_b; \\ e_f t^2 + m_f t + n_f, & T_f - T_b < t \leq T_f. \end{cases} \tag{12}$$

$F(t)$ can be $X(t)$, $Y(t)$, and $Z(t)$, which are the foot's coordinates in X axis, Y axis, and Z axis, respectively. a_f to n_f are the parameters, and they are different for different axes. T_b is the movement time of the parabola phase, and it is the same for $X(t)$, $Y(t)$, and $Z(t)$ in order to guarantee the stabilization of movement acceleration. T_f is the time of the total support phase. Some restricted conditions are set at the same time in (13).

$$\begin{cases} \lim_{t \to T_b^-} F(t) = \lim_{t \to T_b^+} F(t), \\ \lim_{t \to T_b^-} F'(t) = \lim_{t \to T_b^+} F'(t), \\ \lim_{t \to (T_f - T_b)^-} F(t) = \lim_{t \to (T_f - T_b)^+} F(t), \\ \lim_{t \to (T_f - T_b)^-} F'(t) = \lim_{t \to (T_f - T_b)^+} F'(t), \\ F'(0) = 0, \\ F'(T_f) = 0. \end{cases} \tag{13}$$

Take $X(t)$ as an example, as $P_0 = (x_0, y_0, z_0)$ and $P_1 = (x_1, y_1, z_1)$; according to (13), we have the following.

$$\begin{cases} X(0) = c_x = x_0, \\ X(T_f) = e_x T_f^2 + m_x T_f + n_x = x_1. \end{cases} \tag{14}$$

By combining all of the restricted conditions, we have the following.

$$\begin{cases} a_x T_b^2 + b_x T_b + c_x = k_x T_b + d_x, \\ \frac{1}{2} a_x T_b + b_x = k_x, \\ e_x (T_f - T_b)^2 + m_x (T_f - T_b) + n_x = k_x (T_f - T_b) + d_x, \\ \frac{1}{2} e_x (T_f - T_b) + m_x = k_x, \\ b_x = 0, \\ \frac{1}{2} e_x T_f + m_x = 0. \end{cases} \tag{15}$$

For a given robot, when T_b and T_f are set, the equations above can be solved. The same can be performed for $Y(t)$ and $Z(t)$, and the trajectory of the foot in support phase can be obtained and is defined as $G(t) = [X(t), Y(t), Z(t)]$. According to the inverse kinematics model in (11), the angle function of the three joints can be obtained and is defined as $\theta_S(t) = [\theta_1(t), \theta_2(t), \theta_3(t)]$.

4.2. Trajectory Planning in Transfer Phase

Transfer phase is the phase when the foot is in the air and moves from the end point of one movement period to the start point of the next movement period. Compared with support phase, the transfer phase has fewer restrictions. The only requirement is that the foot should complete its trajectory in time T_f.

The trajectory planning in transfer phase is performed in joint-space. Considering computational simplicity as well as the trajectory's reliability, this paper designs trajectory planning to occur in this phase as the superposition of two functions: one is what we call the joint angle reverse-chronological function and the other is the joint angle changes interpolation function.

4.2.1. Joint Angle Reverse-Chronological Function

We named it as reverse-chronological function because the trajectory planned based on this function is symmetrical with the trajectory in the support phase along the time $t = T_f$. As it is planned in joint-space, we provide the joint angle reverse-chronological function as Equation (16).

$$\theta_B(t) = \theta_S(2T_f - t) = [\theta_1(2T_f - t), \theta_2(2T_f - t), \theta_3(2T_f - t)]. \tag{16}$$

4.2.2. Joint Angle Changes Interpolation Function

The joint angle changes interpolation function is defined as $\theta_P(t) = [\theta_{P\_1}(t), \theta_{P\_2}(t), \theta_{P\_3}(t)]$, where $\theta_{P\_j}(t), (j = 1, 2, 3)$ is the angle changes interpolation function for the jth joint. Considering the continuity of the velocity as well as the acceleration, for each joint, $\theta_{P\_j}(t)$ is a quartic polynomial, as described in Equation (17):

$$\theta_{P\_j}(t) = a_4 t^4 + a_3 t^3 + a_2 t^2 + a_1 t + a_0. \tag{17}$$

where a_0 to a_4 are the parameters of $\theta_{P\_j}(t)$ and will be different according to different joints.

To ensure the continuity of the joint angle, the joint angle changes at the start and end time should be zero. To ensure the stability of the joint rotation, the velocity of the joint angle changes at the start, and the end time should also be zero.

$$\begin{cases} \theta_{P\_j}(t)(T_f) = 0, \\ \theta_{P\_j}(t)(2T_f) = 0, \\ \theta_{P\_j}(t)'(T_f) = 0, \\ \theta_{P\_j}(t)'(2T_f) = 0. \end{cases} \tag{18}$$

To realize the symmetry of the angle changes, we hope that $\theta_{P\_j}(t)$ arrives at the set joint angle change $\Delta\theta_j$ at the middle time of this phase; thus, we have the following.

$$\theta_{P\_j}\left(\frac{3}{2}T_f\right) = \Delta\theta_j. \tag{19}$$

Then, the relationship matrices (20) and (21) can be obtained as follows.

$$\begin{bmatrix} 0 \\ \Delta\theta_j \\ 0 \\ 0 \\ 0 \end{bmatrix} = \begin{bmatrix} T_f^4 & T_f^3 & T_f^2 & T_f & 1 \\ (\frac{3}{2}T_f)^4 & (\frac{3}{2}T_f)^3 & (\frac{3}{2}T_f)^2 & \frac{3}{2}T_f & 1 \\ (2T_f)^4 & (2T_f)^3 & (2T_f)^2 & 2T_f & 1 \\ 4T_f^3 & 3T_f^2 & 2T_f & 1 & 0 \\ 4(2T_f)^3 & 3(2T_f)^2 & 2(2T_f) & 1 & 0 \end{bmatrix} \begin{bmatrix} a_4 \\ a_3 \\ a_2 \\ a_1 \\ a_0 \end{bmatrix}. \tag{20}$$

$$\begin{bmatrix} a_4 \\ a_3 \\ a_2 \\ a_1 \\ a_0 \end{bmatrix} = \begin{bmatrix} T_f^4 & T_f^3 & T_f^2 & T_f & 1 \\ (\frac{3}{2}T_f)^4 & (\frac{3}{2}T_f)^3 & (\frac{3}{2}T_f)^2 & \frac{3}{2}T_f & 1 \\ (2T_f)^4 & (2T_f)^3 & (2T_f)^2 & 2T_f & 1 \\ 4T_f^3 & 3T_f^2 & 2T_f & 1 & 0 \\ 4(2T_f)^3 & 3(2T_f)^2 & 2(2T_f) & 1 & 0 \end{bmatrix}^{-1} \begin{bmatrix} 0 \\ \Delta\theta_j \\ 0 \\ 0 \\ 0 \end{bmatrix}. \tag{21}$$

$\Delta\theta_j$ is set according to the real terrain situation. Based on Formula (21), the parameters a_0 to a_4 can be solved, and $\theta_{P\_j}(t)$ is obtained.

By superimposing $\theta_B(t)$ and $\theta_P(t)$, the trajectory function of the transfer phase $\theta_T(t)$ in joint-space is described as follows (22).

$$\theta_T(t) = \theta_B(t) + \theta_P(t) = [\theta_1(2T_f - t) + \theta_{P\_1}(t), \theta_2(2T_f - t) + \theta_{P\_2}(t), \theta_3(2T_f - t) + \theta_{P\_3}(t)]. \tag{22}$$

By combining $\theta_S(t)$ in support phase and $\theta_T(t)$ in transfer phase together, the total trajectory of one joint during a movement period in joint-space is defined as the dotted line in Figure 7. The red dotted line is $\theta_S(t)$. The blue dotted line is $\theta_T(t)$, and it is the superposition of the green line $\theta_B(t)$ and the pink line $\theta_P(t)$.

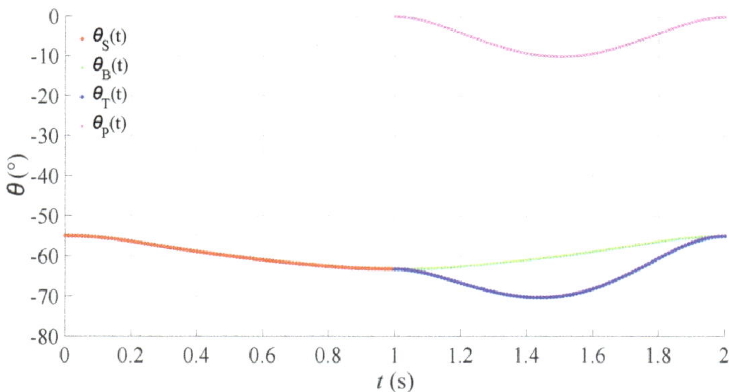

Figure 7. Trajectory in joint-space.

4.3. Verification of the Three-Element Trajectory Determination Method

The trajectory of one foot based on the three-element trajectory determination method during one movement period in Cartesian space is simulated in Figure 8, in which counter clockwise is selected as the positive direction and $T_b = 0.2$, $T_f = 1$.

From Figure 8a, it can be observed that, during the time $t \in (0, T_f)$, the foot's coordinates in X and Z axes remain unchanged, and the robot only moves along the Y axis, which confirms that the robot is moving in a straight line. Figure 8b is the foot's trajectory in space, the straight line is the trajectory in support phase, and the arc line is the trajectory in transfer phase.

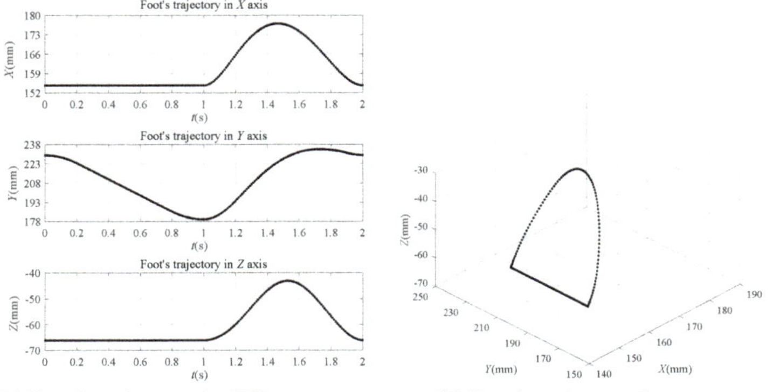

(**a**) Foot's trajectory in different axes.   (**b**) Foot's trajectory in space.

Figure 8. Trajectory in Cartesian space.

Here, we also provide the angle curve of the three joints as in Figure 9. We can observe that, while moving, all of the three angle curves are continuous and smooth, which further demonstrates the stability and practicability of our trajectory planning method for legged robots.

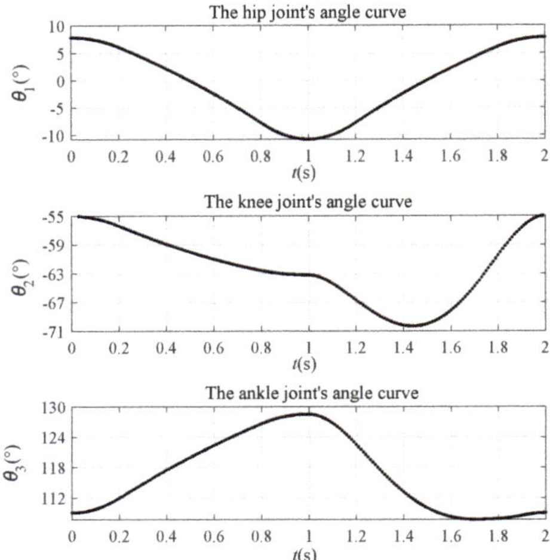

Figure 9. Joints' angle curve.

5. The Three-Element Trajectory Determination Method in Triangle Gait

The advantage of our designed trajectory method is reflected in the realization that the robot's horizontal posture is maintained during movements. For the three elements, $\Delta\theta$ is set according to the feedback information from the foot. What else needs to be performed is to select the start point P_0 and the end point P_1. Supposing the target height of the robot's body is Z , target moving direction is θ_a, and the moving step is D, then it can be divided into three steps for our method. The details are described by combining Figure 10.

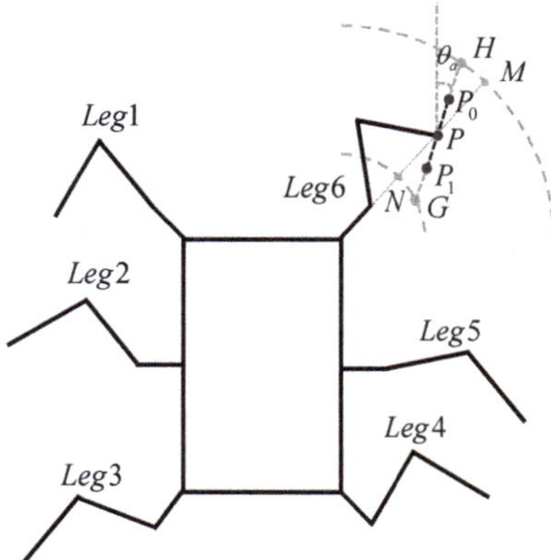

Figure 10. Start and end points selection geometric sketch.

5.1. Stable Point Determination

First of all, to improve our method's stability and flexibility in triangle gait, in this paper, we treat the stable posture for a legged robot as the posture where $link_{i3}$ in Figure 1b is perpendicular to the ground and the hip joint's angle θ_1 in Figure 5 is simultaneously zero. The foot's point is now called a stable point. As in Figure 10, the points on the line $\overline{MN}$ are all stable points. As $link_{i3}$ is perpendicular to the ground, by geometric knowledge, we obtain the following.

$$\theta_3 = \theta_2 + \frac{\pi}{3}. \tag{23}$$

For the foot's inverse kinematics in (11), with the above conditions, there will be only one degree of freedom. To help the hexapod robot adapt to rugged terrains, we set Z as input so that we can obtain the expected stable point P.

5.2. Start Point and End Point Selection

After the stable point P is determined, what also needs to be performed is to select the start point P_s and the end point P_e of the support phase. For a hexapod robot, it is symmetric, and the triangle gait is simultaneously repetitive. Thus, the stable point during movement will be the middle point of the trajectory in the support phase. Supposing that the target moving direction is θ_a, the trajectory in the support phase is along the line $\overline{P_0 P_1}$, as shown in Figure 10. Ideally, $|PP_0| = |PP_1| = \frac{D}{2}$, and the coordinates of P_0 and P_1 can be computed. In reality, the feet can only reach a limited range (as in Figure 10, the foot of *Leg6* can only arrive at the points between the two dotted arc lines); thus, the selection of P_s and P_e should depend on specific conditions.

Extend PP_0 and intersect it with the outer arc at point H. Extend PP_1 and intersect it with the inner arc at point G. If $\min\{|PH|, |PG|\} \geq \frac{D}{2}$, this means both of the points P_0 and P_1 can be reached, and they can be the selected points. Then, the priority is given to movement stability, and all feet move equidistant around their respective stable points. If $|PH| \geq \frac{D}{2}, |PG| < \frac{D}{2}$, this means the point P_1 cannot be reached, and the selected points are P_0 and G. If $|PH| < \frac{D}{2}, |PG| \geq \frac{D}{2}$, this means the point P_0 cannot be reached, then the selected points are H and P_1. If $\max\{|PH|, |PG|\} < \frac{D}{2}$, then the trajectory will be the line $\overline{HG}$. Thus, we have the following.

$$\begin{cases} \text{If } \min\{|PH|, |PG|\} \geq \frac{D}{2}, & P_s = P_0 \text{ and } P_e = P_1, \\ \text{If} |PH| \geq \frac{D}{2}, |PG| < \frac{D}{2}, & P_s = P_0 \text{ and } P_e = G, \\ \text{If} |PH| < \frac{D}{2}, |PG| \geq \frac{D}{2}, & P_s = H \text{ and } P_e = P_1, \\ \text{If } \max\{|PH|, |PG|\} < \frac{D}{2}, & P_s = H \text{ and } P_e = G. \end{cases} \tag{24}$$

Transform the start point $(x_{P_s}, y_{P_s}, z_{P_s})$ and the end point $(x_{P_e}, y_{P_e}, z_{P_e})$ that were obtained above under the world coordinate to the coordinate $\{O_{i0}\}$. Then, the trajectory of each foot can be planned by using the method in Section IV.

6. Experiment on Physical Robot in Real Environment

6.1. Triangle Gait Designed in Practice

Traditionally, the movement of a foot can be divided into support phase in $(0, T_f)$ and transfer phase in $(T_f, 2T_f)$. For the triangle gait [24], when the feet of group A move as the support phase, the feet of group B move as the transfer phase and vice versa. In this paper, to better apply our trajectory method for legged robots in complex environments, we added an adjustment phase (AP) at the end of transfer phase so that the robot can adjust its foot position for trajectory planning in the next period to guarantee the accuracy of the movement. The triangle gait then, in practice, is designed as in Figure 11. For different terrains, the time for the adjustment phase Δt_i can be different. However, when the feet of one group is in the adjustment phase, the feet of the other group will still be in the support phase in order to leave the following planning process unaffected.

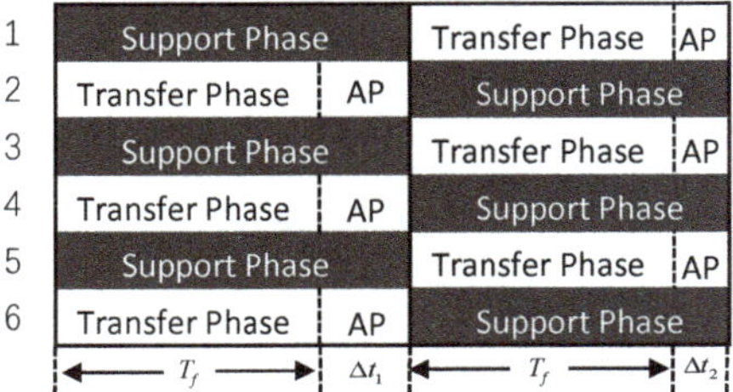

Figure 11. Triangle gait designed in practice.

6.2. Movement Control Flowchart

The movement control flowchart based on the above triangle gait is shown in Figure 12. For any leg, the follow situations may happen. When the leg is in the support phase, judging whether this phase has been completed needs to be performed. If not, keep moving along the planned trajectory; if it is performed, this means that the leg is ready for phase changing. When the foot is not in the support phase but already on the ground, this means that the foot touches the ground or obstacles in advance; here, the foot needs to replan its trajectory based on the ground's height information, which is sensed by the designed foot sensing structure. When the foot is not on the ground, one situation is that the leg is in the transfer phase, then the foot moves along the planned trajectory. The other situation is that the leg has already finished the transfer phase but has not touched the ground, this situation happens when the ground's height is lower than planned. Then, a new height $Z = Z + \Delta Z$ is set, and the foot needs to perform new trajectory planning in the adjustment phase based on Z information. When all legs are ready for phase changing, the phase is changed, and the robot keeps moving.

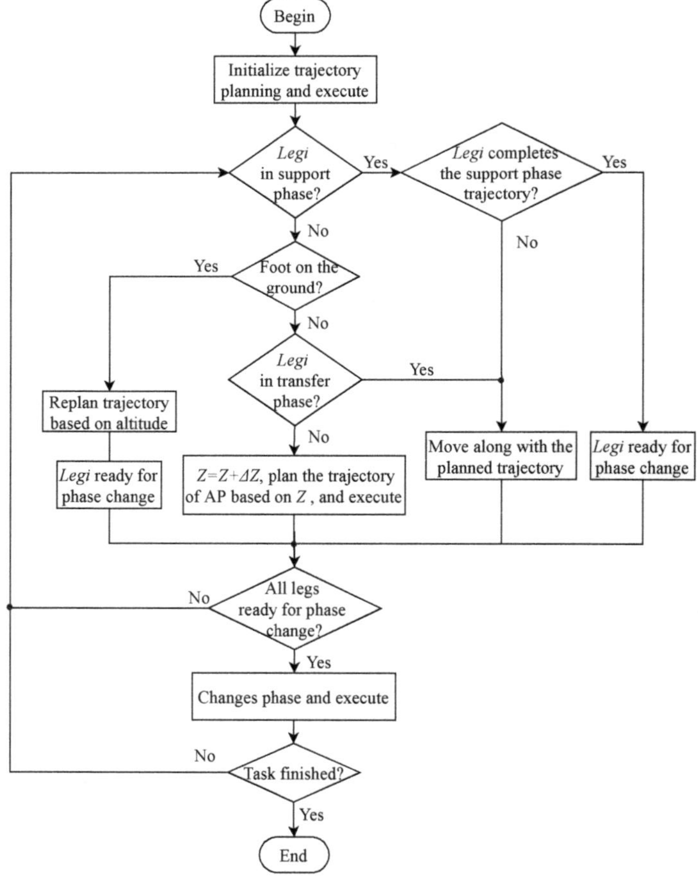

Figure 12. The movement control flowchart.

6.3. Experimental Results and Analysis

To prove the effectiveness of our method in complex environment, we situate the robot in two different real scenes.

6.3.1. Slope Experiment

The first scene that we chose is a slope outside with the angle of 7°. In Figure 13a–c, take *Foot5* as an example; its trajectory planning is given as the red curve. This curve is similar to the simulated trajectory in Figure 8b except that the heights of the start point and the end point in support phase are not the same. At $t = 12$ s, the *Foot5* is on the ground, and it is ready for the transfer phase. At $t = 13$ s, we can observe that *Foot5* arrives at the top of the planned trajectory. At $t = 14$ s, it falls to the ground and enters to the next support phase. This is a typical trajectory planning process. Figure 13d–f show the gait mode. In Figure 13f, at time $t = 27$ s, the *Leg2*, *Leg4*, and *Leg6* are in the air (transfer phase), while the other three legs are on the ground (support phase). At time $t = 34$ s, as in Figure 13e, the phases are switched, the *Leg1*, *Leg3*, and *Leg5* are in the transfer phase, and the *Leg2*, *Leg4*, and *Leg6* are in support phase. The situation in Figure 13f is the same as the one in Figure 13d.

During the robot's slope climbing, we mark its posture with a yellow line. What is worth pointing out is that, during all the movement, the robot can always keep its body horizontal, which is proof of the stability of our method.

Although this experiment is performed on a slope with the angle of 7°, after conducting a lot of tests, it is proved that the robot can adapt to a slope with any angle as long as the friction is sufficient.

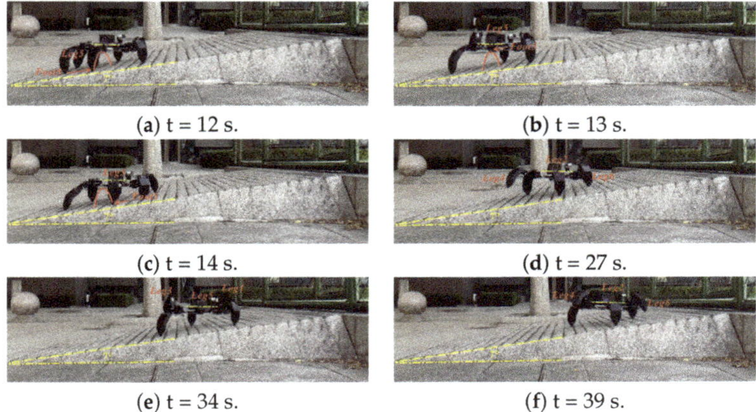

(a) t = 12 s.

(b) t = 13 s.

(c) t = 14 s.

(d) t = 27 s.

(e) t = 34 s.

(f) t = 39 s.

Figure 13. Hexapod robot's movement on a slope.

6.3.2. Obstacle Experiment

To further show the robot's strong movement ability, we situate the robot in an obstacle environment, as shown in Figure 14, where two obstacles with different height are placed inside. The robot's movement is shown as in Figure 14. For the sake of description, we divide *Leg1*, *Leg3*, and *Leg5* as the A group and *Leg2*, *Leg4*, and *Leg6* as the B group.

In Figure 14a–c, we can observe that the A group is in the transfer phase. The *Leg1* lifts up at the time $t = 10$ s and touches the obstacle at time $t = 12$ s. As the ground's height information changes, there is an adjustment process for *Leg1* as stated in the control flowchart in Figure 12. From Figure 14d on, the A group is in support phase, and the B group is in the transfer phase. This process finishes in Figure 14f at time $t = 15$ s. In Figure 14g,h, the *Leg1* moves on the top of obstacle 1. In Figure 14i, *Leg1* starts to walk down the obstacle. Here, the foot tries to sense where the ground is based on the designed foot sensing structure in Figure 2. At the time $t = 34$ s, the *Leg1* touches the ground, and landing information is fed back. Figure 14l–p show the process of leaping over obstacle 2. In Figure 14p, all legs and feet are marked to help better understand the robot's movement.

In this experiment, we also mark the robot's posture. Results show that, although in a very complex environment, the robot with our trajectory planning method can still maintain horizontal posture.

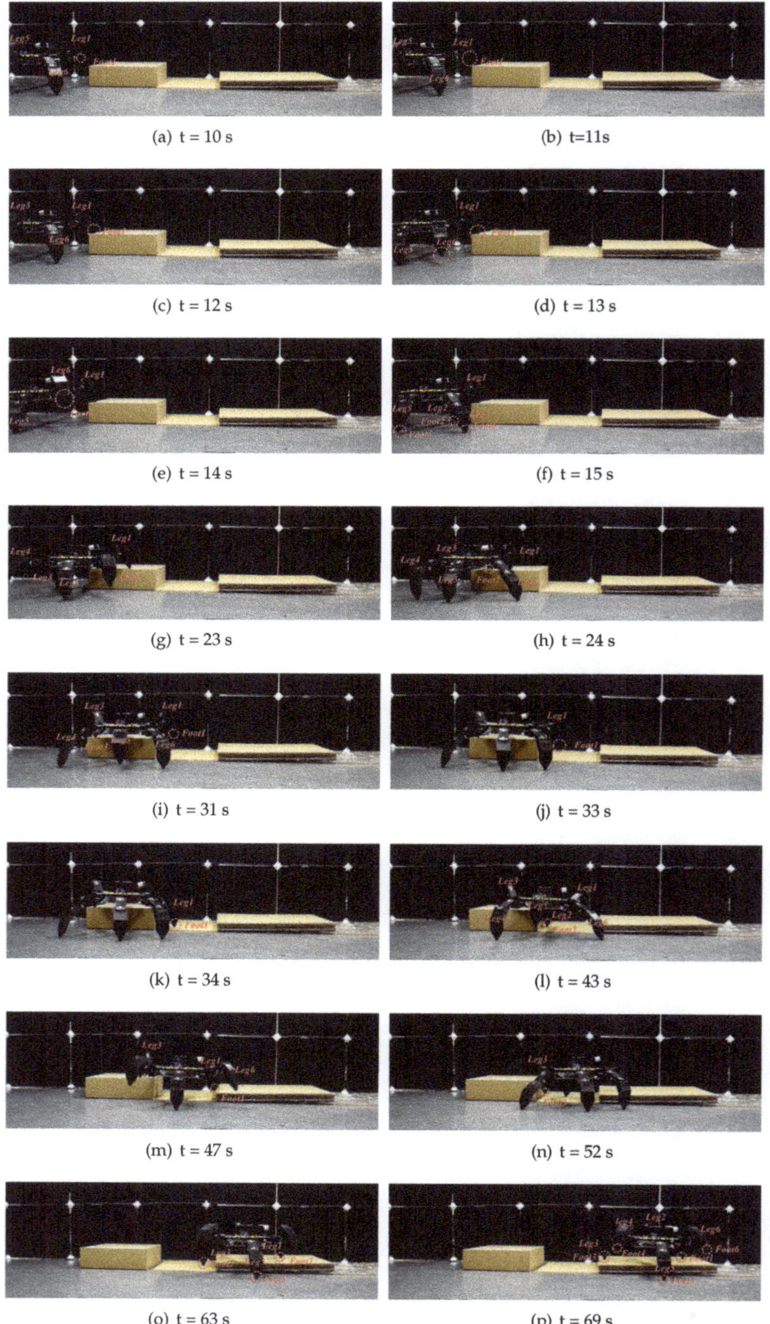

Figure 14. Hexapod robot's movement with obstacles.

7. Conclusions

Legged robots have high movement flexibility, which makes them more suitable for various complex terrains. However, the complex control algorithm and trajectory planning also accompany the robot. Reasonable trajectory planning methods will greatly reduce the difficulty of control algorithm designs. Thus, this paper designs a new trajectory planning method for legged robots named the three-element trajectory determination method. Meanwhile, to realize this method's application on a physical robot, this paper creates a new design of the legged robot's foot sensing structure in order to help obtain landing information. To show how the method works, the details of its application in a triangle gait are given. Indicated by the observation of our experiments, this paper makes an improvement of the traditional triangle gait, where an adjustment phase is added at the end of transfer phase to help the robot adjust its foot position when it deviates from the original planned trajectory because of changing terrain. A control flowchart with the introduced adjustment phase is also given. Experiments are carried out in a slope environment and an obstacle environment. The results prove that our method is effective and stable in practice .

Hexapod robots have extensive prospect and are worth studying. In this paper, the improvement of the movement of hexapod robots mainly focuses on the trajectory planning method. In the future, we plan to take attitude information into account in order to develop novel control algorithm designs.

Author Contributions: Conceptualization, H.X. and X.Z.; methodology, H.X. and X.Z.; software, H.X.; validation, H.X.; formal analysis, H.X. and X.Z.; investigation, H.X.; resources, H.X. and X.Z.; data curation, H.X. and X.Z.; writing—original draft preparation, X.Z. and H.Z.; writing— review and editing, X.Z. and H.Z.; visualization, X.Z.; supervision, X.Z.; project administration, X.Z.; funding acquisition, X.Z. and H.Z. All authors have read and agreed to the published version of the manuscript.

Funding: This research was funded by the National Natural Science Foundation of China grant number 61903006, the Beijing Natural Science Foundation grant number 4204096, the Shaanxi Provincial Natural Science Foundation of China grant number 2021 SF-478, the Beijing Municipal Great Wall Scholar Program grant number CIT&TCD 20190304, the Scientific Research Project of Beijing Educational Committee 201910009008, the Basic Scientific Research Project of Beijing Municipal Education Commission, Youth Yuyou Talent Project of North China University of Technology, and Research Initial Foundation of North China University of Technology.

Conflicts of Interest: The authors declare no conflicts of interest.

References

1. Bayar, G.; Ozturk, S. Investigation of the Effects of Contact Forces Acting on Rollers of a Mecanum Wheeled Robot. *Mechatronics* **2020**, *72*, 1–13. [CrossRef]
2. Amirkhani, A.; Shirzadeh, M.; Shojaeefard, M.H.; Abraham, A. Controlling wheeled mobile robot considering the effects of uncertainty with neuro-fuzzy cognitive map. *ISA Trans.* **2020**, *100*, 454–468. [CrossRef]
3. Wang, W.; Yan, Z.; Du, Z. Experimental study of a tracked mobile robot's mobility performance. *J. Terramechanics* **2018**, *77*, 75–84. [CrossRef]
4. Muscolo, G.G.; Ceccarelli, M. Mechanics of Legged Robots: From Bio-Inspiration to Novel Legged Machines. *Front. Mech. Eng.* **2021**, *7*, 22. [CrossRef]
5. Raiola, G.; Hoffman, E.M.; Focchi, M.; Tsagarakis, N.; Semini, C. A simple yet effective whole-body locomotion framework for quadruped robots. *Front. Robot. AI* **2020**, *7*, 1–17. [CrossRef]
6. Zhou, X.; Bi, S. A survey of bio-inspired compliant legged robot designs. *Bioinspiration Biomim.* **2012**, *7*, 1–20. [CrossRef] [PubMed]
7. Raibert, M.; Blankespoor, K.; Nelson, G.; Playter, R. Bigdog, the rough-terrain quadruped robot. *IFAC Proc. Vol.* **2008**, *41*, 10822–10825. [CrossRef]
8. Konyev, M.; PALlS, F.; Zavgorodniy, Y.; Melnikov, A.; Rudskiy, A.; Telesh, A.; Schmucker, U.; Rusin, V. Walking robot Anton: design, simulation, experiments. In *Advances in Mobile Robotics*; Marques, L., Almeida, A., Eds.; Publishing House: Singapore, 2008; pp. 922–929.
9. Chen, T.; Sun, X.; Xu, Z.; Li, Y.; Rong, X.; Zhou, L. A trot and flying trot control method for quadruped robot based on optimal foot force distribution. *J. Bionic Eng.* **2019**, *16*, 621–632. [CrossRef]

10. Yang, K.; Zhou, L.; Rong, X.; Li, Y. Onboard hydraulic system controller design for quadruped robot driven by gasoline engine. *Mechatronics* **2018**, *52*, 36–48. [CrossRef]
11. Li, J.; You, B.; Ding, L.; Xu, J.; Li, W.; Chen, H.; Gao, H. A novel bilateral haptic teleoperation approach for hexapod robot walking and manipulating with legs. *Robot. Auton. Syst.* **2018**, *108*, 1–12. [CrossRef]
12. Chen, Z.; Gao, F.; Sun, Q.; Tian, Y.; Liu, J.; Zhao, Y. Ball-on-plate motion planning for six-parallel-legged robots walking on irregular terrains using pure haptic information. *Mech. Mach. Theory* **2019**, *141*, 136–150. [CrossRef]
13. Liu, S.; Yang, X. Modeling and simulation analysis of planar biped robot. *J. Am. Med. Inform. Assoc.* **2020**, *37*, 281–285. 352.
14. Rame, L. Dynamic Modelling and Analyzing of a Walking of Humanoid Robot. *Stroj. Cas.-J. Mech. Eng.* **2018**, *68*, 59–76.
15. Ding, C.; Zhou, L.; Li, Y.; Rong, X. A novel dynamic locomotion control method for quadruped robots running on rough terrains. *IEEE Access* **2020**, *8*, 150435–150446. [CrossRef]
16. Manglik, A.; Gupta, K.; Bhanot, S. In Proceedings of the 2016 IEEE 1st International Conference on Power Electronics, Intelligent Control and Energy Systems, Delhi, India, 4–6 July 2016; IEEE: Piscataway, NJ, USA, 2016; pp. 1–7.
17. Vidoni, R.; Gasparetto, A. Efficient force distribution and leg posture for a bio-inspired spider robot. *Robot. Auton. Syst.* **2011**, *59*, 142–150. [CrossRef]
18. Bai, L.; Hu, H.; Chen, X.; Sun, Y.; Ma, C.; Zhong, Y. Cpg-based gait generation of the curved-leg hexapod robot with smooth gait transition. *Sensors* **2019**, *19*, 1–26. [CrossRef] [PubMed]
19. Ouyang, W.; Chi, H.; Pang, J.; Liang, W.; Ren, Q. Adaptive Locomotion Control of a Hexapod Robot via Bio-Inspired Learning. *Front. Neurorobot.* **2021**, *15*, 1–16. [CrossRef] [PubMed]
20. Zha, F.; Chen, C.; Guo, W.; Zheng, P.; Shi, J. A Free Gait Controller Designed for a Heavy Load Hexapod Robot. *Adv. Mech. Eng.* **2019**, *11*, 1–17. [CrossRef]
21. Faial, J.; Čížek, P. Adaptive Locomotion Control of Hexapod Walking Robot for Travesing Rough Terrains with Position Feedback Only. *Robot. Auton. Syst.* **2019**, *116*, 136–147.
22. Lee, Y.H.; Lee, H.; Kang, H.; Lee, J.H.; Park, J.M.; Kang, C.; Lee, Y.H.; Kim, Y.B.; Choi, H.R. Balance Recovery based on Whole-Body Control using Joint Torque Feedback for Quadrupedal Robots. *J. Mech. Robot.* **2021**, *13*, 1–18. [CrossRef]
23. Guo, F.; Cai, H.; Ceccarelli, M.; Li, T.; Yao, B. Enhanced DH: An improved convention for establishing a robot link coordinate system fixed on the joint. *Ind. Robot. Int. J. Robot. Res. Appl.* **2019**, *47*, 197–205. [CrossRef]
24. Khazaee, M.; Sadedel, M.; Davarpanah, A. Behavior-based navigation of an autonomous hexapod robot using a hybrid automaton. *J. Intell. Robot. Syst.* **2021**, *102*, 1–24. [CrossRef]

applied sciences

MDPI

Article

An FPGA Based Energy Efficient DS-SLAM Accelerator for Mobile Robots in Dynamic Environment

Yakun Wu [1], Li Luo [1], Shujuan Yin [2], Mengqi Yu [3], Fei Qiao [3,*], Hongzhi Huang [1], Xuesong Shi [4], Qi Wei [5] and Xinjun Liu [6,*]

[1] School of Electronic and Information Engineering, Beijing Jiaotong University, Beijing 100044, China; 18120025@bjtu.edu.cn (Y.W.); lluo@bjtu.edu.cn (L.L.); 13213037@bjtu.edu.cn (H.H.)
[2] School of Information Science and Technology, Baotou Teachers' College, Baotou 014030, China; 60466@bttc.edu.cn
[3] Department of Electronic Engineering and BNRist, Tsinghua University, Beijing 100084, China; yumq17@mails.tsinghua.edu.cn
[4] Intel Labs China, Beijing 100090, China; xuesong.shi@intel.com
[5] Department of Precision Instrument, Tsinghua University, Beijing 100084, China; weiqi@tsinghua.edu.cn
[6] Department of Mechanical Engineering, Tsinghua University, Beijing 100084, China
* Correspondence: qiaofei@tsinghua.edu.cn (F.Q.); xinjunliu@mail.tsinghua.edu.cn (X.L.); Tel.: +86-138-1035-5024 (F.Q.); +86-135-5279-5847 (X.L.)

Abstract: The Simultaneous Localization and Mapping (SLAM) algorithm is a hotspot in robot application research with the ability to help mobile robots solve the most fundamental problems of "localization" and "mapping". The visual semantic SLAM algorithm fused with semantic information enables robots to understand the surrounding environment better, thus dealing with complexity and variability of real application scenarios. DS-SLAM (Semantic SLAM towards Dynamic Environment), one of the representative works in visual semantic SLAM, enhances the robustness in the dynamic scene through semantic information. However, the introduction of deep learning increases the complexity of the system, which makes it a considerable challenge to achieve the real-time semantic SLAM system on the low-power embedded platform. In this paper, we realized the high energy-efficiency DS-SLAM algorithm on the Field Programmable Gate Array (FPGA) based heterogeneous platform through the optimization co-design of software and hardware with the help of OpenCL (Open Computing Language) development flow. Compared with Intel i7 CPU on the TUM dataset, our accelerator achieves up to $13\times$ frame rate improvement, and up to $18\times$ energy efficiency improvement, without significant loss in accuracy.

Keywords: visual semantic SLAM; semantic segmentation; FPGA; hardware acceleration; mobile robots

Citation: Wu, Y.; Luo, L.; Yin, S.; Yu, M.; Qiao, F.; Huang, H.; Shi, X.; Wei, Q.; Liu, X. An FPGA Based Energy Efficient DS-SLAM Accelerator for Mobile Robots in Dynamic Environment. *Appl. Sci.* **2021**, *11*, 1828. https://doi.org/10.3390/app11041828

Academic Editor: Pavol Božek

Received: 10 December 2020
Accepted: 15 February 2021
Published: 18 February 2021

1. Introduction

Mobile robots are not only widely used in fields such as industry, agriculture, medical and services [1,2], but are also well used in dangerous situations such as urban security [3], national defense [4] and space detection [5,6]. With the development of artificial intelligence technology, mobile robots perform complex algorithms to complete more advanced tasks. The insufficient computing capability and limited energy of traditional computing platforms has become the bottleneck of intelligent robot development [7,8].

Mobile robots not only need to perceive the information of the surrounding environment but to also clearly determine their position, which leads to the SLAM (Simultaneous Localization and Mapping) algorithm. Nowadays, the visual SLAM framework is quite mature, mainly composed of critical parts such as visual odometry, optimization, loop closing and mapping. However, visual SLAM has some problems with the dynamic environment and the demand for advanced tasks, such as the lack of understanding about environmental information and the stability of the system in dynamic environments [9].

Appl. Sci. **2021**, *11*, 1828. https://doi.org/10.3390/app11041828

Semantic information introduced through deep learning is more and more tightly integrated with SLAM to solve these problems. On the one hand, semantic SLAM can improve human-computer interaction ability and help robots understand the surrounding environment as human beings do. On the other hand, the result of semantic segmentation can help to remove additional feature points and improve the accuracy of mapping. As a new development direction, there is a series of semantic SLAM represented by DS-SLAM (Semantic SLAM towards Dynamic Environment) [9]. DS-SLAM introduces SegNet [10] network for semantic segmentation in the front-end visual odometry to solve the problems of location caused by moving objects in the environment. Compared with ORB (Oriented Fast and Rotated Brief)-SLAM2 [11], the accuracy of the localization in dynamic scenes has been improved by an order of magnitude.

However, the addition of a semantic segmentation module increases the computational complexity of the system and cannot be processed on the CPU (Central Processing Unit) in real-time. When paired with the high-performance GPU (Graphics Processing Unit) for computing acceleration, the system consumes a lot of energy and increases the hardware costs, which limits the actual deployment on the terminal devices and mobile robots with limited battery capacity. The FPGA (Field Programmable Gate Array) platform has been widely used in Convolutional Neural Network (CNN) acceleration because it is reconfigurable, energy-efficient, and parallelizable.

For the design of FPGA, traditional development is mainly based on the RTL (Register Transfer Level) [12,13], which has a long development cycle and complicated debugging. This article uses the high-level synthesis tool OpenCL [14] to realize algorithm acceleration. OpenCL works with the host-device model and its code is divided into two parts—the host program and device kernel. The host program is used to obtain device information, set memory objects and execute kernels. In order to control the device, users need to build the host programming code, which meets the OpenCL development specification. The device kernel mainly writes functions that need to be accelerated and these functions are compiled into the accelerator. In the OpenCL development environment, FPGA compiler and synthesis and kernel simulation and debugging are used to get the image for FPGA execution. It can automatically compile from the high-level programming language (C/C++) to the low-level RTL design, provide a faster hardware development cycle and be easily integrated with the applications.

As an industry standard, OpenCL can program heterogeneous computing platforms supported by many suppliers and is not limited by the hardware device. The processing platform used in this article is shown in Figure 1, which is composed of the Turtlebot2 robot, the Kincet camera and the HERO (Heterogeneous Extensible Robot Open) platform. Turtlebot2 is a small, low-cost, fully programmable, ROS (Robot Operating System) based mobile robot with simple operation and strong scalability. HERO is the heterogeneous extensible robot open platform provided by Intel [15]. It has high-quality features such as low power consumption, high performance and small size (18 cm × 14 cm × 7.5 cm). The whole of the HERO platform is composed of the host system and the external FPGA acceleration card. In the whole hardware system, CPU has rich IO (Input/Output) functions, provides various high-speed and low-speed interfaces, and its performance is enough to meet that of the simple algorithms. FPGA integrates abundant logic and computing units, storage and routing resources, which ensures the performance acceleration of FPGA in neural network inference. Besides, there is an 8-channel PCIe Gen 3 interface that provides a bandwidth of 7.88 GB/s to perform transmission and communication [15].

The main contributions of this work are summarized below:

1. The quantization strategy of combining convolution and batch normalization into one operation and grouping convolution filters with different data distribution is proposed to solve the problem of the large model and long execution time of a semantic segmentation network.

2. A function configurable pipeline design method and a multi-level memory access optimization scheme for the convolutional Encoder-Decoder architecture of semantic segmentation networks is put forward.
3. The high energy-efficient hardware acceleration solution of the DS-SLAM algorithm on a heterogeneous computing platform is proposed, which can be utilized for the functional module to provide mobile robots with autonomous localization.

Figure 1. The processing platform.

The remainder of this paper is organized as follows—Section 2 provides an overview of the current implementation of different SLAM algorithms on the embedded platform. In Section 3, we give a brief description of DS-SLAM algorithms and the SegNet semantic segmentation network, as well as our proposed quantization strategy. Section 4 presents the framework of our FPGA accelerator and some optimization methods. Subsequently, Section 5 provides the experimental results of the proposed energy-efficient DS-SLAM implementation. Finally, a brief conclusion is provided in Section 6.

2. Related Work

As the SLAM algorithm of mobile robots becomes more and more complex, the general processors on traditional mobile platforms cannot meet the requirements of real-time and energy consumption. The hardware acceleration for the SLAM algorithm has become vital research in the field of robotics.

In [16], an accelerator implemented in FPGA can expand the actual EKF-SLAM (Extended Kalman Filter SLAM) system and observe and correct up to 45 landmarks under the real-time constraints. However, because the complexity of the EKF-SLAM algorithm increases with conversion landmarks, it is not suitable for large-scale environments. Most of the operations are executed sequentially; thus, the parallel architecture's advantages cannot be reflected.

For the LSD-SLAM (Large Scale Direct monocular SLAM) algorithm, Konstantinos Boikos et al. propose a hardware implementation based on an FPGA SoC (System-on-a-Chip) [17]. The design shows that an FPGA SoC can introduce a more intensive SLAM algorithm into embedded low-power devices, and has a higher performance than other solutions. However, the SLAM algorithm implemented in this work is relatively simple, which limits the actual deployment of the system in complex scenarios.

M. Abouzahir et al. use OpenCL to implement the large-scale parallel processing on the FPGA to achieve the Fast-SLAM system [18]. Compared with the GPU-based implementation, the processing time is accelerated by 7.5 times. The research shows that real-time SLAM systems can be implemented on heterogeneous embedded platforms using software and hardware co-design methods.

The eSLAM is proposed by Runze Liu et al. [19]. On a heterogeneous platform based on the ARM (Advanced RISC Machine) processor and FPGA board, a highly energy-efficient ORB-SLAM (Oriented Fast and Rotated Brief SLAM) system is implemented. This work puts the most time-consuming feature extraction and matching process on the FPGA for acceleration, while the remaining threads are executed on the ARM processor. The eSLAM proposes an improved rotationally symmetric BRIEF (Binary Robust Independent Elementary Features) descriptor. By reducing the operation of rotating test locations, the calculation and storage consumption are significantly reduced. Simultaneously, through the pipeline design, the throughput is further improved and the memory is reduced.

It can be seen that the SLAM algorithm implemented on embedded platforms is relatively simple and the related work is still lacking on semantic SLAM hardware implementation. So, this paper proposes a method to achieve the semantic SLAM algorithm based on the HERO platform.

3. Algorithm Implementation

3.1. DS-SLAM Overview

The DS-SLAM algorithm introduces semantic information through deep learning, which endows the robot with learning ability to understand the surrounding environment and reduces the impact of dynamic objects. As can be seen from the algorithm in Figure 2, the DS-SLAM algorithm is a five-thread parallel architecture, including tracking, semantic segmentation, local mapping, loop closure and dense mapping creation. The core idea is to extract ORB feature points from the tracking thread and then check these feature points to get potential dynamic points through a moving mobile consistency check. At the same time, the segmentation results are obtained in parallel with a semantic segmentation thread. Finally, dynamic objects are removed by combining the results of two threads. Table 1 shows the processing time of some major modules in the corresponding thread. The results show that the semantic segmentation thread, whether running on CPU or GPU, is the bottleneck that affects the system's real-time performance. Figure 2 shows the overall framework of the DS-SLAM algorithm based on the HERO platform. Firstly, the images and model parameters are read from CPU. The computationally complex semantic segmentation thread can be accelerated on FPGA with the help of Intel OpenCL SDK (Software Development Kit) and BSP (Board Support Package). Finally, the output is transmitted back to CPU to complete the other threads, enough for real-time processing.

Table 1. Execution time of different modules in the Semantic Simultaneous Localization and Mapping towards Dynamic Environment (DS-SLAM) algorithm [9].

Module	ORB Feature Extraction	Moving Consistency Check	Semantic Segmentation	
Thread	Tracking	Tracking	Semantic segmentation	
Platform	Intel Core i7-6800K CPU	Intel Core i7-6800K CPU	Intel Core i7-8750H CPU	NVIDIA Quadro P4000 GPU
Time (ms)	9.37	29.5	2582	37.6

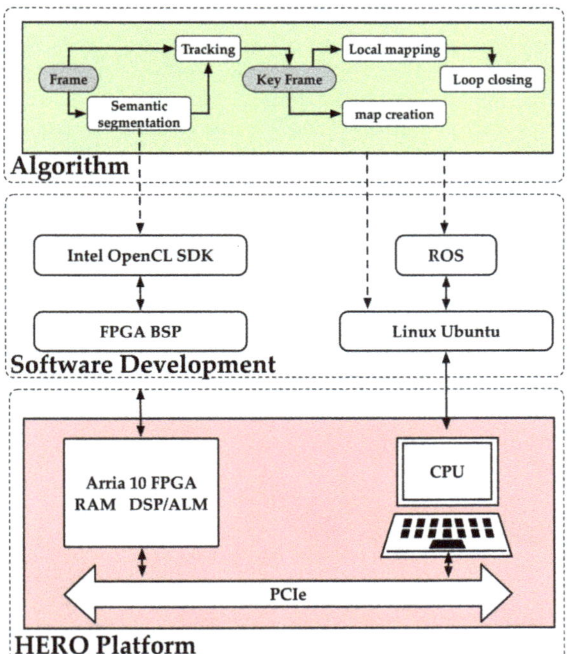

Figure 2. The DS-SLAM framework on Heterogeneous Extensible Robot Open (HERO).

The extraction of semantic information is realized by the semantic segmentation network. In order to reduce the complexity of the whole system and ensure the accuracy and real-time, the SegNet algorithm is selected to achieve semantic segmentation. It is a typical convolutional encoder-decoder architecture.

3.2. Quantization Strategy

The model size of SegNet is 112.3 MB, which should be compressed by pruning, quantization, coding and other means [20] before running on the embedded platform to reduce the computational cost and storage consumption. The method used in this paper is linear quantization, and the overall process is shown in Figure 3.

First, convolution and Batch Normalization (BN) are combined into one operation, because the model parameters are fixed during inference. For SegNet, each convolution layer is followed by a BN layer. Merging these two layers can reduce the use of computing resources and precision loss and improve the inference speed. Next, the data distribution of filters in a convolution layer is calculated. Filters with similar data distribution are merged into one group, corresponding to the same scale factor f. The f determines the decimal point position of the quantized dynamic fixed-point number. Equations (1) and (2) take the weight as an example to introduce the calculation of the f. The w and w' represent the weight before and after quantization, and N represents the bit width of the fixed-point number. f is designed as the integer power of 2, so that the shift can be used instead of the multiplication, which is well implemented on FPGA. Load the preprocessed input image for convolution, and the generated output activations are arranged in the channel dimension. Use the same method described above to calculate the decimal position of the output activation f_o, and the f_o of the current layer is the f_i (decimal position of the input data) of the next layer. To ensure the following correct calculation, a set of 1×1 filter is used to complete the reordering of feature maps. Subsequent pooling and unpooling operations are unchanged. After the final convolution layer, the quantized model and the corresponding scale factor are exported. The whole quantization operation is processed in the host CPU in advance, and will not increase the inference latency on FPGA. The 8-bit quantized SegNet algorithm is evaluated on the Pascal VOC dataset [21]. The global accuracy is 87.3%, the class accuracy is 69.4%, and the mean IoU is 51.5%, which is 0.8%, 1.1%, and 1.6% lower than the original model. The quantized model dramatically reduces the storage requirements and computational complexity without significant precision loss.

$$f_w = ceil(log_2 \frac{2^N - 1}{max(|w|)}) \tag{1}$$

$$w' = round(\frac{w}{2^{-f_w}}). \tag{2}$$

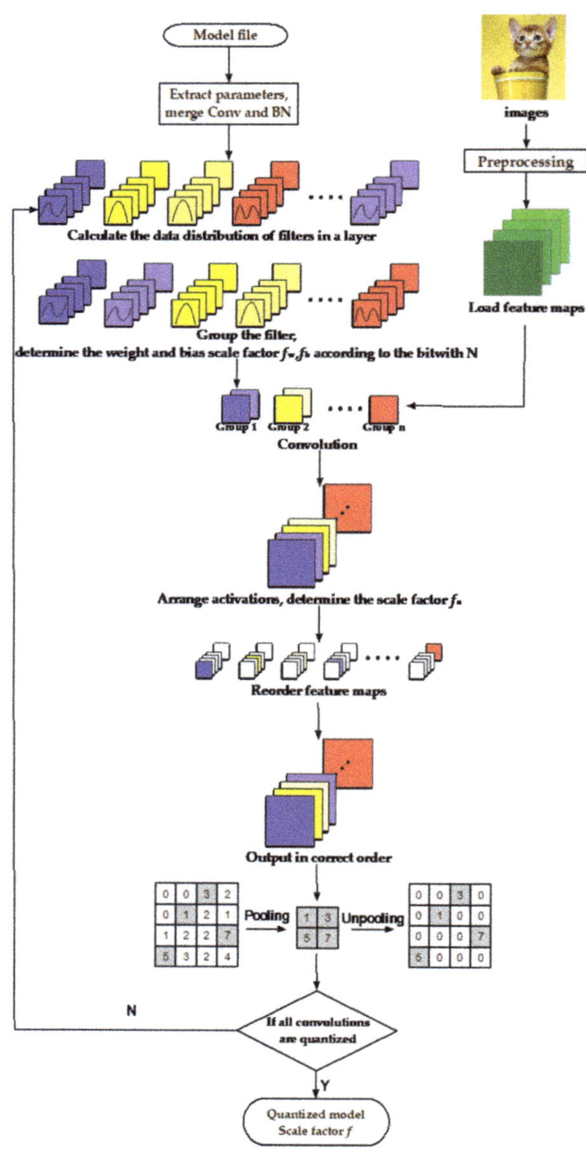

Figure 3. The process of quantization.

4. Proposed Method

The semantic segmentation thread implemented by SegNet is the most time-consuming part of the DS-SLAM algorithm and needs to be accelerated. In [9], SegNet runs on GPU but its high power consumption is not suitable for actual deployment on battery-powered mobile robots. Therefore, this paper will accelerate the inference of semantic segmentation network on FPGA to improve the real-time performance and energy efficiency of the DS-SLAM system.

4.1. Arria 10 FPGA Description

The Arria 10 GX 1150 FPGA board used in this paper adopts TSMC 20 nm technology and has rich design resources. The adaptive logic module (ALM) is the basic component of the FPGA logic structure and 427,200 ALMs can be equivalent to 1,150,000 logic elements (LE) [22], so this FPGA board is named GX 1150. One thousand five hundred and eighteen precision adjustable DSPs (Digital Signal Processings) have encapsulated as hardcore IP (intellectual property) for developers to call, supporting high-performance multiplication and addition operations. Memory can be divided into two types; one is the M20k module, the size of a single module is 20 KB; the other is memory logic array block (MLAB), the size of a single module is 0.64 kb. The M20k module is the best choice for larger memory arrays and provides a large number of independent ports. MLAB is the best choice for the wide and shallow memory array, which can realize shift register and a small FIFO (first in first out) buffer.

4.2. Accelerator Architecture

Through the OpenCL development process, the overall architecture [23,24] of the Seg-Net accelerator implemented on Arria 10 GX 1150 FPGA is shown in Figure 4. The storage module is mainly divided into two parts. One is off-chip DRAM as the global memory, which is used to store model parameters (weights, bias, and decimal position generated by quantization), pooling indices caused by maximum value pooling and operation data of input and output. Another part is the on-chip local buffer as the local memory, which is used to store the input feature map and weights of the current calculation layer, as well as the output activation value. On-chip memory can be further subdivided into channels and registers, which, together with off-chip DRAM, form a multi-level memory access structure. The channel extension defined by OpenCL is widely used between local buffer and kernel, as well as between kernel and kernel. It acts like FIFO and allows the transfers on chip to reduce latency. The dataflow determines what data is read into which level of memory structure and when it is processed. Compared with on-chip memory, DRAM consumes more energy per access. Therefore, we should increase the data reuse of on-chip memory to reduce access to high-cost memory. In this paper, each weight is read from DRAM to local buffer and stays stationary for data reused and the input feature maps are broadcast to different CUs, then the partial sums are accumulated. Weight reuse minimizes the energy consumption of reading weights. Currently, the DRAM access bandwidth is 1234 MB/s.

OpenCL uses a host-device model, and the design is divided into two parts—the host program and the device kernel. The host program includes an API (Application Programming Interface) to define and control the platform, which is used to obtain device information. Besides, the host program operates memory objects and creates a command queue and executes kernel queues. The design of the device kernel is based on the standard C language with some restrictions and extensions. The kernel program mainly includes the function that needs to be accelerated and these functions are compiled into the accelerator. For the SegNet network, this paper mainly designs convolution kernel, pooling kernel and unpooling kernel for SegNet, and the ReLU activation function is integrated into the convolution kernel. Through a 2-bit control flag, the different combinations of kernels realize a function configurable pipeline. There are four modes in total, shown in Figure 5.

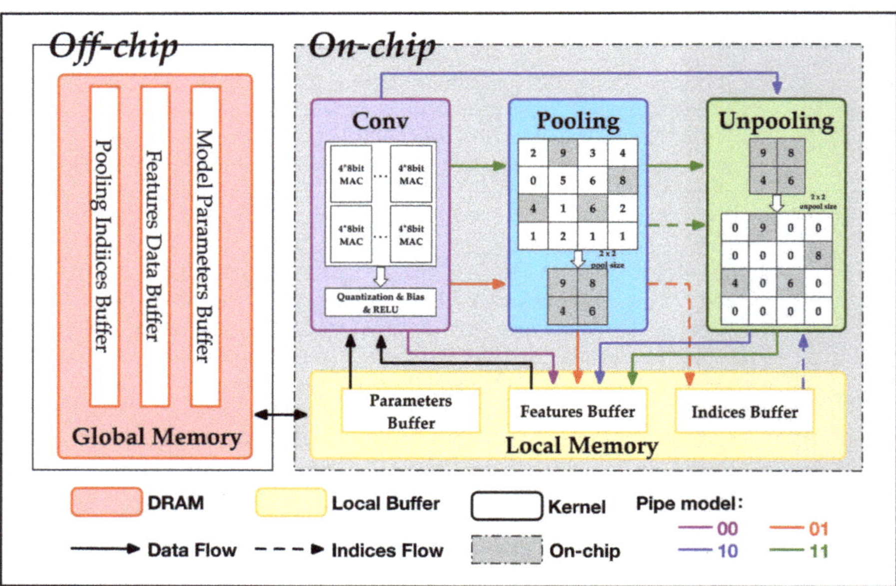

Figure 4. The architecture of the SegNet accelerator.

The 00 mode pipeline only executes convolution operation, shown in Figure 5a. The weight and decimal point of one layer are read from DRAM to the local buffer at one time and then they are pressed into the channel continuously and transferred to the convolution kernel for operation. However, the input feature map of the convolution layer is read in batches. The on-chip read buffer improves data transmission efficiency by Ping-Pong operation. Two buffers of Ping-Pong operation are designed on the chip. When one buffer transmits characteristics, the other buffer reads the input characteristic group of the next stage from the off-chip DRAM at the same time, so as to improve the throughput. Convolution operations have different computing units to execute in parallel. The result is transferred back to off-chip DRAM, through the quantification and ReLU (Rectified Linear Unit). Since there is no dependency between the result of off-chip DRAM, multiple work items are executed in parallel to write the result back.

In the 01 mode pipeline, convolution and pooling operations are executed in parallel. Part of the convolution results are passed to the pooling kernel through the channel, avoiding the data transmission delay caused by multiple access to DRAM to maintain a high-speed processing efficiency. The pooling results and pooling indices are passed to the local buffer through the channel and finally written back to the off-chip DRAM. The input and output memory are used alternately and the output memory of the current operation will become the input memory of the next operation. The data flow is shown in Figure 5b.

The 10 mode pipeline is a combination of convolution and unpooling, shown in Figure 5c. For the unpooling operation, the input not only has the calculation results of the previous kernel but also has the pooling indices with location information. Pooling indices need to be read from the off-chip DRAM to on-chip local buffer, which will cause unnecessary delay. Therefore, we prepare the required pooling indices in advance through data pre-reading. When the convolution results are transmitted to the unpooling kernel with the help of the channel, the operation can be executed directly without waiting time. This design method is beneficial for the encoder-decoder network, with a connection between the front and back layers. As shown in Figure 5d, the 11 mode pipeline performs all of the kernels, including convolution, pooling, and unpooling operations. The pooling indices required by unpooling are directly obtained from the last kernel through the channel without accessing off-chip DRAM.

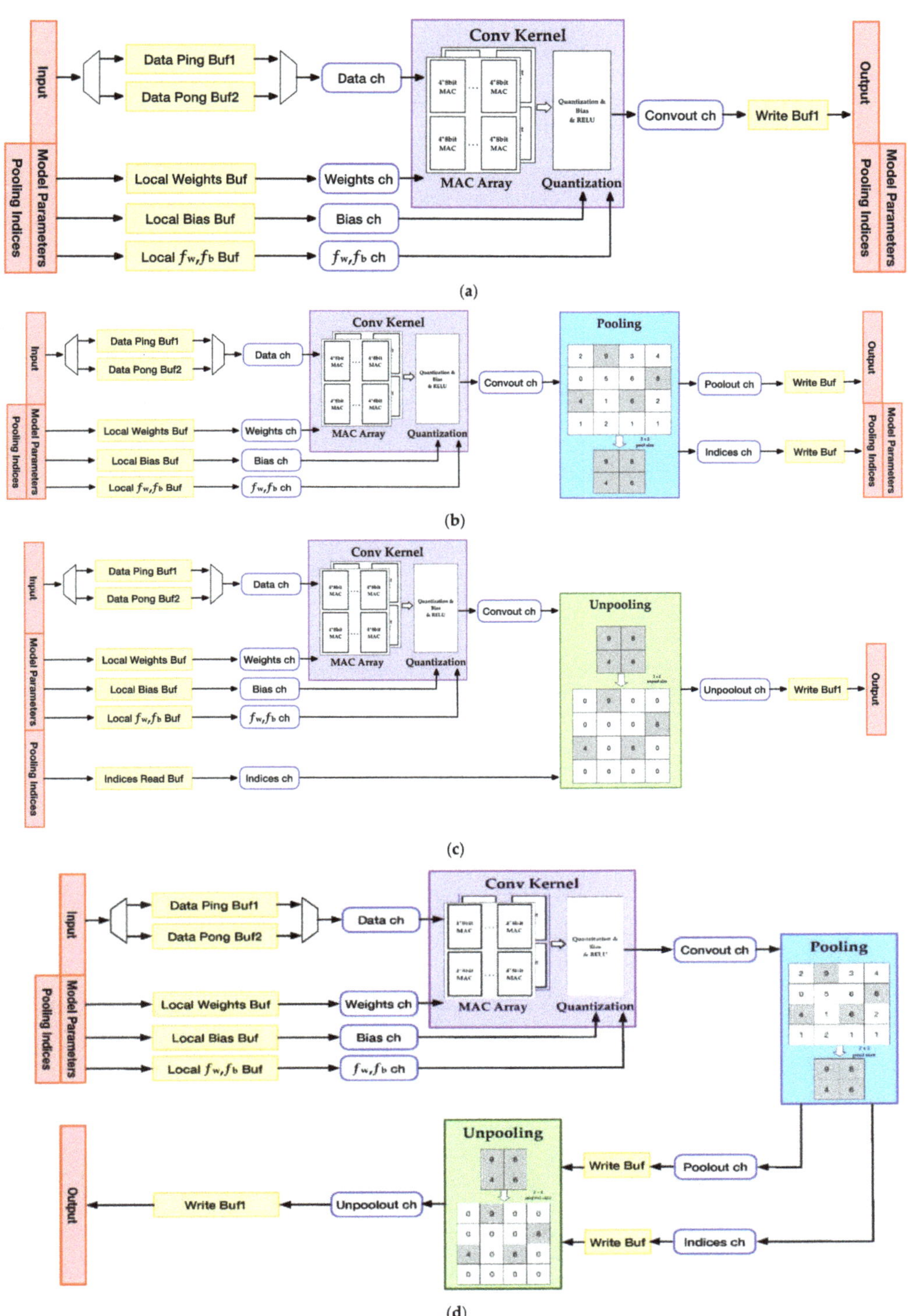

Figure 5. Four pipeline modes with configurable functions. (**a**) Mode 00: convolution; (**b**) Mode 01: convolution + pooling; (**c**) Mode 10: convolution + unpooling; (**d**) Mode 11: convolution + pooling + unpooling.

Each pipeline mode completes an iteration of "data reading in, computing, and writing back." Then multiple iterations are combined in an orderly way to realize the whole network. For example, to implement SegNet, the combination is in sequence of "00-01-00-01-00-00-01-00-00-01-00-00-11-00-00-10-00-00-10-00-00-10-00-10-00-00". The benefit of this design approach is that using multiple pipeline loops in different patterns allows us to implement different network frameworks flexibly. This kind of function configurable pipeline architecture has a certain expansibility. By designing and replacing different kernels, we can flexibly implement other convolutional neural networks. In the process of implementing other neural networks, we should evaluate the boards' resources in advance and focus on the use of local memory. We must pay attention to the kernel design and follow the coding rules of OpenCL, so as to improve the efficiency of the hardware implementation.

4.3. Softmax Optimization

In addition to the inference part of FPGA, the result of segmentation also needs to be accelerated on CPU in the implementation of the semantic segmentation thread. Softmax operation, as the last layer of the SegNet network, is used to realize multi classification. It maps the output of multiple neurons to the (0~1) range. The sum of the mapped values is 1, which can be regarded as the probability value. In the calculation of softmax, the maximum probability is taken as the final prediction result, as shown in Figure 6. This operation only normalizes the result and does not change the numerical relationship, so the operation can be omitted. Softmax is relatively complex because of its exponential operation, which affects the real-time performance of the system. After removing softmax, the speed of semantic segmentation is improved by nearly 77 ms.

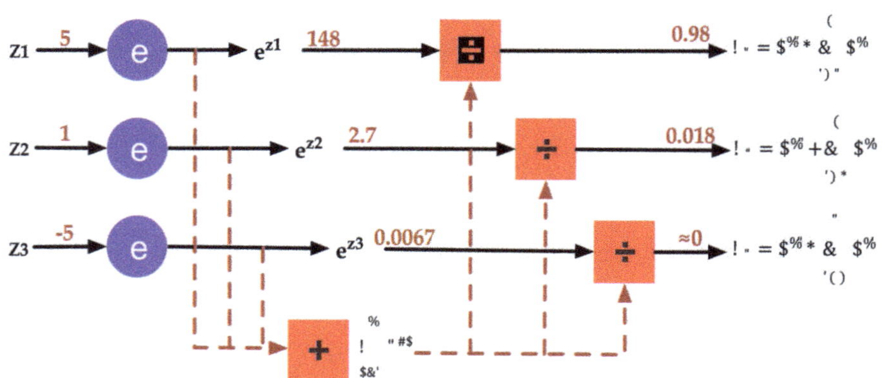

Figure 6. Softmax operation diagram.

5. Experimental Results

5.1. HERO Platform Description

The proposed DS-SLAM accelerator is implemented on the HERO platform, which contains an Intel i5-7260U CPU as a host side and an Arria 10 GX 1150 FPGA as an acceleration board. The semantic segmentation thread that needs to be accelerated is deployed on FPGA, and the others run on CPU. We use the OpenCL development process to implement the SegNet network through the Intel FPGA SDK. The clock frequency of the FPGA is 202 MHz and the hardware resource utilization is shown in Table 2. It can be seen that the utilization rate of DSP reaches 100%, almost all of which is used to realize convolution operation.

Table 2. Hardware resource utilization of Arria 10 Field Programmable Gate Array (FPGA).

	ALMS	**Registers**	**M20Ks**	**DSPs**
Utilization	24%	12%	63%	100%

The energy-efficient DS-SLAM system implemented on a heterogeneous computing platform is evaluated on the TUM RGB-D dataset [25]. The TUM dataset is mainly for office scenes. Its "walking" sequence contains dynamic objects, which are two moving people.

5.2. Performance Evaluation

In Table 3, the execution time of semantic segmentation thread and the whole DS-SLAM algorithm, frame rate, the power consumption and the energy efficiency of the proposed method in this paper are compared with the software implementations on the Intel i7-8750H CPU and the NVIDIA P4000 GPU. In contrast with CPU, our acceleration platform achieves up to 13× frame rate improvement and up to 18× energy efficiency improvement. The GPU server has more computing resources, thus achieves a higher real-time performance than ours. An important application of mobile robots is to provide indoor services. In order to ensure human safety, the moving speed is relatively slow, so the real-time performance of 5.3 fps can meet the application requirements. In [9], the experimental platform is Intel i7-6800K CPU with NVIDIA P4000 GPU and the overall power consumption is 245 W, which is unacceptable for energy-limited applications. Mobile robots are usually powered by batteries, so it is essential to reduce the power consumption of the processor. The HERO is a low power platform, and the proposed accelerator implemented on it is 2.2 times more energy-efficient than GPU.

Table 3. Execution time, frame rate, power consumption, and energy efficiency comparison results.

	Software Implementations		**Our Work**	**Improvements (vs. CPU)**
	i7-8750H CPU	i7-6800K CPU + P4000 GPU	HERO	-
Platform	i7-8750H CPU	i7-6800K CPU + P4000 GPU	HERO	-
Power	45W	245 W	35 W	22.2%
Execution time of semantic segmentation	2582 ms	37.6 ms	155 ms	94.0%
Execution time of the whole DS-SLAM	2600 ms	59.4 ms	190 ms	92.7%
Frame rate	0.38 fps	16.8 fps	5.3 fps	13×
Energy efficiency	0.008 fps/W	0.068 fps/W	0.151 fps/W	18×

The proposed high energy efficiency accelerator provides the feasibility of deploying the DS-SLAM algorithm on mobile robots. The SegNet network, which is used by the semantic segmentation thread, can be used for 21 kinds of objects including indoor and outdoor scenes, and well supports the semantic segmentation of the datasets used in the SLAM algorithm. The evaluation results on the TUM dataset are shown in Figure 7 and the objects are well segmented. As the dynamic object, people will be eliminated in the final mapping to improve the localization accuracy.

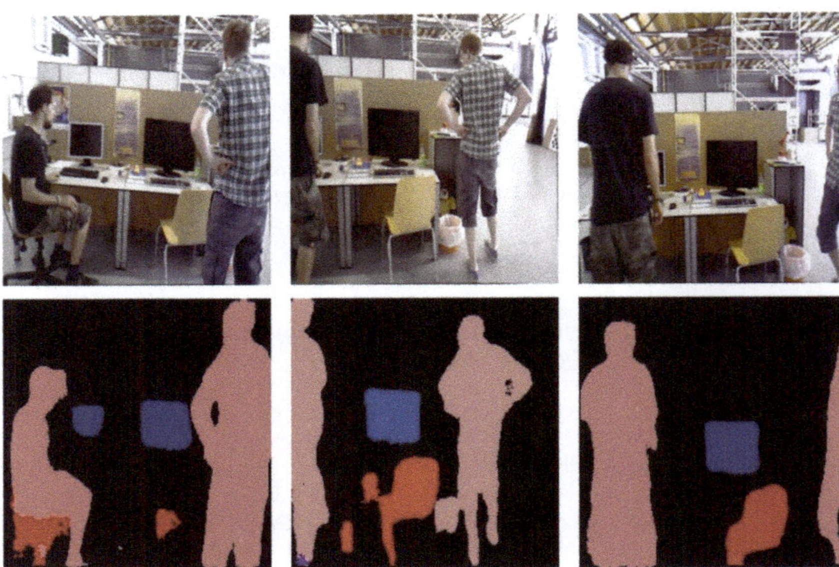

Figure 7. The result of semantic segmentation on the TUM dataset.

5.3. Accuracy Analysis

The TUM dataset provides the ground truth of the real trajectory for the convenience of comparison with the experimental results. The indicators to evaluate the accuracy of DS-SLAM implemented in this paper are ATE (Absolute Trajectory Error) and RPE (Relative Pose Error). ATE is the direct difference between estimated pose and real pose, which can directly reflect the accuracy of the algorithm and the global consistency of the trajectory. RPE is used to evaluate the drift of the system. In this paper, we present the values of Root Mean Squared Error (RMSE), Mean Error, Median Error, and Standard Deviation (S.D.). The experimental results are shown in Tables 4 and 5, comparing the accuracy of ORB-SLAM2, DS-SLAM software algorithm and hardware implementation DS-SLAM proposed in this paper under different data sequences. Figure 8 shows ATE and RPE plots from ORB-SLAM2 and hardware implementation DS-SLAM in high dynamic sequence (fr3_walking_xyz) respectively. The ATE results show that black is the true value of the trajectory, blue is the estimated trajectory, and red is the error between the two.

Table 4. The results of metric absolute trajectory error (ATE).

Sequences	ORB-SLAM2				DS-SLAM				Our Work			
	RMSE	Mean	Median	S.D.	RMSE	Mean	Median	S.D.	RMSE	Mean	Median	S.D.
Fr3_walking_xyz	0.7521	0.6492	0.5857	0.3759	0.0247	0.0186	0.0151	0.0161	0.0316	0.0176	0.0129	0.0262
Fr3_walking_static	0.3900	0.3554	0.3087	0.1602	0.0081	0.0073	0.0067	0.0036	0.0078	0.0069	0.0063	0.0035
Fr3_walking_rpy	0.8705	0.7425	0.7059	0.4520	0.4442	0.3768	0.2835	0.2350	0.5031	0.4276	0.3066	0.2646
Fr3_walking_half	0.4863	0.4272	0.3964	0.2290	0.0303	0.0258	0.0222	0.0159	0.0292	0.0250	0.0214	0.0151
Fr3_sitting_static	0.0087	0.0076	0.0066	0.0043	0.0065	0.0055	0.0049	0.0033	0.0060	0.0052	0.0046	0.0030

Table 5. The results of metric translational drift (RPE).

Sequences	ORB-SLAM2				DS-SLAM				Our Work			
	RMSE	Mean	Median	S.D.	RMSE	Mean	Median	S.D.	RMSE	Mean	Median	S.D.
Fr3_walking_xyz	0.4124	0.3110	0.2465	0.2684	0.0333	0.0238	0.0181	0.0229	0.0459	0.0275	0.0198	0.0367
Fr3_walking_static	0.2162	0.0905	0.0155	0.1962	0.0102	0.0091	0.0082	0.0048	0.0115	0.0104	0.0099	0.0048
Fr3_walking_rpy	0.4249	0.2825	0.1487	0.3166	0.1503	0.0942	0.0457	0.1168	0.1903	0.1076	0.0436	0.1570
Fr3_walking_half	0.3550	0.2161	0.0774	0.2810	0.0297	0.0256	0.0226	0.0152	0.0414	0.0363	0.0335	0.0198
Fr3_sitting_static	0.0095	0.0083	0.0073	0.0046	0.0078	0.0068	0.0061	0.0038	0.0089	0.0079	0.0071	0.0042

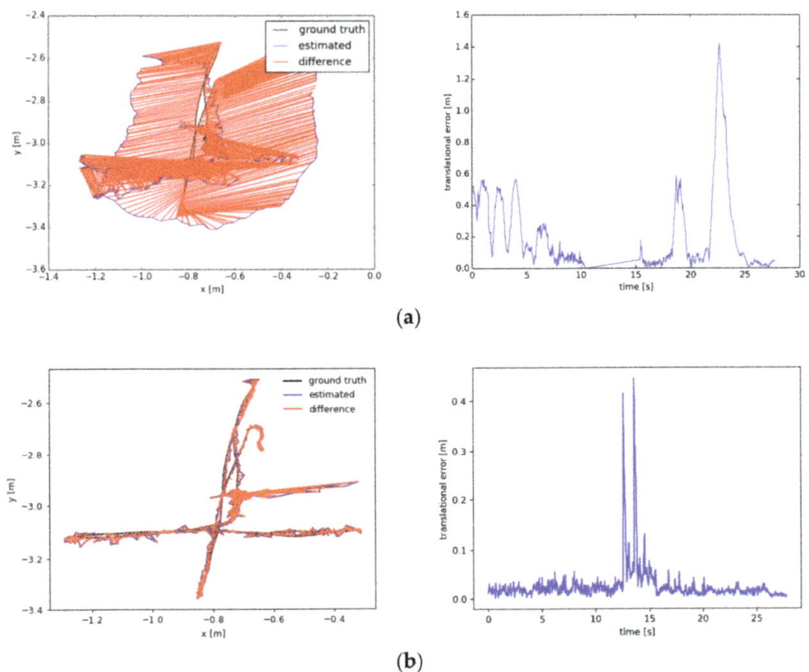

Figure 8. Visual experimental results: (**a**) ATE and RPE from ORB-SLAM2, (**b**) ATE and RPE from our work.

Compared with ORB-SLAM2, the introduction of semantic information in DS-SLAM can eliminate the interference of moving objects and significantly improve the accuracy in thr dynamic environment. The accuracy of our work is in the same order of magnitude as DS-SLAM, but it is slightly reduced because of model quantization. We have made a trade-off between accuracy and energy consumption, sacrificing a little precision to improve the energy efficiency of the system, providing a solution for the actual deployment of DS-SLAM on mobile robots.

6. Conclusions

In this paper, an energy-efficient DS-SLAM based semantic SLAM system is proposed on the HERO heterogeneous platform. The model of SegNet for extracting semantic information is first quantized to the 8-bits dynamic fixed-point number. The proposed quantization strategy reduces the storage requirements and computational complexity, and only loses 0.8% of the global accuracy on the Pascal VOC dataset. Meanwhile, as the most time-consuming part, the semantic segmentation thread is accelerated on the FPGA to reduce the latency. A function configurable pipeline architecture with expansibility is proposed. By designing and replacing kernels, other convolutional neural networks can be implemented flexibly. Compared with Intel i7 CPU, our DS-SLAM accelerator on HERO platform achieves up to 13× frame rate improvement and 18× energy efficiency improvement. Moreover, there is a minute localization accuracy loss on the TUM dataset. The proposed energy- efficient DS-SLAM accelerator fundamentally solves the dependence of the algorithm on high energy consumption GPU. It increases the working time of battery-powered robots, which can be utilized as a functional module to provide mobile robots with autonomous localization.

Author Contributions: Conceptualization, Y.W., M.Y., H.H. and F.Q.; data curation, Y.W., H.H. and M.Y.; formal analysis, Y.W.; funding acquisition, F.Q., Q.W. and X.L.; investigation, Y.W. and M.Y.; methodology, Y.W., M.Y., H.H. and X.S.; project administration, F.Q., S.Y. and L.L.; resources, X.S., F.Q., Q.W. and X.L.; software, M.Y. and X.S.; supervision, F.Q., S.Y. and L.L.; validation, Y.W. and X.S.; visualization, Y.W.; writing—original draft, Y.W.; writing—review & editing: M.Y. All authors have read and agreed to the published version of the manuscript.

Funding: This research was funded by National Key R&D Program of China under grant No. 2018YFB1702500.

Informed Consent Statement: Informed consent was obtained from all subjects involved in the study.

Data Availability Statement: No new data were created or analyzed in this study. Data sharing is not applicable to this article.

Acknowledgments: The authors would like to acknowledge supports from National Key R&D Program of China under grant No. 2018YFB1702500. The authors would also acknowledge support from Beijing Innovation Center for Future Chips, Tsinghua University.

Conflicts of Interest: The authors declare no conflict of interest.

References

1. Song, S.; Choi, D.; Hur, J.; Lee, M.; Park, Y.J.; Kim, J. Development and application of Mobile Robot system for Marking Process in LNGC cargo tanks. In Proceedings of the 2009 ICCAS-SICE, Fukuoka, Japan, 18–21 August 2009; pp. 2636–2638.
2. Saitoh, M.; Takahashi, Y.; Sankaranarayanan, A.; Ohmachi, H.; Marukawa, K. A mobile robot testbed with manipulator for security guard application. In Proceedings of the 1995 IEEE International Conference on Robotics and Automation, Nagoya, Japan, 21–27 May 2002; Volume 3, pp. 2518–2523.
3. Murphy, R. Human–Robot Interaction in Rescue Robotics. *IEEE Trans. Syst. Man. Cybern. Part C Appl. Rev.* **2004**, *34*, 138–153. [CrossRef]
4. Dunbabin, M.; Marques, L. Robots for Environmental Monitoring: Significant Advancements and Applications. *IEEE Robot. Autom. Mag.* **2012**, *19*, 24–39. [CrossRef]
5. Bapna, D.; Rollins, E.; Murphy, J.; Maimone, E.; Whittaker, W.; Wettergreen, D. The Atacama Desert Trek: Outcomes. In Proceedings of the 1998 IEEE International Conference on Robotics and Automation (Cat. No. 98CH36146), Leuven, Belgium, 20 May 2002; Volume 1, pp. 597–604.
6. Rollins, E.; Luntz, J.; Foessel, A.; Shamah, B.; Whittaker, W. Nomad: A Demonstration of the Transforming Chassis. In Proceedings of the IEEE International Conference on Robotics and Automation, Leuven, Belgium, 20 May 1998; pp. 611–617.
7. Kimon, P.V. Classification of UAVs. In *Handbook of Unmanned Aerial Vehicles*; Kimon, P.V., George, J.V., Eds.; Springer: Dordrecht, The Netherlands, 2015.
8. Mei, Y.; Lu, Y.H.; Hu, Y.C. A case study of mobile robot's energy consumption and conservation techniques. In Proceedings of the International Conference on Advanced Robotics, Seattle, WA, USA, 18–20 July 2005.
9. Yu, C.; Liu, Z.; Liu, X. Ds-slam: A semantic visual slam towards dynamic environments. In Proceedings of the 2018 IEEE/RSJ International Conference on Intelligent Robots and Systems (IROS), Madrid, Spain, 1–5 October 2018; pp. 1168–1174.
10. Badrinarayanan, V.; Kendall, A.; Cipolla, R. SegNet: A Deep Convolutional Encoder-Decoder Architecture for Image Segmentation. *IEEE Trans. Pattern Anal. Mach. Intell.* **2017**, *39*, 2481–2495. [CrossRef] [PubMed]
11. Mur-Artal, R.; Tardos, J.D. Orb-slam2: An open-source slam system for monocular, stereo and rgb-d cameras. *IEEE Trans. Robot.* **2017**, *33*, 1255–1262. [CrossRef]
12. Zhang, C.; Li, P.; Sun, G. Optimizing FPGA-based Accelerator Design for Deep Convolutional Neural Networks. In Proceedings of the ACM/SIGDA International Symposium on Field-Programmable Gate Arrays, Monterey, CA, USA, 22–24 February 2015; pp. 161–170.
13. Qiu, J.; Wang, J.; Yao, S. Going Deeper with Embedded FPGA Platform for Convolutional Neural Network. In Proceedings of the ACM/SIGDA International Symposium on Field-Programmable Gate Arrays, Monterey, CA, USA, 21–23 February 2016; pp. 26–35.
14. Khronos Group. OpenCL-The Open Standard for Parallel Programming of Heterogeneous Systems. Available online: http://www.khronos.org/opencl (accessed on 10 December 2020).
15. Shi, X.; Cao, L.; Wang, D. HERO: Accelerating Autonomous Robotic Tasks with FPGA. In Proceedings of the 2018 IEEE/RSJ International Conference on Intelligent Robots and Systems (IROS), Madrid, Spain, 1–5 October 2018.
16. Tertei, D.T.; Piat, J.; Devy, M. FPGA design of EKF block accelerator for 3D visual SLAM. *Comput. Electr. Eng.* **2016**, *55*, 123–137. [CrossRef]
17. Boikos, K.; Bouganis, C.-S. Semi-dense SLAM on an FPGA SoC. In Proceedings of the 2016 26th International Conference on Field Programmable Logic and Applications (FPL), Lausanne, Switzerland, 29 August–2 September 2016; pp. 1–4.

18. Abouzahir, M.; Elouardi, A.; Latif, R.; Bouaziz, S.; Tajer, A. Embedding SLAM algorithms: Has it come of age? *Robot. Auton. Syst.* **2018**, *100*, 14–26. [CrossRef]
19. Liu, R.; Yang, J. Eslam: An energy-efficient accelerator for realtime orb-slam on fpga platform. In *DAC*; ACM: New York, NY, USA, 2019; p. 193.
20. Han, S.; Mao, H.; Dally, W.J. Deep compression: Compressing deep neural networks with pruning, trained quantization and huffman coding. *arXiv* **2015**, arXiv:1510.00149.
21. Everingham, M.; Eslami SM, A.; Van Gool, L. The pascal visual object classes challenge: A retrospective. *Int. J. Comput. Vis.* **2015**, *111*, 98–136. [CrossRef]
22. Intel Corporation. Intel Arria 10 Device Overview. Available online: https://www.intel.com/content/www/us/en/programmable/documentation/sam1403480274650.html (accessed on 10 December 2020).
23. Yu, M.; Huang, H.; Liu, H.; He, S.; Qiao, F.; Luo, L.; Xie, F.; Liu, X.-J.; Yang, H. Optimizing FPGA-based Convolutional Encoder-Decoder Architecture for Semantic Segmentation. In Proceedings of the 2019 IEEE 9th Annual International Conference on CYBER Technology in Automation, Control, and Intelligent Systems (CYBER), Guangzhou, China, 29 July–2 August 2019; pp. 1436–1440.
24. Huang, H. EDSSA: An Encoder-Decoder Semantic Segmentation Networks Accelerator on OpenCL-Based FPGA Platform. *Sensors* **2020**, *20*, 3969. [CrossRef] [PubMed]
25. Sturm, J.; Engelhard, N.; Endres, F. A benchmark for the evaluation of rgb-d slam systems. In Proceedings of the 2012 IEEE/RSJ International Conference on Intelligent Robots and Systems, Vilamoura-Algarve, Portugal, 7–12 October 2012; pp. 573–580.

applied sciences

MDPI

Article

Estimation of Grey-Box Dynamic Model of 2-DOF Pneumatic Actuator Robotic Arm Using Gravity Tests

Monika Trojanová [1], Tomáš Čakurda [1], Alexander Hošovský [1] and Tibor Krenický [2,*]

[1] Department of Industrial Engineering and Informatics, Faculty of Manufacturing Technologies with a Seat in Prešov, Technical University of Košice, Bayerova 1, 080 01 Prešov, Slovakia; monika.trojanova@tuke.sk (M.T.); tomas.cakurda@tuke.sk (T.Č.); alexander.hosovsky@tuke.sk (A.H.)

[2] Department of Technical Systems Design and Monitoring, Faculty of Manufacturing Technologies with a Seat in Prešov, Technical University of Košice, Štúrova 31, 080 01 Prešov, Slovakia

* Correspondence: tibor.krenicky@tuke.sk; Tel.: +421-55-602-6337

Abstract: This article describes the dynamics of a manipulator with two degrees of freedom, while the dynamic model of the manipulator's arm is derived using Lagrangian formalism, which considers the difference between the kinetic and potential energy of the system. The compiled dynamic model was implemented in Matlab, taking into account the physical parameters of the manipulator and friction term. Physical parameters were exported from the 3D CAD model. A scheme (model) was compiled in the Simulink, which was used for the subsequent validation process. The outputs of the validations were compared with measured data of joint angles from the system (expected condition) obtained by using gravity tests. For obtaining better results were parameters of the model optimizing by using the Trust Region Algorithm for Nonlinear Least Squares optimization method. Therefore, the aim of the research described in the article is the comparison of the model with the parameters that come from CAD and its improvement by estimating the parameters based on gravitational measurements. The model with estimated parameters achieved an improvement in the results of the Normal Root Mean Square Error compared to the model with CAD parameters. For link 1 was an improvement from 28.49% to 67.93% depending on the initial joint angle, and for link 2, from 63.84% to 66.46%.

Keywords: planar robotic arm; pneumatic artificial muscle (PAM); gravity test; dynamic model; validation; optimization; trust region algorithm

Citation: Trojanová, M.; Čakurda, T.; Hošovský, A.; Krenický, T. Estimation of Grey-Box Dynamic Model of 2-DOF Pneumatic Actuator Robotic Arm Using Gravity Tests. *Appl. Sci.* **2021**, *11*, 4490. https://doi.org/10.3390/app11104490

Academic Editor: Angelo Luongo

Received: 23 April 2021
Accepted: 12 May 2021
Published: 14 May 2021

Publisher's Note: MDPI stays neutral with regard to jurisdictional claims in published maps and institutional affiliations.

1. Introduction and Related Works

Robotics has been an area of research that has attracted the attention of scientists around the world for many decades, but in recent years has enjoyed enormous interest, especially in the safe interaction between humans and robots [1]. The robotic system may be assigned tasks that a person cannot perform in some respects, e.g., if the operations are physically demanding for a person, or if the activity requires high accuracy of movement, or last but not least, the operating and working conditions are not suitable for human work in the vicinity of such a mechanism. All these facts give researchers from different fields many actuations to progress in the research and development of robots and manipulators, whether in the process of designing [2,3], production or the actual application of robots [4–7]. Ref. [4] describes the application of robots, and thus, in particular, the topic of self-employed robots so that its interaction is safe in a dynamic environment. In [5], are described, several types of robots, robot drives, sensors, control methods, as well as industrial applications are described in several sections. Ref. [6], using Lab Activities, helps to understand the parallel between the theory and the application of industrial robotics. Advances in robotics are reflected in every single application area of robotics. This is the case with rehabilitation robotics too, which is described in [7]. Current advances in robotics (with a focus on explaining new concepts) are also described in more detail in [8].

Behind the effective movement of the robot or manipulator are many aspects that have to be mastered in motion control (detailed characteristics of manipulator properties, complex process of motion simulation, correct derivation of kinematic and dynamic models, or last but not least, correct identification of control algorithm). Therefore, the manipulator control will not be correct until the system described in the form of a model is correct. In the publications [9–11], which are manuals in the field of robotics, it is in some chapters to the introduction of kinetic and dynamic methods of modeling such systems. For example, in [9], Chapter 3 is devoted to Inverse Kinematics, and Chapters 4 and 5 are oriented to Newton–Euler Dynamics and Lagrangian Dynamics. Ref. [10], in the introductory parts, describes the basic topics of robotics, and then, among other things, one part is devoted to various areas of application of robotics. Another more comprehensive publication is [11], which explains the constructions, drives, principles of operation of robots and applications of robots, so that the work helps not only researchers but also potential enthusiasts and students of robotics. Publication [12] summarizes the most basic information on the design, but also the control of mechatronic systems, focusing on the needs of engineers and designers who develop and apply their machines to the industrial environment. The part of the work is devoted to robotics, including the dynamic modeling of systems.

Modeling of kinematics and dynamics of manipulators plays an important role in the control accuracy of these devices. For example, in the publication [13], the kinematics and dynamics of the Hyundai HS165 robot are described, which is one of the industrial robots used for spot welding. A large part of the article is devoted to the modeling of kinematics and dynamics, where the load capacity of the manipulator is monitored. Another article [14], in turn, describes a mathematical model of a mobile manipulation robot, where a kinematic model is used for the draft regulation, and a dynamic model is used to monitor (compensate for errors) the speed of the robot. Additionally, the publication [15] describes dynamic modeling, but for a manipulator, which is designed for underwater work and is driven by an electric motor. The manipulator has heterogeneous arms, while each joint of the arm behaves differently within the kinematics. By identifying the minimum number of parameters, the model is experimentally validated on an underwater hybrid remote-controlled vehicle Ifremer Ariane. In [16] is described the dynamic characteristics of the arm of a manipulator that is driven by pneumatic artificial muscles. In addition to the dynamics of the manipulator, the article also deals with the dynamics of the drive. It is validated the model using measured data from a real device. The created complex model of system dynamics is to be used for control.

Dynamic modeling is generally used to describe the basic relationship between two parameters of the system, namely between force and motion. The dynamic model of the system can be derived within the dynamic modeling based on two basic principles, which are described in the publication [17]. The first is the Euler–Lagrange formulation. This method assesses the system as a whole based on the total kinetic and potential energy (or the difference between these energies). On the contrary, the Newton–Euler formulation describes each member of the system separately. It is necessary to take into account when choosing a method, e.g., number of joints in the manipulator (number of degrees of freedom), location of joints and other mechanical parts, and others. In [18], the dynamic modeling of a robotic manipulator is described using the Euler–Lagrange formulation. The robot has four degrees of freedom, and the location of the base is oriented to the ground/floor (horizontal mounting upwards). In addition to analyzing the properties of this model, they suggest using a PID controller to control the position of the manipulator. In [19], a combination of the Lagrange equation and the Morison formula is used to derive a dynamic model of a manipulator with 3 DOF, which can be used in a hydrostatic environment, while during modeling, they had to take into, for example, other forces acting on the joints when working with the manipulator underwater. The next publication [20] used the Euler–Lagrange method to derive dynamic equations for the prototypes of a four-legged robot—Mini-Bot—where they tracked the position and angular velocity of each leg in the Cartesian coordinate system. They used the Matlab-Mupad software tool as a tool for deriving dynamic equations. The Newton–Euler formulation was used in the publication [21] to derive object equations, while a recursive Gibbs–Appell formulation was used to obtain the dynamic equations of a flexible, cooperative manipulator. The manipulator consists of two arms, which have two flexible links. Two limitations had to be taken into account in the simulations, namely dynamic and kinematic limitations resulting from the placement of the object on a non-holonomic mobile platform. In the article [22], the research of the derivation of a dynamic model using a Newton–Euler formulation for a rigid manipulator is described, which consists of two links.

Some papers also present new methods in their publications that can be used for the dynamic modeling of manipulators. For example, [23] proposes a new method for derivate a dynamic model of a manipulator with 6 DOF. The hysteresis effect of friction in the joints is described using a centrosymmetric static friction model. The dynamic parameters of the six joints were identified using a hybrid whale optimization algorithm and a genetic algorithm. One of the most recent publications [24], contains another new approach to deriving equations of motion, called Pseudo-Symbolic Dynamic Modeling. It is a numerical algorithm whose task is to create functions that form the basis for the inverse dynamics of the manipulator. The result is simplified dynamic models, while this algorithm has been implemented in Matlab and can be applied to manipulators up to 7 degrees of freedom of movement.

To achieve better results of the modeling, models can be optimized using numerical iteration methods, for example, algorithms as, e.g., the Levenberg–Marquardt method or the Trust Region Algorithm. In [25], the Levenberg–Marquardt method for the unconstrained optimization problem is studied, whereas the method of convergence technique (using line search) is proposed. In one of the publication's chapters [26], the Levenberg–Marquardt method is used to determine the minimum volume of the longitudinal cooling fin. The Trust Region Algorithm is also used to solve nonlinear problems. At present, its popularity is higher than with the previously mentioned method. Many researchers describe algorithms that are derived from the Trust Region Algorithm. The aim is to increase its reliability and convergence properties. For example, in [27], two trust-region algorithms are proposed that examine trust outside the specified area. In [28] is described the application of the algorithm to the online estimation of tire/road friction parameters in real-time. In addition, a new LuGre model-based nonlinear least-squares parameter estimation algorithm is presented in this work.

The aim of the research described in the article is to use an optimization method to estimate parameters of the model (improve the results of validation) and the comparison of the model with the parameters that come from CAD and its improvement by estimating the parameters based on gravitational measurements. The article describes the dynamics of a manipulator with two degrees of freedom, while the dynamic model of the manipulator's arm is derived using Lagrangian formalism. The compiled model was implemented in Matlab, taking into account the physical parameters of the manipulator. A scheme (model) was compiled in the Simulink, which was used for the subsequent validation process. The article is divided some parts, where the introduction is followed by the dynamic model of the kinematic structure of a planar arm. Next follows the characteristics of the experimental manipulator. Data measured from the system using gravity test are described in chapter Measured Data. A simulation model was developed for the validation process, and the results of the validation with CAD parameters are described in the fifth chapter. To achieve better results were estimated the parameters of the model. The sixth chapter described the optimization algorithm used to estimate parameters of the model (Trust Region Algorithm for Nonlinear Least Squares), and in the next chapter are summarized results of validation with estimated parameters. The last part consists of a discussion, in which the results of validation and estimation of parameters, based on statistical indicators MAE and NRMSE.

2. Dynamic Model of the System

The experimental system represents one of the most basic kinematic structures (a planar arm with two rotating joints moving in the x-y plane). The schematic diagram of the kinematic structure of the manipulator shown in Figure 1 serves to show the basic elements, their location within the whole system, but also the basic parameters. Two rigid arms, lengths l_1 and l_2, have a mass of m_1 and m_2. The distance from the axis of rotation of the joint to the center of gravity of the arm is denoted as c_1 and c_2. I_1 and I_2 represent the moments of inertia of the bodies (arms) concerning an axis that passes through the center of gravity of the body and is parallel to the horizontal axis x. The deflection of the arm from the zero position (the angle of rotation of the joint is not equal to 0°) is indicated by the angles φ_1, φ_2, which represent the degrees of freedom of the manipulator. The deflection of the arm in the positive direction from the vertical y-axis is expressed by the angle of rotation of the joint φ_1. The angle φ_2 indicates the angle of rotation of the second joint, but concerning the axis of the first deflected link.

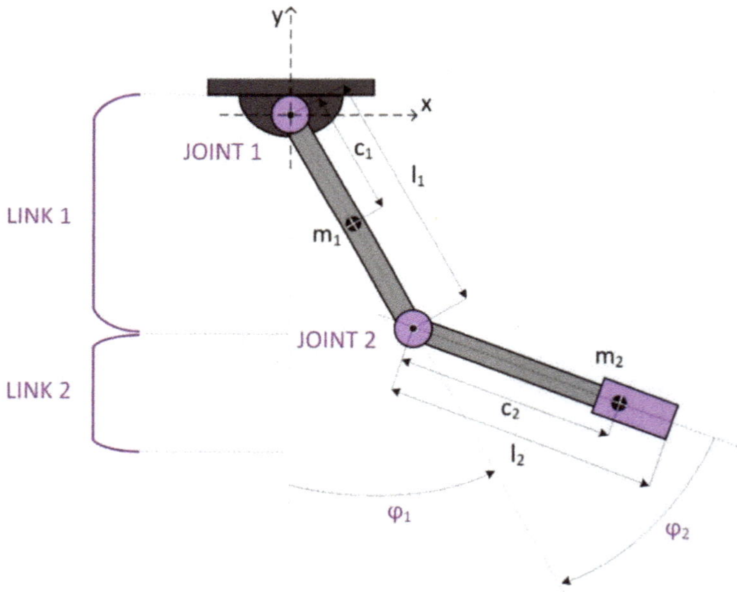

Figure 1. Scheme of kinematic structure of experimental system.

Lagrange's equations were used to derivate a dynamic model of the described system, which has two degrees of freedom [29]. The general formulation for a robot with 2 degrees of freedom is given in Equation (1), where L is Lagrangian, φ_i is joint positions, derivation of the joint position is joint velocities, and τ_i is force term.

$$\frac{d}{dt}\left[\frac{\partial L}{\partial \dot{\varphi}_i}\right] - \frac{\partial L}{\partial \varphi_i} = \tau_i \qquad i = 1,2 \tag{1}$$

The Lagrangian is given by the difference of kinetic and potential energy, and therefore Equation (2) applies to its determination. Member $E_k\,(\varphi, \dot{\varphi})$ expresses the kinetic energy of the manipulator and is formed by the sum of individual kinetic energies of members 1 and

2 (E_{k1} and E_{k2}). Member E_p (φ) characterizes the energy of the manipulator concerning its position and is given by the sum of individual potential energies of arms 1 and 2 (E_{p1} and E_{p2}).

$$L(\varphi, \dot{\varphi}) = E_k(\varphi, \dot{\varphi}) - E_p(\varphi) \tag{2}$$

Determination Lagrange's Equations

By calculating the Lagrangian and applying Equation (1), two nonlinear differential equations are obtained, which represent the equations of the manipulator dynamics. For link 1, the force component τ_1 is determined in the final form:

$$\begin{aligned}
\tau_1 = \ & \left[m_1 \cdot c_1{}^2 + m_2 \cdot c_2{}^2 + m_2 \cdot l_1{}^2 + 2m_2 \cdot c_2 \cdot l_1 \cdot \cos \varphi_2 + I_1 + I_2\right] \cdot \ddot{\varphi}_1 \\
& + \left[m_2 \cdot c_2{}^2 + m_2 \cdot c_2 \cdot l_1 \cdot \cos \varphi_2 + I_2\right] \cdot \ddot{\varphi}_2 - 2m_2 \cdot c_2 \cdot l_1 \cdot \sin \varphi_2 \cdot \dot{\varphi}_1 \cdot \dot{\varphi}_2 \\
& - m_2 \cdot c_2 \cdot l_1 \cdot \sin \varphi_2 \cdot \dot{\varphi}_2{}^2 + m_2 \cdot g \cdot c_2 \cdot \sin(\varphi_1 + \varphi_2) \\
& + \left[m_1 \cdot c_1 + m_2 \cdot l_1\right] \cdot g \cdot \sin(\varphi_1)
\end{aligned} \tag{3}$$

and for link 2, the force component τ_2 is defined:

$$\begin{aligned}
\tau_2 = \ & \left[m_2 \cdot c_2{}^2 + m_2 \cdot c_2 \cdot l_1 \cdot \cos \varphi_2 + I_2\right] \cdot \ddot{\varphi}_1 + \left[m_2 \cdot c_2{}^2 + I_2\right] \cdot \ddot{\varphi}_2 \\
& + m_2 \cdot c_2 \cdot l_1 \cdot \sin \varphi_2 \cdot \dot{\varphi}_1{}^2 + m_2 \cdot g \cdot c_2 \cdot \sin(\varphi_1 + \varphi_2)
\end{aligned} \tag{4}$$

The Lagrangian dynamic model expressed by Equations (3) and (4) can be written in a compact form, and when extended by the friction component F then has the form of Equation (5).

$$M(\varphi)\ddot{\varphi} + C(\varphi, \dot{\varphi})\dot{\varphi} + G(\varphi) + F(\dot{\varphi}) = \tau \tag{5}$$

The friction component F can be defined by Equation (6), where ω represents the angular velocity in [rad/s]; Tc is the Coulomb friction magnitude is given in [N·m]; the Viscous friction coefficient β has given in units [N·m·s/rad]. This whole expression on the right side of the equation is also called Coulomb and Viscous friction [30].

$$F = sgn(\omega) \cdot Tc + \beta \cdot \omega \tag{6}$$

3. Experimental System—Planar Robotic Arm

The object of the research was a planar robotic arm of the manipulator, which is driven by an unconventional type of actuators (pneumatic artificial muscles) and has two degrees of freedom. The manipulator is shown in Figure 2. The construction of the manipulator consists of metal profiles, which ensure the stability of the manipulator and the attachment of the upper arm of the manipulator to the upper base of the structure. A load weighing 3.34 kg is attached to the other arm of the manipulator. The type of pneumatic drive was chosen for the manipulator drive-fluid muscles from the manufacturer FESTO. If these muscles are pressurized, their activation occurs, as a muscle contraction, which causes the translational tensile force of the muscle. However, since the muscles within the manipulator are antagonistically involved, and also the chain-sprocket and joint subsystems are present within the system, the resulting movement is rotational (the subsystems ensure the transmission of torque). During the research, the angle of rotation of the joint was monitored as an output parameter. An incremental sensor was installed in the manipulator to be able to read this parameter. To achieve the required outputs with the manipulator, the pressure in the muscles must be regulated—electronic pressure regulators (one for each muscle) are used for this. In addition to the possibility of pressure coordination, the regulators contain built-in pressure sensors. An AMD Athlon X2 computer with a 5 GB RAM processor was used as a control unit to control the manipulator. The software support of the stand consisted of the use of the Matlab and Matlab Simulink programs, while the control signals were operated with a Humusoft MF624 card.

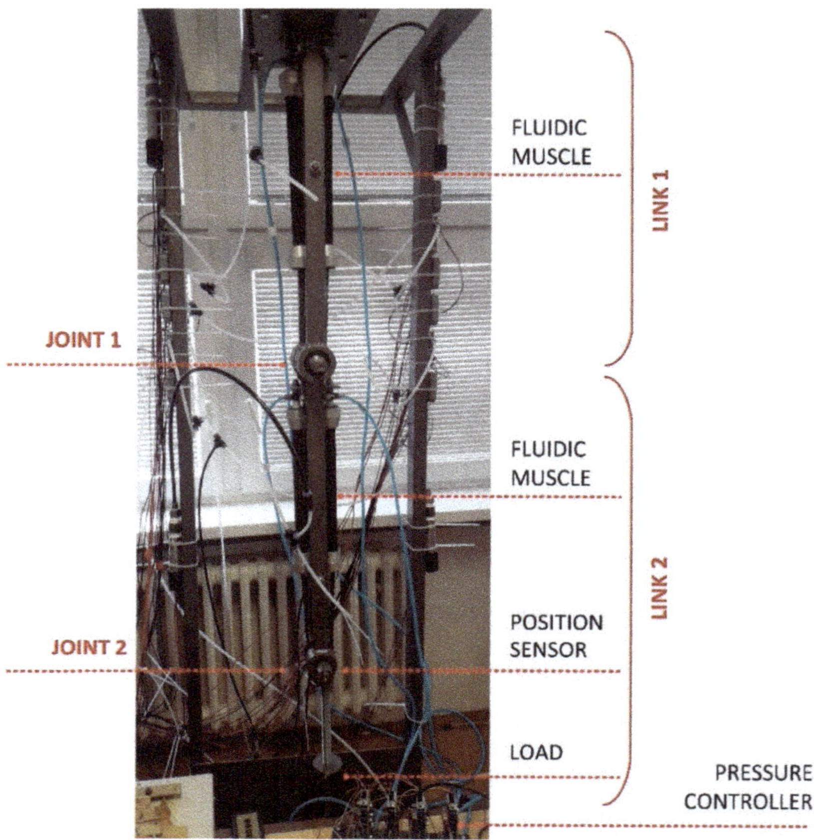

Figure 2. Experimental system—planar robotic arm with 2-DOF and actuated with PAMs.

Table 1 contains an overview of the main components used in the experimental system. In addition to the name of the component, its basic parameters are also given. A total of 4 fluid muscles of the same size and diameter were used. Each muscle had one electronic pressure regulator (a total of four were used). An incremental sensor was used to read the output variable. The mechanical parts of the stand (except the structure) consist of two arms, two joints, two chains and two sprockets with a diameter of 0.07 m.

Table 1. The list of main components used in experimental system.

Component	Parameter Name	Value	Component	Parameter Name	Value
Fluid muscle FESTO MAS-20-250N-AA-MC-K	Diameter Length Max. muscle force Max. working pressure	0.02 m 0.25 m 1200 N 6 bar	Incremental encoder KUBLER 3610	Power supply Max. resolution	5–30 V DC 2500 pulse/rev.
Electronic pressure regulator MATRIX EPR50	Input signal Input pressure Max. flow Power supply Output pressure Response Reaction time	4–20 mA 1–8 bar 60 L/min 24 V DC 0–7 bar 60–100 ms less than 5 ms	Compressor FIAC LEONARDO	Max. pressure Flow Power	8 bar 105 L/min 750 W

4. Measured Data

Measurements were performed by using gravity experimental tests, the task of which was to obtain the waveforms of the output monitored parameter—the joint angle of the manipulator's arm. The first measurement started at a joint rotation angle of 10°, i.e., the arm was moved from the zero position by 10°. Then, it was released, and the values of the joint rotation angle were recorded by incremental sensors until the moment when the arm stopped. Thus, measurements were subsequently performed for the joint angle 20°, 30° and 40° too. The total number of measurements for one initial angle was 20, with the tests being performed in both directions. During the processing, all waveforms were converted to one side based on a simplified assumption of symmetry of the arm dynamics. The tests were performed independently for link 1 and link 2. The total measurement time was different for each initial joint angle, but the sampling was the same at 0.005 s.

4.1. Measured Data—Link 1

Figure 3 shows 4 boxplots (when viewed from left to right for an initial joint angle of 10°–40°), which show how the initial values of the joint angle have changed in 4 different measurement cases. Each boxplot was processed based on the measured initial joint angles in 20 measurements. It can be seen from the graphs that the mean of the 20 measured initial values (red line) approaches the required input at most in the case of an angle of 20° (the second graph from the left with mean value 20.016°), where it is not possible to see the 25th and 75th percentiles (blue rectangle) is present. Only one different value is also a remote value of 19.872° (shown in the second graph on the left by a red cross). It is possible also to observe a remote measured value of 40.464° in the fourth graph from the left when the initial angle was set to the reference value of 40°. The largest difference between the minimum and maximum set input value (black horizontal line) can be observed for an initial joint angle of 10° (first graph from the left).

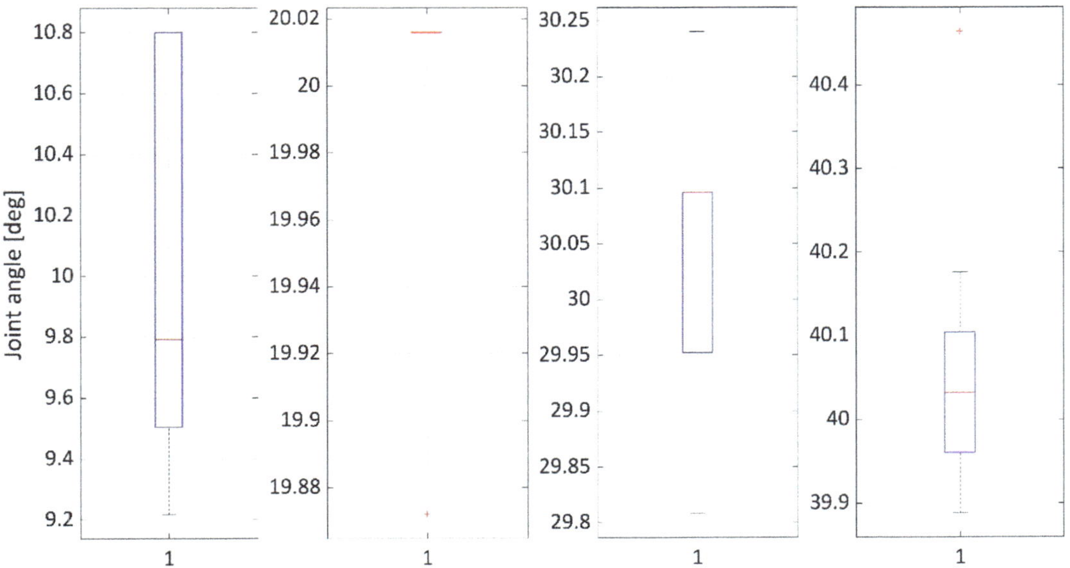

Figure 3. Boxplots of initial values of joint angles—link 1.

From the 20 measured time courses of the joint angle, where the initial joint angle is 10°, the angles were set for individual measurements in the interval <9.216°; 10.08°>. The maximum value of the interval represents the statistic parameter Mode. Figure 4 represents the mean response (red curve) and standard deviation (the grey area) from measured data.

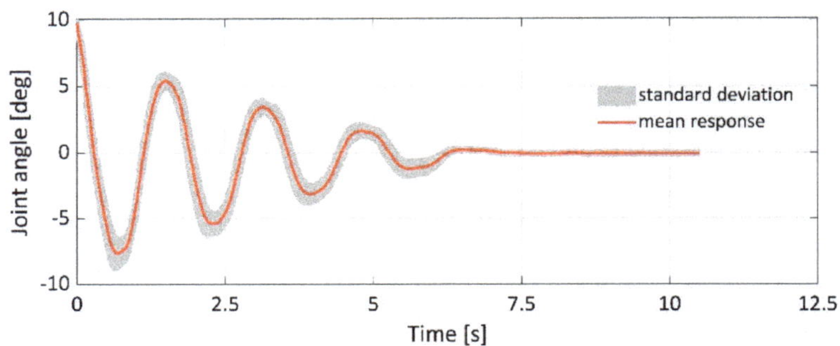

Figure 4. The mean response and standard deviation for the joint angle 10°—link 1.

The interval of the initial angles of rotation of the joint with a reference value of 20° was <19.872°; 20.016°>. Such a small range was caused by the fact that out of 20 measurements, the value 20.016° was set 19 times. Waveforms of standard deviation and mean response of data are shown in Figure 5.

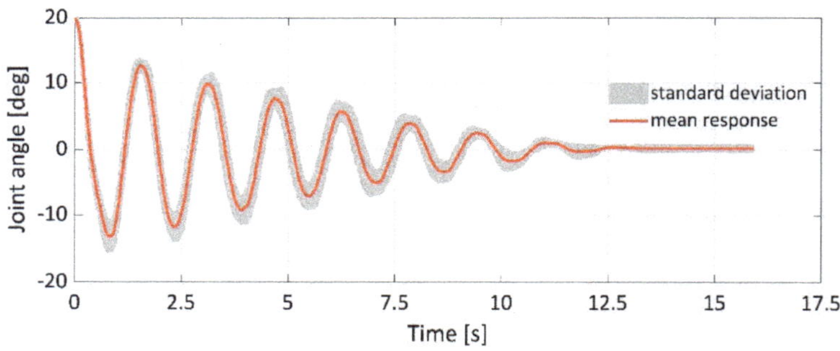

Figure 5. The mean response and standard deviation for the joint angle 20°—link 1.

Figure 6 shows the waveform of mean response at a joint angle of 30° for link 1, where the initial values of the joint angle were set in the range <29.808°; 30.24°>. More than half of the measurements started from 30.096°. Figure 6 shows a more pronounced standard deviation also (compared to the deviation for angles of 10 ° and 20 °).

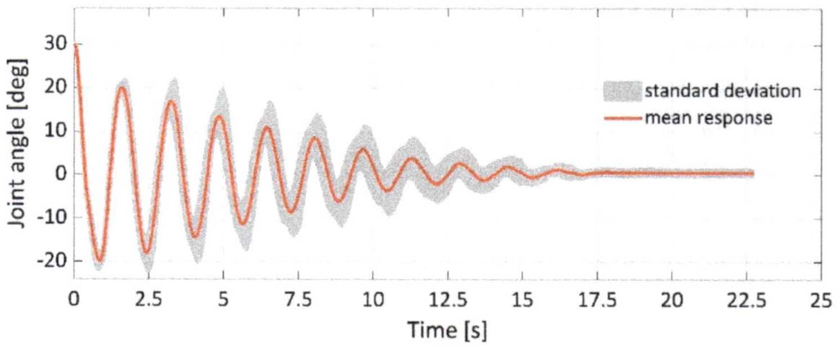

Figure 6. The mean response and standard deviation for the joint angle 30°—link 1.

The last group within link 1 is closed by the curve mean response and standard deviation if the initial angle was 40°, as shown in Figure 7 (respectively, in reality, the initial angles were set in the interval <39.888°; 40.464°> during the measurements). Half of the measurements were started at a joint rotation angle of 40.032°.

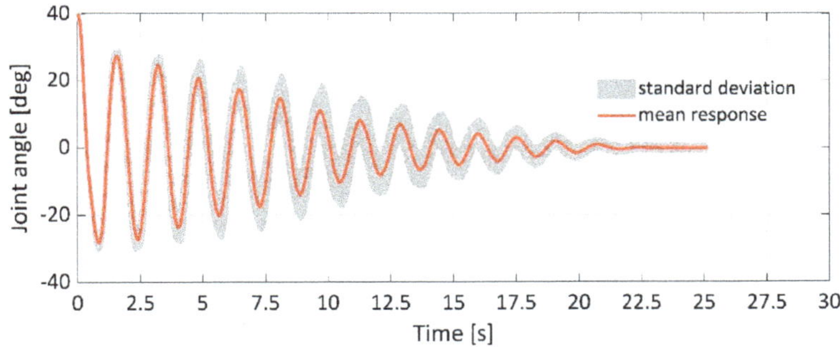

Figure 7. The mean response and standard deviation for the joint angle 40°—link 1.

4.2. Measured Data—Link 2

Figure 8 shows four boxplots for link 2 (when viewed from left to right for an initial joint angle of 10° to 40°). In these graphs, in comparison with the graphs for link 1, it can be seen that the minimum and maximum values of the set initial angles are much further apart, which results in a more significant shift of the mean value from the desired initial angle (especially for the angle −40°, fourth graph from the left). The remote value was measured only for an angle of 20° (red crosses, second graph from the left), two times (−17.28° and −14.832°).

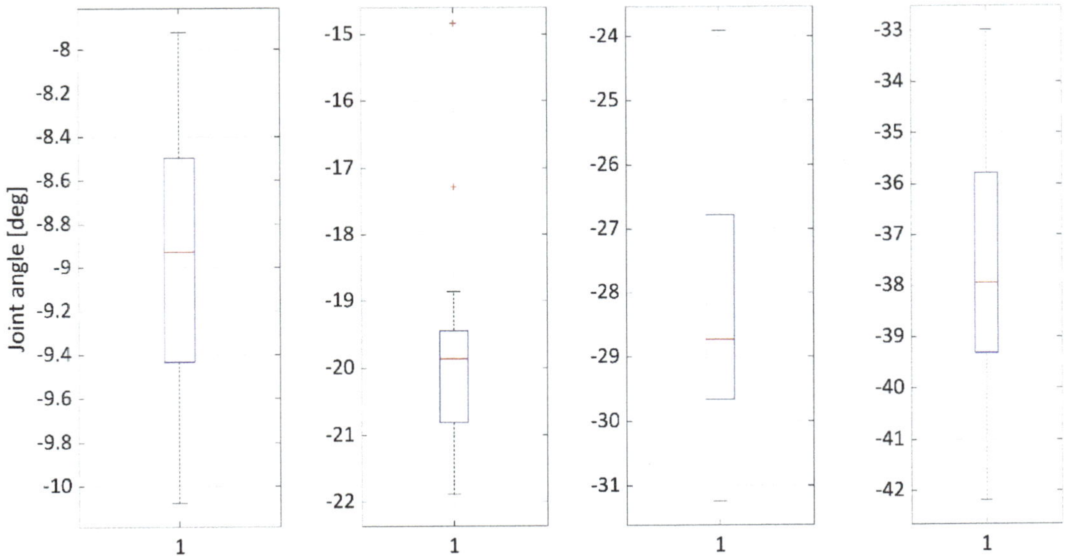

Figure 8. The boxplots of initial values of joint angles—link 2.

While for link 1, the initial angles were set in the range of up to 1°, the setting of the initial joint angle for the measurements of link 2 was more difficult in terms of constraints in the construction of the system. Figure 9 graphically presents the mean response (red curve) and standard deviation (grey area) when the initial angle was set to −10°. The total interval of the initial values of the angle of rotation was <−7.92°; −10.08°>, the Mode from the twenty initially set angles was −8.928°.

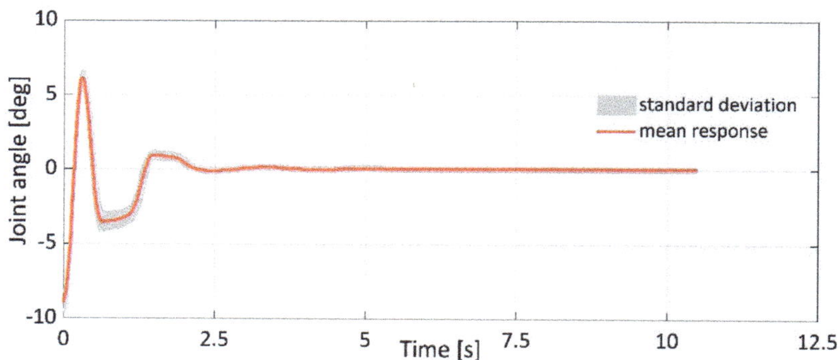

Figure 9. The mean response and standard deviation for the joint angle 10°—link 2.

As the joint angle for link 2 increases, the range of the intervals of the initial values of the joint angle also increases (it is also visible in the area of the standard deviation parameter in Figure 10). The interval for the initial value of the joint rotation angle −20° is <−14.832°; −21.888°>, where the difference between the minimum and maximum value of the interval already exceeds 7°. Nevertheless, the value of the Mode parameter (−20.736°) is close to the reference value.

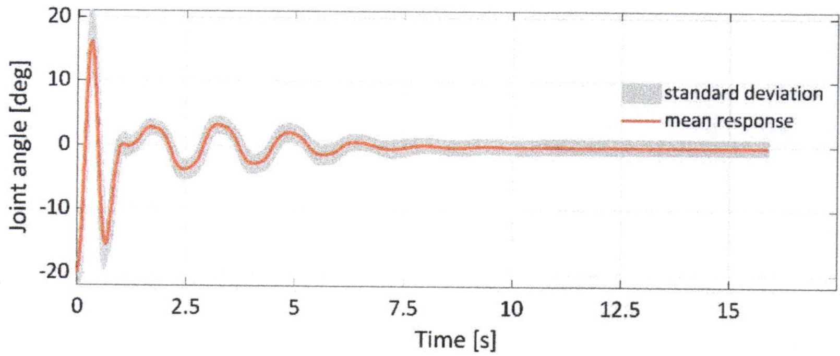

Figure 10. The mean response and standard deviation for the joint angle 20°—link 2.

The interval for the reference value −30° for link 2 was <−23.904°; −31.248°>, where from the range of initial values, it is possible to monitor a more significant value of the parameter Mode (−24.048°) in comparison with the reference value. The shift of the beginning of the mean response curve (Figure 11) and the standard deviation area itself is caused by one of the measurements when the initially set value was −23.904 °.

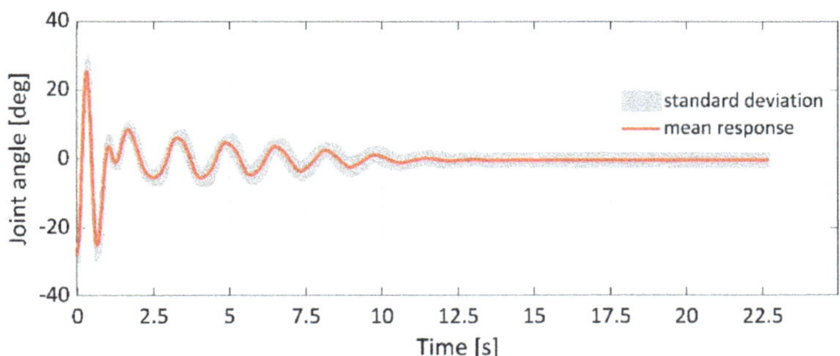

Figure 11. The mean response and standard deviation for the joint angle 30°—link 2.

The last measured data was for the initial joint angle was $-40°$. The total interval of the initial values of the joint angle was $<-32.976°; -42.192°>$, the Mode of these twenty values was $-38.736°$. In Figure 12, it is possible to see the fact that the variance in the measurements is the most expressive, which of course, was reflected in the standard deviation.

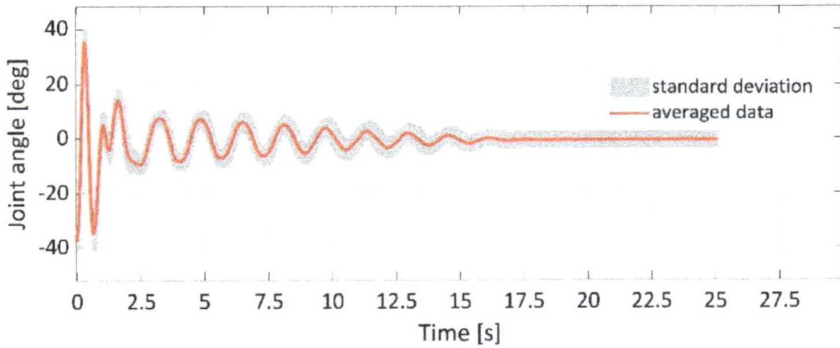

Figure 12. The mean response and standard deviation for the joint angle 40°—link 2.

Gravity tests showed that within link 1, the time course of mean response of the joint angle was more uniform with gradual stabilization to the value of the zero joint angle. For link 2, it is clear from the graphic waveforms that the course is different compared to link 1. A more pronounced stabilization for link 2 for each initial joint angle begins only after a certain amplitude has been exceeded. Additionally, in the case of graphical dependencies, link 2 can see more significant friction, which can affect the validation process. Further, from the sample mean responses and standard deviations for link 1 and link 2, it can be presented that for initial joint angles of 10° and 20°, the standard deviation range is narrower than for 30° and 40° for both axes, which means that at higher initial joint angles, the deviation from the mean response is more pronounced. This could be caused by the stronger manifestation of unmodeled effects, such as additional oscillations of the structure, which was not tightly fastened to the floor (itself slightly uneven). In addition, the larger dispersion of the initial angles had an impact on the wider standard deviation band for link 2 compared to link 1. At the same time, it can be seen from the mean response that the arm stabilization was not completed at 0° in any case due to the effect of friction being more pronounced for this axis. The most visible deviation is for a joint rotation of 20° and 30° for link 1 (Figures 5 and 6). After the phase of implementation of gravity tests and analysis of measured data, the stage of validation in the Matlab and Simulink followed.

In research were used only experiments (not simulation) obtained through gravity tests as a validating agent. However, in the first case, are used parameters values that could be obtained from CAD, and thus could directly be used for simulation without the need for experiments (nevertheless, even in this case, this was validated using the measurements—Chapter 5). In the second case, these parameter values were used as initial values for parameter optimization to obtain an improved model that includes the effects not incorporated in the model with CAD parameters—Chapter 7 (such as fluidic muscles, chains, etc.).

5. Validation

Based on the dynamic model presented in the section Dynamic Model of The System, a simulation scheme in the Simulink was created for the validation process. The basic physical parameters of the system (mass, length, a moment of inertia and distance from the axis of rotation of the joint to the center of gravity for link 1 and link 2) were exported from the CAD software, which was used to make a 3D model of the system. The coefficients viscous friction coefficient and Coulomb friction magnitude were determined experimentally. An overview of these parameters (with parameter designation, value and unit) is summarized in Table 2.

Table 2. The list of input parameters of the Simulink model.

Notation	Description	Value and Unit
m_1	Mass—link 1	5.914 kg
m_2	Mass—link 2	3.437 kg
l_1	Length—link 1	0.615 m
c_1	Distance to the center of mass—link 1	0.293 m
c_2	Distance to the center of mass—link 2	0.19 m
I_1	Moment of inertia—link 1	0.357 kg·m^2
I_2	Moment of inertia—link 2	0.013 kg·m^2
f_{c1}	Viscous friction coefficient—link 1	0.35 N·m·s·rad^{-1}
f_{c2}	Viscous friction coefficient—link 2	0.35 N·m·s·rad^{-1}
f_{d1}	Coulomb friction magnitude—link 1	0.35 N·m
f_{d2}	Coulomb friction magnitude—link 2	0.35 N·m

The validation outputs were evaluated based on the MAE (Mean Absolute Error) and NRMSE (Normal Root Mean Square Error) statistical indicators. The indicators are used to compare the output of the model with the outputs of the system. The MAE indicator by a numerical value represents the error that occurred during the validation compared to the measured data. The general mathematical formulation of MAE is given in Equation (7), where y_j represents the output of the system on the k-th sample and $\bar{y}_j$ represents the output of the model on the k-th sample. The NRMSE, unlike the MAE, compares the degree of agreement of the simulated output with the measured data (Equation (8)). The desired value is for NRMSE = 100% when the model output corresponds 100% to the system output.

$$MAE = \frac{1}{n}\sum_{j=1}^{n}\left|y_j - \hat{y}_j\right| \tag{7}$$

$$F_{gof} = \left(1 - \frac{\sqrt{\sum_{j=1}^{n}\left[y_j - \hat{y}_j\right]^2}}{\sqrt{\sum_{j=1}^{n}\left[y_j - \frac{1}{n}\sum_{j=1}^{n}y_j\right]^2}}\right) \times 100\% \tag{8}$$

5.1. Results of Validation—Link 1

The validation results based on the selected MAE and NRMSE criteria for link 1 are summarized in Table 3. The model output for link 1 achieved the best result based on both statistical indicators at an initial joint angle of 10° (MAE = 1.0932 and NRMSE = 35.47%). However, even with the best model, the results represent that the agreement of the model output with the measured data is low and insufficient.

Table 3. Results of validation for link 1.

Angle of Rotation	MAE	NRMSE
10°	1.0932	35.47%
20°	2.9180	20.29%
30°	3.2550	34.37%
40°	6.1717	16.69%

Figures 13–16 show the model outputs for the joint 10°–40° for link 1. It can be seen from the waveforms that the model outputs (blue solid line) are largely unable to simulate an expected output represent by mean responses (red dashed line) at the specified input parameters. The error (black dotted line) for a joint angle of 10° has a decreasing tendency with a gradual approach to zero (Figure 13). The error waveforms for the joint angles 20–40° (Figures 14–16) are initially smaller but then increase over time, and at the moment when the mean response and thus the output model approaches zero, the value of the error is minimized.

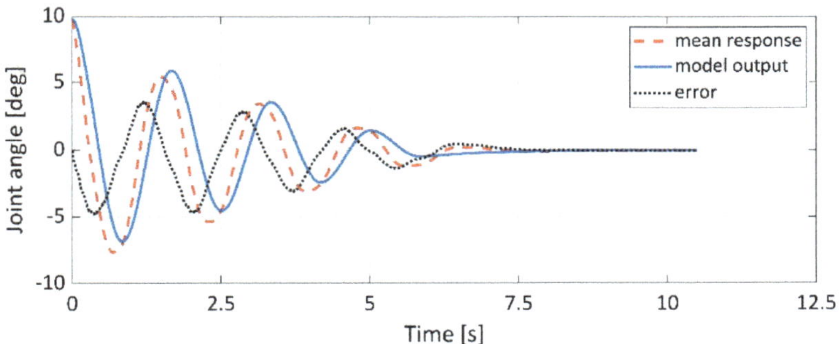

Figure 13. Model output of validation for the joint angle 10°—link 1.

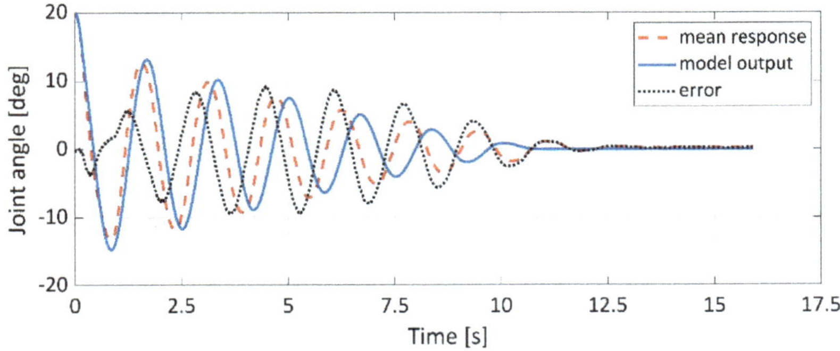

Figure 14. Model output of validation for the joint angle 20°—link 1.

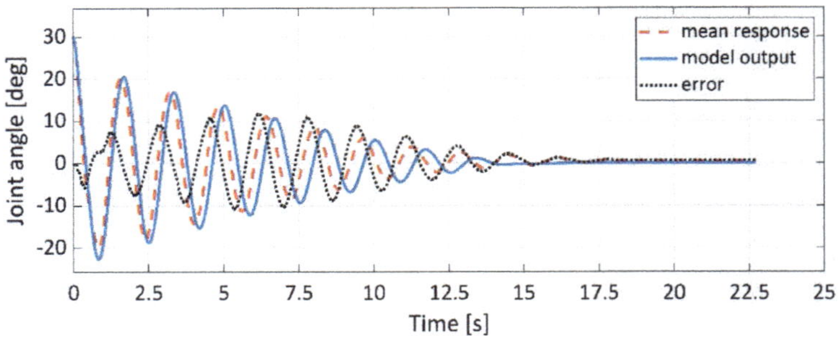

Figure 15. Model output of validation for the joint angle 30°—link 1.

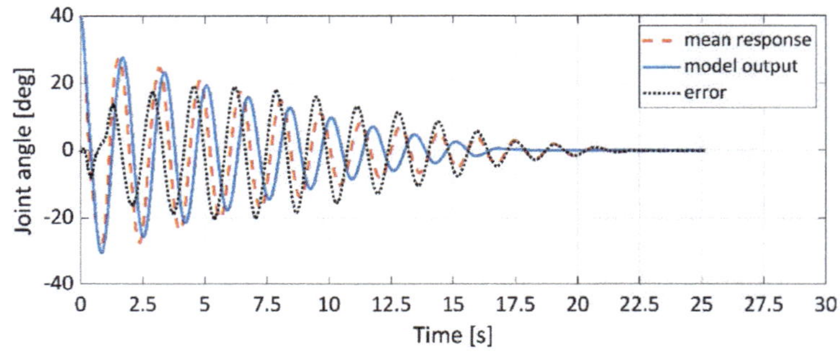

Figure 16. Model output of validation for the joint angle 40°—link 1.

5.2. Results of Validation—Link 2

From the analysis of the validation results for link 2 (Table 4) based on the above indicators, it is clear that the values of the indicators achieve even worse results than in the case of link 1 (this may be due to oscillations that originate from the interconnection of individual parts of the system for link 2 as for link 1). The highest value of the indicator NRMSE = 18.14% was achieved for the model when the input joint angle was 30°, but MAE achieved the best result at the initial joint angle of 10° (MAE = 0.5655).

Table 4. Results of validation for link 2.

Angle of Rotation	MAE	NRMSE
10°	0.5655	14.94%
20°	1.1078	16.36%
30°	1.7524	18.14%
40°	2.8270	16.88%

Since the course of the measured data for link 2 has a different character compared to the courses of link 1, and at the same time, a larger oscillation is present, these facts are also reflected in the actual outputs of the model. In Figures 17–20, where the model outputs of validation for the joint angle 10–40° for link 2 are presented, it is possible to see to what extent over time the model output (blue solid line) was able to approach an average of the measured values (red dashed line). Since the output of the model moves many times close to the zero value of the joint angle, in all cases, the shape of the error (black dotted line) is similar to the course of the mean response. Similarly, for both link 1 and link 2, at the moment when the mean response and thus also the output model approaches zero, the error value is minimized.

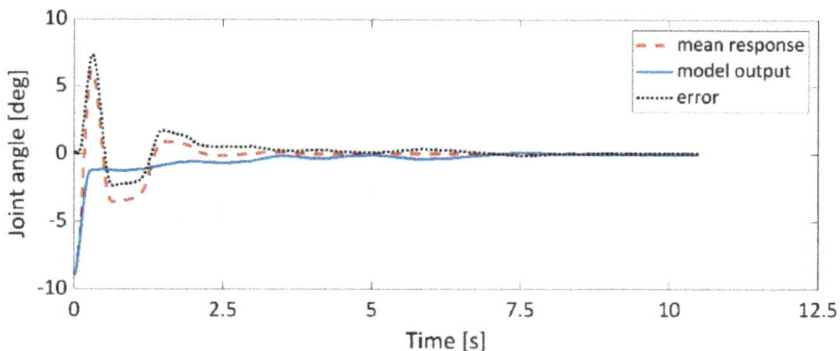

Figure 17. Model output of validation for the joint angle 10°—link 2.

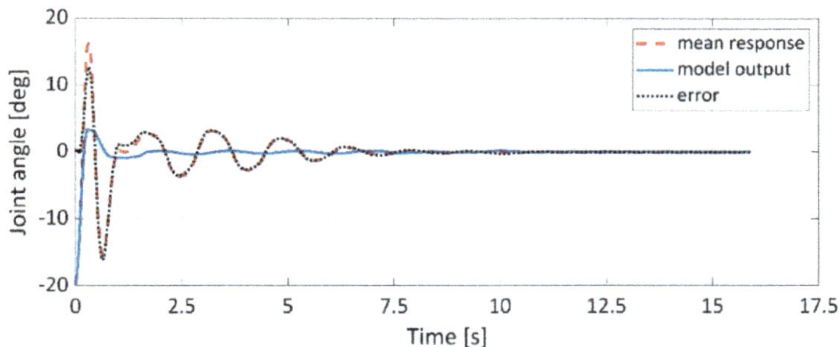

Figure 18. Model output of validation for the joint angle 20°—link 2.

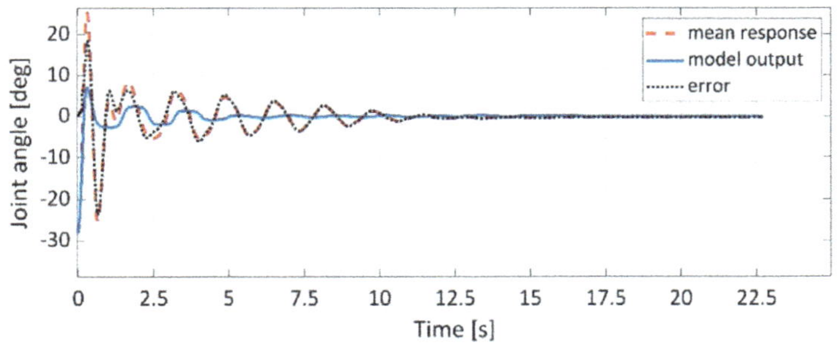

Figure 19. Model output of validation for the joint angle 30°—link 2.

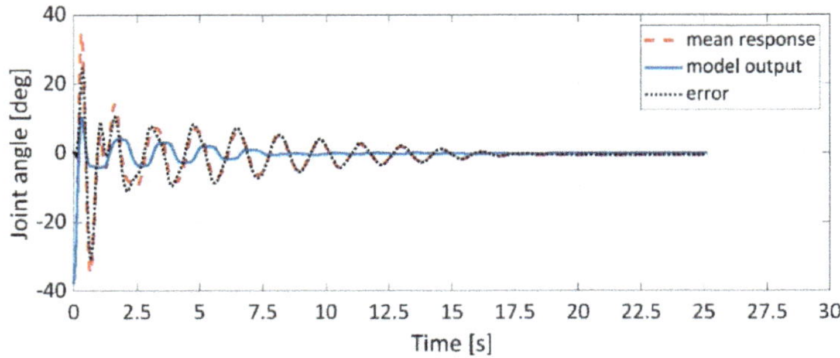

Figure 20. Model output of validation for the joint angle 40°—link 2.

Based on the insufficient results obtained in the validation phase and presented either in the form of MAE and NRMSE criteria parameters or by time dependences of the model outputs alone, compared to the average of measured values and the error between model output and real setpoint, was as next step selected estimation of model parameters. Parameter estimation is an efficient tool that uses mathematical models to accurately describe the behavior of systems. This phase is described in the following two sections.

6. Parameter Estimation of 2-DOF Planar Arm

To estimate the parameters of the model given in Equations (9) and (10), one has to formulate this as an optimization problem. Based on [31], the parameter estimation problem can be formulated as follows: assuming it is a vector of measured values of joint angle $\lambda \in \Re^{n_\lambda}$ consisting of K samples, a vector of model parameters $\bar{\theta} \in \Re^{n_\theta}$ has to be found using an optimization algorithm and properly defined objective function. The objective function used in the experiments was a sum-squared type of error function defined as:

$$(\lambda) = \sum_{k=1}^{K} \left[\bar{\theta}(k) - \theta(k, \lambda) \right]^{T} \left[\bar{\theta}(k) - \theta(k, \lambda) \right] \tag{9}$$

The purpose of the optimization algorithm is to find the values of parameter vector $\hat{\lambda} \in \Re^{n_\lambda}$ so that

$$\hat{\lambda} \in arg \min_{\lambda \in L}(\lambda) \tag{10}$$

where $L \subset \Re^{n_\lambda}$ is the parameter space.

Trust Region Algorithm for Nonlinear Least Squares

Nonlinear optimization is the minimization of a function according to the following formulation:

$$\min_{x \in R^n} f(x) \tag{11}$$

subject to $c_i(x) = 0$ for $i = 1,2, \dots ,m_e$ and $c_i(x) \geq 0$ for $i = m_e + 1, \dots ,m$. Functions $f(x)$ and $c_i(x)$ hold that they are real, defined in R^n; at least one of them is nonlinear; for the coefficients m and m_e, it holds that are nonnegative integers, and $m \geq m_e$. It holds that, if $m = m_e = 0$, the function (Equation (11)) presents an unconstrained optimization problem. If this condition does not apply, then Equation (11) is a constrained optimization problem.

Optimization problems are solved by numerical iteration methods, which means that the approximate solution of the function at the point x_k is in the k-th iteration. The new point (approach to the solution) x_{k+1} is determined computationally based on an algorithm. Such approximation to the solution is applied until the solution is acceptable. An important aspect in the direction of the search, whereby solving subtasks it is verified whether the points are approaching, resp. do not come close to the solution. One of the algorithms that can be applied is the Trust Region Algorithm [32–35]. The Trust Region Algorithm for Nonlinear Least Squares is similar to the method Levenberg–Marquardt [25,36] used to solve nonlinear functions. Although it is a newer algorithm (compared to Levenberg–Marquardt), it brings advantages such as robustness, reliability and convergence properties. In this algorithm, the model is said to be trusted if the area is close to the solution. Such an area is called a trust region and is modified at each subsequent iteration. This means that if in a certain iteration the approximate model is not good enough (it, therefore, contains a bad point), the area of confidence will not be increased but, on the contrary, will decrease.

After a nonlinear optimization problem was defined using Equation (12), if this problem were unconstrained, the algorithm model could be constructed under the condition that the trial step in the k-th iteration would be determined as [34]:

$$\min_{d \in R^n} g_k^T d + \frac{1}{2} d^T b_k d = \varphi_k(d) \tag{12}$$

$$\text{Subject to } \|d\|_2 \leq \Delta_k \tag{13}$$

where b_k is a symmetric matrix of dimension $n \times n$ approaching the Hessian of the function $f(x)$; g_k represents the gradient at the current iteration x_k; Δ_k is greater than zero and represents the trust-region radius.

If Equation (13) is subtracted from Equation (12), then the difference is S_k. Subsequently, if the difference $\varphi_k(0) - \varphi_k(S_k)$ is determined, the result is P_k, which represents the predicted reduction. For P_k, holds it is positive if x_k is a stationary point and b_k is a positive semi-definite. Actual reduction A_k (reduction in objective function) can be determined as Algorithm 1 [34]:

$$A_k = f(x_k) - f(x_k + S_k) \tag{14}$$

If the ratio is the actual reduction A_k and the predicted reduction P_k, it is obtained an important ratio r_k, which firstly expresses whether the test step is acceptable and secondly determines the adaptation of the new trust-region radius.

Algorithm 1 [34]

---

1. Step: Given: $x_1 \in R^n$, $\Delta_1 > 0$, $\epsilon \geq 0$, $b_1 \in R^{n \times n}$;

$$0 < \tau_3 < \tau_4 < 1 < \tau_1,\ 0 \leq \tau_0 \leq \tau_2 \langle 1,\ \tau_2 \rangle 0,\ k1$$

2. Step: If $\|g_k\|_2 \leq \epsilon$ then stop;

   Solve is S_k

3. Step: Calculated $r_k = A_k / P_k$

$$x_{k+1} = \begin{cases} x_k & \textit{if } r_k \leq \tau_0 \\ x_k + S_k & \textit{otherwise} \end{cases}$$

Choose Δ_{k+1} that satisfies

$$\Delta_{k+1} \in \begin{cases} [\tau_3\|g_k\|_2,\ \tau_4\Delta_k] & \textit{if } r_k < \tau_2 \\ \Delta_k,\ \tau_1\Delta_k & \textit{otherwise} \end{cases}$$

4. Step: Update

   b_{k+1}; $k := k+1$;   go to 2. Step.

τ_i is chosen by the user (usually $\tau_0 = 0$, $\tau_1 = 2$, $\tau_2 = \tau_3 = 0.25$, $\tau_4 = 0.5$). Normally, τ_0 is chosen to be zero, but if $\tau_0 > 0$, count of $\lim\limits_{k \to \infty} \inf \|g_k\|_2 = 0$ will be stronger. As a satisfactory optimal criterion for terminating the algorithm at point x_k is $\|g_k\|_2 \leq \epsilon$.

7. Parameter Estimation of Simulink Model

Within the estimation of model parameters, attention, in this case, was focused on the parameter estimation of the Simulink model, where the task is to process the input/output data used for the testing phase. In the validation phase, tests were performed separately for link 1 and link 2 and at the same time for each axis at four different initial joint angles. However, when estimating the model parameters, the estimation takes place simultaneously for link 1 and link 2 for the specified joint angle. To create the experiment, it was necessary to select the measured outputs—the matrix of the required measured outputs depending on time. Specifically, in this case, it was the time dependence of the mean response obtained from the measured data. Another important step in configuring the experiment was to choose which parameters to estimate. In the performed experiment, all physical parameters of the system were selected for estimation, which are listed in Table 5. At the same time, it was necessary to determine the limits of the set parameter. The limits were set to ±30% of the original parameter values obtained from the CAD software (excluding friction). The basis of parameter estimation is an optimization method and algorithm that determine the sequence of steps in which the estimation takes place and what criteria for parameter estimation are chosen. Based on what was mentioned in Section 4, the Trust Region Algorithm for Nonlinear Least Squares was chosen.

Table 5. The list of the number of samples and the parameters' spaces.

Angle of Rotation	K	Notation	Parameter Estimation Bounds
10°	2100	m_1	5–7 kg
20°	3182	m_2	3–5 kg
30°	4546	l_1	0.5–0.7 m
40°	5028	c_1	0.2–0.4 m
		c_2	0.13–0.25 m
		I_1	0.25–0.45 kg·m^2
		I_2	0.009–0.017 kg·m^2
		f_{c1}	0.00001–20 N·m·s·rad^{-1}
		f_{c2}	0.00001–20 N·m·s·rad^{-1}
		f_{d1}	0.00001–20 N·m
		f_{d2}	0.00001–20 N·m

The result of parameter estimation is new estimated values of selected model parameters, which were reused for validation to obtain better model results. An overview of these parameters is summarized in the list of model parameters after parameter estimation in Table 6. The simulation scheme in Simulink was modified with new detected parameters, and the process of output validation based on measured data was performed as in section Validation. The results of the validation after estimating the parameters are divided into two parts—for link 1 and link 2.

Table 6. The list of model parameters after parameter estimation.

Notation	Value and Unit	Notation	Value and Unit
m_1	6.813 kg	I_1	0.264 kg·m^2
m_2	3.361 kg	I_2	0.012 kg·m^2
l_1	0.552 m	f_{c1}	0.335 N·m·s·rad^{-1}
c_1	0.346 m	f_{c2}	0.279 N·m·s·rad^{-1}
c_2	0.212 m	f_{d1}	0.203 N·m
		f_{d2}	0.076 N·m

7.1. Results of Validation after Parameter Estimation—Link 1

The results of validation after the estimation of parameters were assessed based on the same statistical criteria as in the validation before estimation. Table 7 provides an overview of the MAE and NRMSE indicators for link 1. From the results, it can be seen that based on the MAE indicator, the model at the initial joint angle of 10° (MAE = 0.6076) is the best, while based on the NRMSE indicator, it is the model at an initial joint angle of 40°, with NRMSE = 84.62%.

Table 7. Results of validation after parameter estimation for link 1.

Angle of Rotation	MAE	NRMSE
10°	0.6076	63.96%
20°	0.9420	75.27%
30°	1.0382	81.61%
40°	1.1045	84.62%

The improvement of the model results is obvious also from the graphic curves. Figures 21–24 show the model outputs after parameter estimation for the joint 10°–40° for link 1. The model output (blue solid line) approaches the required mean response (red dashed line) with increasing initial joint angle, which also indicates an increase in the value of the NRMSE indicator in Table 7. Before estimating the parameters, the model output had the problem of achieving mean response amplitudes and also the peaks of the model output amplitudes that were shifted for the time axis.

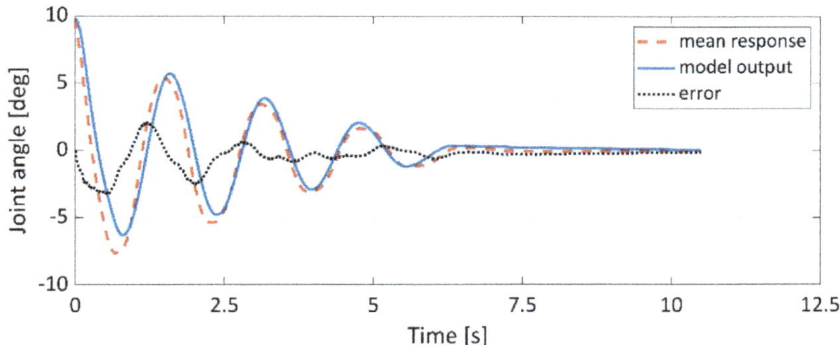

Figure 21. Model output after parameter estimation for the joint angle 10°—link 1.

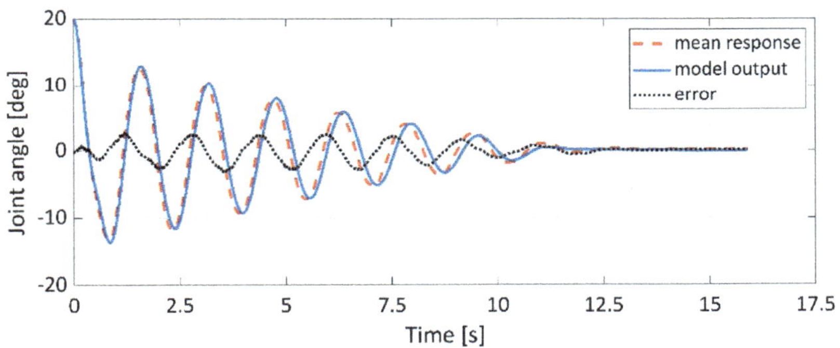

Figure 22. Model output after parameter estimation for the joint angle 20°—link 1.

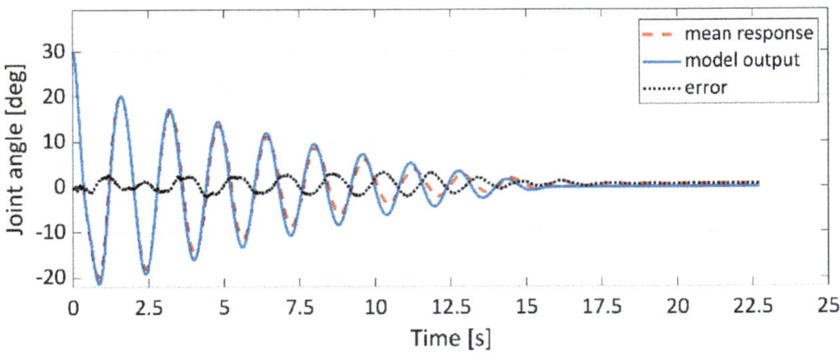

Figure 23. Model output after parameter estimation for the joint angle 30°—link 1.

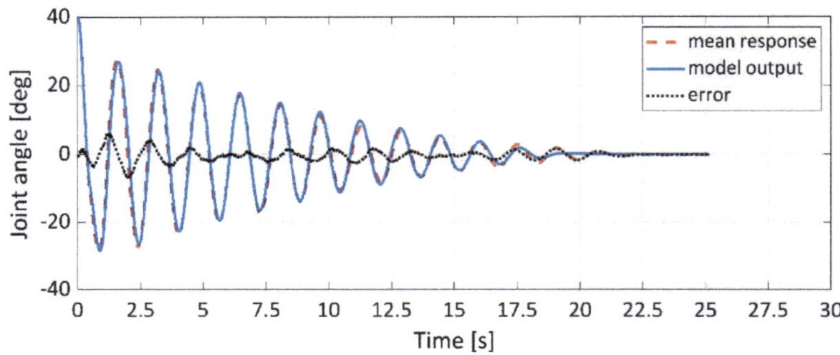

Figure 24. Model output after parameter estimation for the joint angle 40°—link 1.

7.2. Results of Validation after Parameter Estimation—Link 2

The results of the validation before estimating the parameters were significantly worse for link 2 compared to the results for link 1. However, after estimating the parameters and performing the validation again, the models were able to produce much better model outputs, as evidenced by the MAE and NRMSE values in Table 8. The MAE indicator had a value of less than one for each initial joint angle, with the best output of the model obtained at an initial joint angle of 10° (MAE = 0.2407). The value of the NRMSE indicator was around 80% at all initial joint angles examined, with the model achieving the best result at an angle of 20° (NRMSE = 82.82%).

Table 8. Results of validation after parameter estimation for link 2.

Angle of Rotation	MAE	NRMSE
10°	0.2407	79.64%
20°	0.3401	82.82%
30°	0.5159	81.98%
40°	0.7207	82.69%

This significant difference in the improvement of the model results after parameter estimation is especially visible in Figures 25–28. The model before parameter estimation had the biggest problem for link 2 with data in the time zone 0 to 2.5 s when there are significant fluctuations of the manipulator's arm. However, these jumps are followed by more pronounced damping. However, the model after estimation can simulate the output (blue solid line) at this time with high accuracy to the mean response (red dashed line) for each initial joint angle, which is of course also reflected by the change in the curve error (black dotted line).

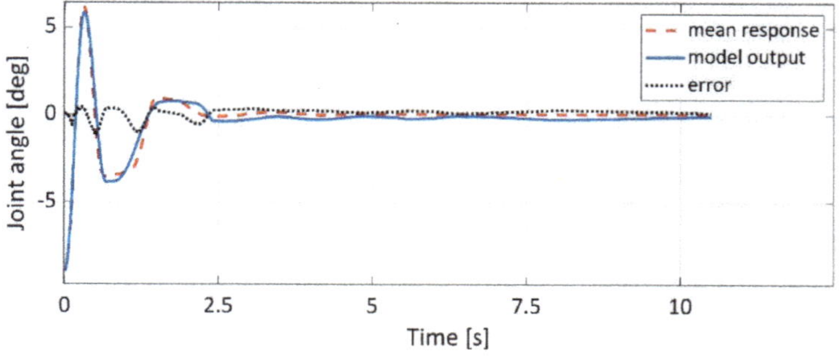

Figure 25. Model output after parameter estimation for the joint angle 10°—link 2.

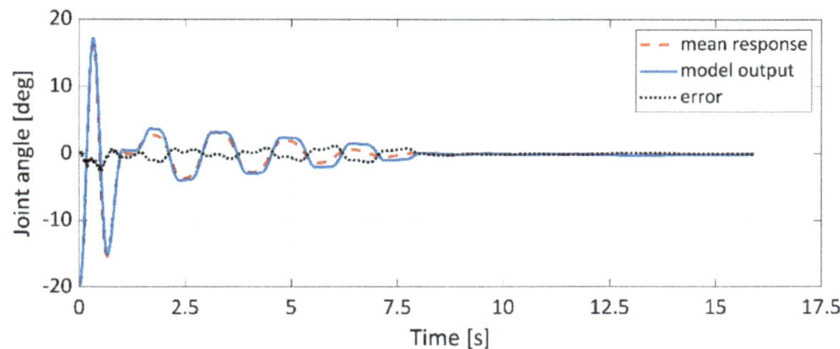

Figure 26. Model output after parameter estimation for the joint angle 20°—link 2.

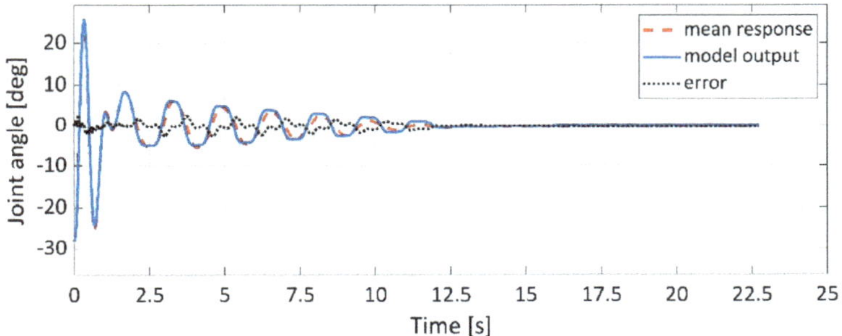

Figure 27. Model output after parameter estimation for the joint angle 30°—link 2.

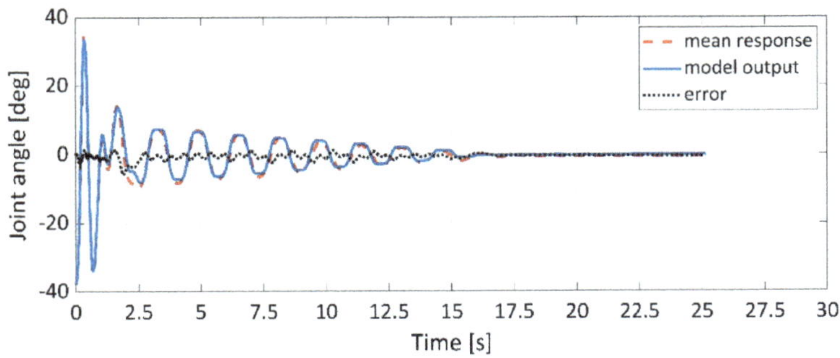

Figure 28. Model output after parameter estimation for the joint angle 40°—link 2.

8. Discussion

The estimation of model parameters using the Trust Region Algorithm for Nonlinear Least Squares optimization method for the known structure of the planar kinematic system model derived based on Lagrange formalism was intended to improve the validation results. As mentioned above, the limits of the estimated parameters were set to ±30% of the original value of each parameter (excluding friction). This value was set because deviations could occur within the parameters specified by the CAD software, e.g., the resulting weights can be skewed, as an element, such as a chain or muscles, is not included in the CAD. CAD, therefore, determined the parameters only based on the modeled construction. A summary and comparison of validation results before and after parameter estimation is given in Table 9 for link 1 and in Table 10 for link 2. It is clear from the indicators that the estimation of parameters significantly helped to achieve better model outputs in the process validation. If the indicators for link 1 are compared, the most significant improvement occurred at the initial joint angle 40°, when the original output of the model reached MAE = 6.1717 and NRMSE = 16.69%, while after estimation, the output of the model was obtained at MAE = 1.1045 and NRMSE = 84.62%. For link 2, the results in the initial validation were even worse than for link 1. As mentioned above, the cause could be a more pronounced oscillation in this axis. Therefore, this fact was expected to play a role in the results even after optimization. However, it is clear from the results that the optimization within the parameter estimation helped the model so much that the outputs not only achieved much better results than before the estimation, but within the NRMSE indicator, they were also similar in value (stable) at different initial rotation angles However, the most significant change after the determination occurred at an initial angle of 40° (taking into account the values of both indicators).

Table 9. Comparison of results validation before and after parameter estimation for link 1.

Angle of Rotation	Before		After	
	MAE	NRMSE	MAE	NRMSE
10°	1.0932	35.47%	0.6076	63.96%
20°	2.9180	20.29%	0.9420	75.27%
30°	3.2550	34.37%	1.0382	81.61%
40°	6.1717	16.69%	1.1045	84.62%

Table 10. Comparison of results validation before and after parameter estimation for link 2.

Angle of Rotation	Before		After	
	MAE	NRMSE	MAE	NRMSE
10°	0.5655	14.94%	0.2407	79.64%
20°	1.1078	16.36%	0.3401	82.82%
30°	1.7524	18.14%	0.5159	81.98%
40°	2.8270	16.88%	0.7207	82.69%

Since the process of estimating the parameters took place simultaneously for both axis 1 and axis 2, from the results of the estimation, it was necessary to select the model that achieved the best results as a whole (not separately for axis 1 and axis 2). As already mentioned at the beginning of this chapter, the most significant progress in the results occurred at the initial angle of rotation of the joint of 40°, so this model was chosen as the best. Figure 29 shows the time courses of the joint angle for both axes at the initial joint angle of 40° before the model optimization process. For both axes, you can see that the output model (blue solid curve) is different compared to the required mean response (red dashed curve). After optimizing the model, however, significant differences (improvements) for both axes can be seen in the model outputs in Figure 30. The output of the model no outrun the required output (this phenomenon was visible especially for axis 1 before estimation). The model also achieves the amplitude of the measured values much better (especially with axis 2 before estimation, this problem is visible).

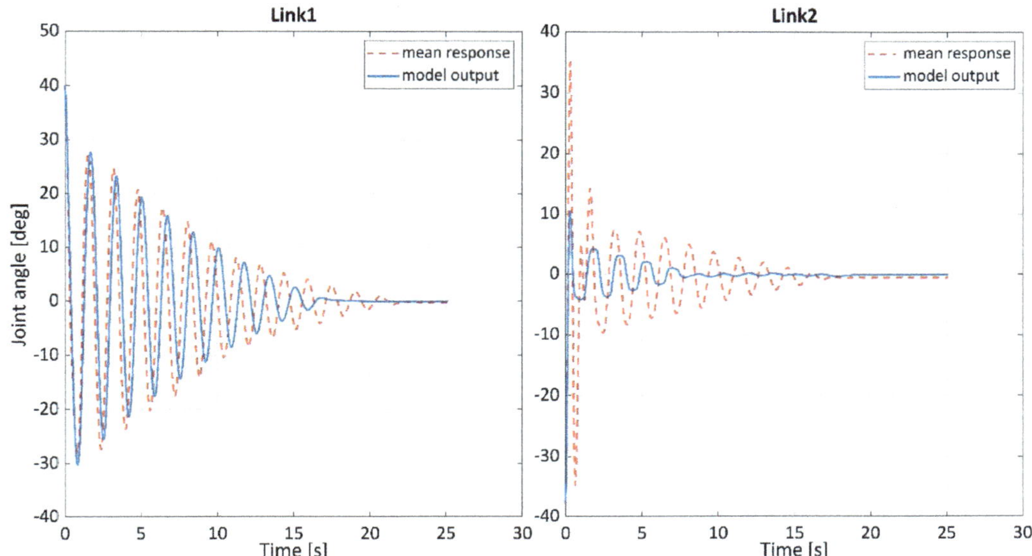

Figure 29. The best model output before parameter estimation for the joint angle 40°.

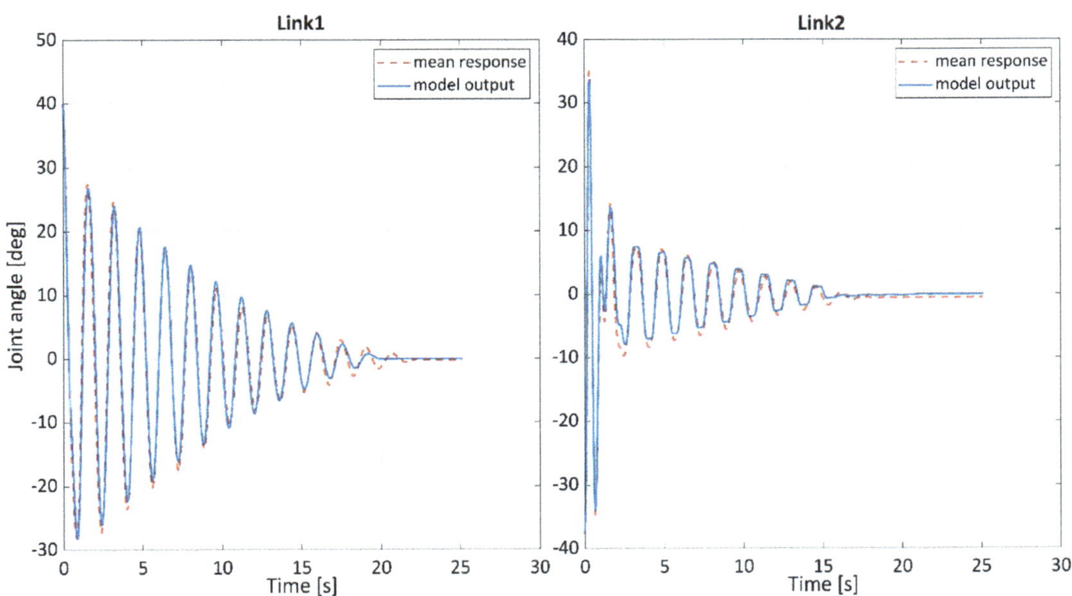

Figure 30. The best model output after parameter estimation for the joint angle 40°.

Since the essence of the optimization of the model was the estimation of the model parameters using the chosen algorithm, it is important to analyze the estimated parameters themselves, which ensured the improvement of the results. These were the parameters of the manipulator together with the coefficients found in the friction components. In connection with this, it should be noted that since the goal was set in the validations—the best possible model outputs and results within the indicators. Therefore the arm parameters, such as length, weight, etc., were also included in the estimation. Table 11 shows a comparison of the values of these parameters before estimation and after estimation.

Table 11. Comparison of parameters before and after parameter estimation.

Notation	Value and Unit before Estimation	Value and Unit after Estimation
m_1	5.914 kg	6.813 kg
m_2	3.437 kg	3.361 kg
l_1	0.615 m	0.552 m
c_1	0.293 m	0.346 m
c_2	0.19 m	0.212 m
I_1	0.357 kg·m^2	0.264 kg·m^2
I_2	0.013 kg·m^2	0.012 kg·m^2
f_{c1}	0.35 N·m·s·rad^{-1}	0.335 N·m·s·rad^{-1}
f_{c2}	0.35 N·m·s·rad^{-1}	0.279 N·m·s·rad^{-1}
f_{d1}	0.35 N·m	0.203 N·m
f_{d2}	0.35 N·m	0.076 N·m

The weights of the arm parts m_1 and m_2 were changed by the estimation—the weight of the upper part of the arm was estimated to 0.9 kg extra, while the weight of the lower part of the arm was estimated to be 0.076 kg less. The parameter, Distance to the center of mass, for both links were increased by a value of 0.053 m for c_1 and a value of 0.022 m for c_2. The length of the arm on axis 1 is less by 0.063 m. Due to these changes, of course, the values of the moments of inertia I_1 and I_2 have also changed. The Viscous friction coefficient and Coulomb friction magnitude were set experimentally before the validations. However, after optimization, the values of the Viscous friction coefficient and Coulomb friction magnitude are smaller than the initial value of 0.35 (especially the value of f_{d2}).

The parameters that were the subject of estimation and used for the subsequent optimization of the model were estimated in the Parameter Estimator. During the estimation process, this tool (among other things) offers the possibility to monitor how the values of the parameters are estimated (changed) depending on the iterations until all the values of the estimated parameters are stabilized. The result of this phase from the given toolbox can be seen in Figures 31 and 32. During the estimation, only one graph could be viewed, where the waveforms of all 11 estimated parameters were. However, since the masses m_1 and m_2 are represented by much higher values than the remaining parameters, the mass curves were exported to the second graph for better clarity. Based on the waveforms of Figure 32, it can be seen that the smallest changes in the parameter estimation occurred in the parameters c_2 (Distance to the center of mass for link 2) and Iz_2 (Moment of inertia for link 2). On the contrary, significant changes took place in the values of Coulomb friction magnitude f_{d1}, f_{d2}. In Figure 31, it is possible to observe how the weight estimate has changed. From the course of estimating the mass m_2, it can be seen how the weight changed in the first iterations, but the result of the estimation in the last iteration was, in the end, very similar to the original value.

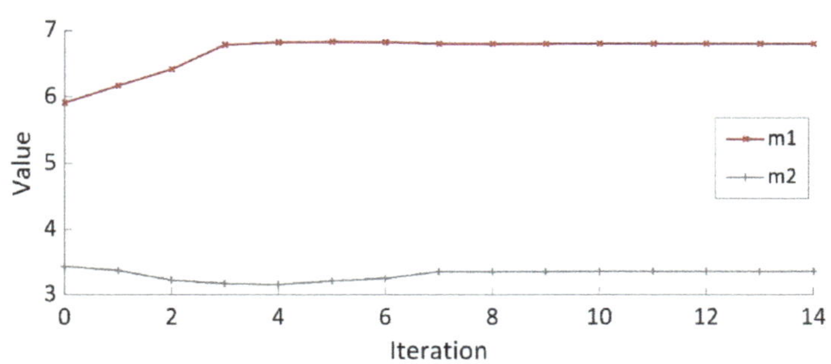

Figure 31. Estimation parameters—part 1.

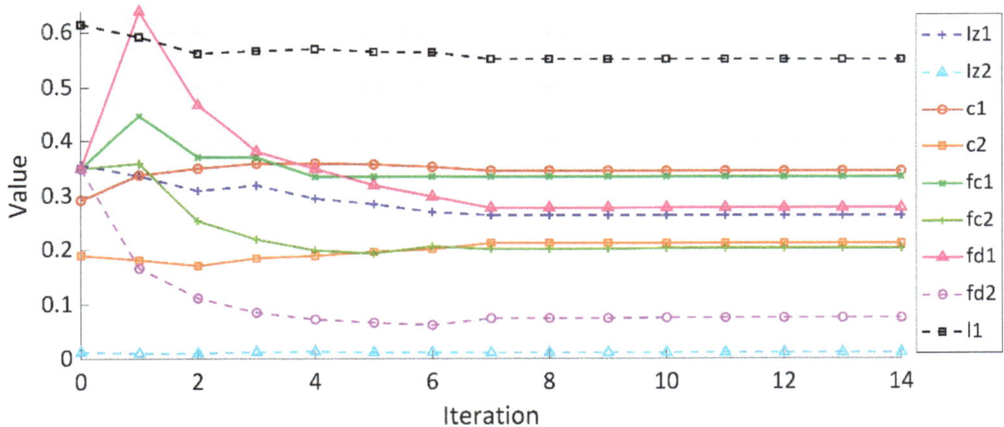

Figure 32. Estimation parameters—part 2.

Significant estimates and changes in some parameters can be evaluated based on the following hypotheses:

- The weights m_1 and m_2 that were exported from the CAD software did not take into account the total weight of the arms even with connected muscles, chain, etc. Based on which the resulting weight could increase by estimation (this phenomenon is visible especially for axis 1—weight m_1).
- Given the first hypothesis, it is possible to follow up. By changing the weight and shape of the arm, its length, the estimation of the parameter Distance to the center of mass for link 1 and 2—c_1 and c_2 (a visible increase of these parameters during estimation) is influenced.
- Changes in the estimation of the moment of inertia parameters I_{z1} and I_{z2} may also be closely related to the change in mass and the shift of the center of gravity.
- In terms of the use of a chain and a sprocket, it was assumed that the friction would be more pronounced (than if another gear had been used), so the value of the coefficients was set to 0.35. However, it is clear from the estimation that the experimentally determined initial values of the friction coefficients were oversized, as evidenced by the courses of the estimated parameters f_{c1}, f_{c2} and f_{d1}, f_{d2}.

9. Conclusions

The research described in this article aimed to compare the model with the parameters that come from CAD and its improvement by estimating the parameters based on gravitational measurements. An improvements model could be subsequently applied to the control phase.

The study described in this article can be summarized in four stages:

- Obtain a representative set of joint angle data by performing gravity tests.
- Creation of a simulation model in the Matlab and Simulink, which presented the dynamics of a real system—a planar robotic arm with two degrees of freedom (taking into account the friction component).
- The validation of a dynamic model built based on Lagrange formalism. However, at this stage, the expected results based on the MAE and NRMSE indicators were not achieved.
- The estimation of parameters—as the initial results of validation were insufficient, the research was extended to the estimation of model parameters using the Trust Region Algorithm, which helped to significantly improve the values of MAE and NRSME.

Based on the results of parameter validation and estimation, it can be stated that the values of system parameters, which serve as input data for the model, significantly affect the validation results, which was demonstrated after estimating these parameters significantly affected the results. Within the course of the estimated parameters, a certain connection between the parameters was seen, which gives rise to the assumption that the values of the parameters not only influence the results of the validation but also interact with each other. The more accurate the system parameters are obtained in the step, the more accurate the validation results will be. By reducing the oscillations (especially for link 2 and at higher initial rotation angles) which occur in the system due to the individual structural bonds, it would be possible to achieve a more accurate stabilization of the arm movement to zero, and thus to improve the validation results.

In further research, the authors would like to extend this research to the following parts:

- Accurate identification of system parameter values, including drive parameters.
- The inclusion of drive dynamics, which would create a more complex dynamic system model.
- The addition of a part of the drives to the simulation scheme in Simulink, the use of other tools, such as SimMechanics.

Author Contributions: Conceptualization, A.H. and M.T.; methodology, A.H.; model design, M.T. and T.Č.; measuring data, A.H. and T.K.; validation in Matlab, T.Č.; estimation in Matlab, T.Č.; formal analysis, M.T.; investigation, M.T.; writing—original draft preparation, M.T.; writing—review and editing, M.T., A.H., T.K. and T.Č.; visualization, M.T. and T.Č.; project administration, T.K.; funding acquisition, A.H. All authors have read and agreed to the published version of the manuscript.

Funding: The project was supported by the Project VEGA no. 1/0393/18—Research of Methods for Modeling and Compensation of Hysteresis in Pneumatic Artificial Muscles and PAM-actuated Mechanisms to Improve the Control Performance Using Computational Intelligence and by the Project of the Structural Funds of the EU, ITMS code: 26220220103.

References

1. Liu, C.; Tang, T.; Lin, H.-C.; Tomizuka, M. *Designing Robot Behavior in Human-Robot Interactions*, 1st ed.; CRC Press: Boca Raton, FL, USA, 2019; ISBN 978-0-367-17969-4.
2. Gasparetto, A.; Ceccarelli, M. Mechanisms and Machine Science. In *Mechanism Design for Robotics: Proceedings of the 4th IFToMM Symposium on Mechanism Design for Robotics*; Springer: Cham, Switzerland, 2019; ISBN 978-3-030-00364-7.
3. Kroff, J. *Modern Robotics: Designs, Systems and Control*; Willford Press: Forest Hills, NY, USA, 2019; ISBN 978-1-68285-676-5.
4. Yang, C.; Ma, H.; Fu, M. *Advanced Technologies in Modern Robotic Applications*; Springer: Singapore, 2016; ISBN 978-981-10-0829-0.
5. Miller, R.; Miller, M. *Robots and Robotics: Principles, Systems, and Industrial Applications*, 1st ed.; McGraw-Hill Education: New York, NY, USA, 2017; ISBN 978-1-259-85978-6.

6.   Ross, L.T.; Fardo, S.W.; Walach, M.F. *Industrial Robotics Fundamentals: Theory and Applications*, 3rd ed.; Goodheart-Willcox Publ: Tinley Park, IL, USA, 2017; ISBN 978-1-63126-941-7.
7.   Colombo, R.; Sanguineti, V. Rehabilitation Robotics. Elsevier Science Publishing Co., Inc.: Amsterdam, The Netherlands, 2018; ISBN 978-0-12-811995-2.
8.   Barrett, L. *Recent Advances in Robotics*; NY Research Press: Forest Hills, NY, USA, 2020; ISBN 978-1-63238-723-3.
9.   Kurfess, T.R. *Robotics and Automation Handbook*; CRC Press: Boca Raton, FL, USA, 2018; ISBN 978-1-4200-3973-3.
10. Siciliano, B.; Khatib, O. *Springer Handbook of Robotics*; Springer: Heidelberg, Berlin, 2016; ISBN 978-3-319-32552-1.
11. Barrett, L. *Handbook of Robotics*; Willford Press: Forest Hills, NY, USA, 2020; ISBN 978-1-68285-776-2.
12. Hurmuzlu, Y.; Nwokah, O.D.I. *The Mechanical Systems Design Handbook: Modeling, Measurement, and Control*; CRC Press: Boca Raton, FL, USA, 2017; ISBN 978-1-4200-3674-9.
13. Nezhad, M.N.; Korayem, A.H. Dynamic Modeling of Industrial Manipulator Hyundai HS165 in Order to Determine the Dynamic Load-Carrying Capacity for a Specified Trajectory. In Proceedings of the 7th International Conference on Robotics and Mechatronics (ICRoM), Teheran, Iran, 20–21 November 2019; pp. 538–543.
14. Varela-Aldás, J.; Andaluz, V.H.; Chicaiza, F.A. Modelling and Control of a Mobile Manipulator for Trajectory Tracking. In Proceedings of the International Conference on Information Systems and Computer Science (INCISCOS), Quito, Ecuador, 13–15 November 2018; pp. 69–74.
15. Leborne, F.; Creuze, V.; Chemori, A.; Brignone, L. Dynamic Modeling and Identification of an Heterogeneously Actuated Underwater Manipulator Arm. In Proceedings of the IEEE International Conference on Robotics and Automation (ICRA), Brisbane, Australia, 21–25 May 2018.
16. Hošovský, A.; Piteľ, J.; Židek, K.; Tóthová, M.; Sárosi, J.; Cveticanin, L. Dynamic Characterization and Simulation of Two-Link Soft Robot Arm with Pneumatic Muscles. *Mech. Mach. Theory* **2016**, *103*, 98–116. [CrossRef]
17. Spong, M.W.; Hutchinson, S.; Vidyasagar, M. *Robot Modeling and Control*; John Wiley & Sons: Hoboken, NJ, USA, 2020; ISBN 978-1-119-52399-4.
18. Al-Qahtani, H.M.; Mohammed, A.A.; Sunar, M. Dynamics and Control of a Robotic Arm Having Four Links. *Arab. J. Sci. Eng.* **2017**, *42*, 1841–1852. [CrossRef]
19. Yuguang, Z.; Fan, Y. Dynamic Modeling and Adaptive Fuzzy Sliding Mode Control for Multi-Link Underwater Manipulators. *Ocean Eng.* **2019**, *187*, 106202. [CrossRef]
20. Naing, S.Y.; Rain, T. Analysis of Position and Angular Velocity of Four-Legged Robot (Mini-Bot) from Dynamic Model Using Euler-Lagrange Method. In Proceedings of the International Conference on Industrial Engineering, Applications and Manufacturing (ICIEAM), Sochi, Russia, 25–29 March 2019; pp. 1–4.
21. Korayem, M.H.; Dehkordi, S.F. Dynamic Modeling of Flexible Cooperative Mobile Manipulator with Revolute-Prismatic Joints for the Purpose of Moving Common Object with Closed Kinematic Chain Using the Recursive Gibbs–Appell Formulation. *Mech. Mach. Theory* **2019**, *137*, 254–279. [CrossRef]
22. Hasnaa, E.H.; Mohammed, B. Robust Control of Two Link Rigid Manipulator with Nonlinear Dynamic Model. In Proceedings of the International Conference on Electrical and Information Technologies (ICEIT), Rabat, Morocco, 15–18 November 2017; pp. 1–6.
23. Zhang, L.; Wang, J.; Chen, J.; Chen, K.; Lin, B.; Xu, F. Dynamic Modeling for a 6-DOF Robot Manipulator Based on a Centrosymmetric Static Friction Model and Whale Genetic Optimization Algorithm. *Adv. Eng. Softw.* **2019**, *135*, 102684. [CrossRef]
24. Lloyd, S.; Irani, R.; Ahmadi, M. A Numeric Derivation for Fast Regressive Modeling of Manipulator Dynamics. *Mech. Mach. Theory* **2021**, *156*, 104149. [CrossRef]
25. Izmailov, A.; Kurennoy, A.; Stetsyuk, P. Levenberg–Marquardt Method for Unconstrained Optimization. *Tambov Univ. Rep. Ser. Nat. Tech. Sci.* **2019**, 60–74. [CrossRef]
26. Quan, N.; Nguyen-Hoai, S.; Chuong-Thiet, T.; Lam-Phat, T. Optimization of the Longitudinal Cooling Fin by Levenberg-Marquardt Method. In *ACOME 2017, Proceedings of the International Conference on Advances in Computational Mechanics 2017, Phu Quoc, Vietnam, 2–4 August 2017*; Lecture Notes in Mechanical Engineering; Springer: Singapore, 2018; pp. 217–227. ISBN 978-981-10-7148-5.
27. Rezapour, M.; Asaki, T. Adaptive Trust-Region Algorithms for Unconstrained Optimization. *Optim. Methods Softw.* **2019**, 1–23. [CrossRef]
28. Sharifzadeh, M.; Akbari, A.; Timpone, F.; Daryani, R. Vehicle Tyre/Road Interaction Modeling and Identification of Its Parameters Using Real-Time Trust-Region Methods. *IFAC Pap.* **2016**, *49*, 111–116. [CrossRef]
29. Kelly, R.; Davila, V.S.; Loría, A. Introduction to Adaptive Robot Control. In *Advanced Textbooks in Control and Signal Processing*; Springer: London, UK, 2005; ISBN 978-1-85233-999-9.
30. Papageorgiou, D.; Blanke, M.; Niemann, H.H.; Richter, J.H. Online Friction Parameter Estimation for Machine Tools. *Adv. Control Appl.* **2020**, *2*, e28. [CrossRef]
31. Tulsyan, A.; Barton, P.I. PERKS: Software for Parameter Estimation in Reaction Kinetic Systems. In *Computer Aided Chemical Engineering*; 26 European Symposium on Computer Aided Process Engineering; Kravanja, Z., Bogataj, M., Eds.; Elsevier: Amsterdam, The Netherlands, 2016; Volume 38, pp. 25–30.
32. Byrd, R.H.; Schnabel, R.B.; Shultz, G.A. A Trust Region Algorithm for Nonlinearly Constrained Optimization. *SIAM J. Numer. Anal.* **1987**, *24*, 1152–1170. [CrossRef]

33. Santos, S.A. Trust-Region-Based Methods for Nonlinear Programming: Recent Advances and Perspectives. *Pesqui. Oper.* **2014**, *34*, 447–462. [CrossRef]
34. Yuan, Y. Recent Advances in Trust Region Algorithms. *Math. Program.* **2015**, *151*. [CrossRef]
35. Zhu, B.; Zhang, X.; Zhang, H.; Liang, J.; Zang, H.; Li, H.; Wang, R. Design of Compliant Mechanisms Using Continuum Topology Optimization: A Review. *Mech. Mach. Theory* **2020**, *143*, 103622. [CrossRef]
36. Kanzow, C.; Yamashita, N. *Levenberg-Marquardt Methods for Constrained*; Universität Würzburg: Würzburg, Germany, 2002.

applied
sciences

Article

Chimney Sweeping Robot Based on a Pneumatic Actuator

Peter Ján Sinčák [1], Ivan Virgala [1], Michal Kelemen [1,*], Erik Prada [1], Zdenko Bobovský [2] and Tomáš Kot [2]

[1] Department of Mechatronics, Faculty of Mechanical Engineering, Technical University of Košice, 04200 Košice, Slovakia; peter.jan.sincak@tuke.sk (P.J.S.); ivan.virgala@tuke.sk (I.V.); erik.prada@tuke.sk (E.P.)
[2] Faculty of Mechanical Engineering, VSB-Technical University of Ostrava, 708 00 Ostrava, Czech Republic; zdenko.bobovsky@vsb.cz (Z.B.); tomas.kot@vsb.cz (T.K.)
* Correspondence: michal.kelemen@tuke.sk

Abstract: The need of improving the quality of professions led to the idea of simplification of processes during chimney sweeping. These processes have been essentially the same for tens of years. The goal of this paper is to bring an automation element into the chimney sweeping process, making the job easier for the chimney sweeper. In this paper, an essentially in-pipe robot is presented, which uses brushes to move while simultaneously cleaning the chimney or pipeline. The problem of the robot motion was reduced using an in-pipe robot due to the environments and obstacles that the robot has to face. An approach of using a pneumatic actuator for motion is presented along with the mechanical design. The next part of this paper is focused on the mathematical model of the robot motion, as well as its simulation and testing in the experimental pipeline. The simulations were compared with the experimental measurements and a few analyses were conducted describing the simulation model and its differences with the real robot, as well as considering certain parameters and their impact on the performance of the robot. The results are discussed at the end of the paper.

Keywords: chimney; cleaning; in-pipe robot; pneumatic actuator; simulation

Citation: Sinčák, P.J.; Virgala, I.; Kelemen, M.; Prada, E.; Bobovský, Z.; Kot, T. Chimney Sweeping Robot Based on a Pneumatic Actuator. *Appl. Sci.* **2021**, *11*, 4872. https://doi.org/10.3390/app11114872

Academic Editor: Alessandro Gasparetto

Received: 15 April 2021
Accepted: 24 May 2021
Published: 26 May 2021

Publisher's Note: MDPI stays neutral with regard to jurisdictional claims in published maps and institutional affiliations.

1. Introduction

The increasing trend of automation, the introduction of robotic systems into various professions, is on its way across all fields. However, one of the professions that has remained unchanged since its very beginning is the chimney-cleaning profession. The process of chimney sweeping has been essentially the same since the 18th century. Nowadays, the main concerns of the chimney-cleaning process are the safety of the chimney cleaner and the fact that it is a physically demanding process. This knowledge led to the idea of designing a system for locomotion inside the chimney as well as cleaning it, thereby improving the whole process. To start off, the problem was generalized into an in-pipe robotic system, due to the similarities of the environments that the robot is going to move in.

In the following sections of this paper, a brief overview of the related works that inspired our design process is presented. These works represent innovations in the area of in-pipe robots. In the subsequent section, the paper focuses on the principle of motion that is used for the actuation of this robotic system. Further, the mechanical design is developed, as well as a mathematical model that describes the forces involved in the actuation process of this robot. To prove the concept of this robot, a computer simulation was created, determining travel distance at certain frequencies of the actuation. To evaluate the created simulations, the verification of the proposed robotic system was carried out, and the results were evaluated in the conclusion of this paper.

2. Related Works

The problem of robots that are able to move inside of a chimney is very similar to the problem of in-pipe robots. Throughout the years, there have been many different approaches to the design of these kinds of robots.

Previous research in our Cognitics Lab in this area was focused on monitoring and inspection tasks in pipelines. On the one hand, we developed an in-pipe bristle-based mechanisms [1] for narrow pipelines. On the other hand, we developed several snake robots for a more complex type of pipeline, such as [2], where the travelling wave locomotion was experimentally analyzed. The concertina locomotion of the snake robot was experimentally analyzed using the digital image correlation method in [3], for the pipes with a rectangular and circular cross-section. Considering in-pipe locomotion, snake robots have many advantages in comparison with wheeled or bristled mechanisms, due to their kinematic redundancy [4], which may be used for optimization tasks [5].

One of the notable in-pipe robots is AIRO-II, which stands for multi-link articulated inspection robot, as presented in [6]. Its unique design consists of four units that are connected by spring joints, along with five sets of wheels, of which two are hemispherical and not actuated and three are so-called omni-directional wheels. Two sets of omni-directional wheels are actuated, and one set is passive. Because of the spring joints, the whole robot is in a zigzag shape, which pushes the wheels against the walls of the pipeline, allowing the robot to move both horizontally and vertically. AIRO was designed to be used in dust-, gas-, and water-filled environments; therefore, it is well protected against outer influences. The robot is also equipped with various sensors and cameras that are used for the inspection of the pipeline. Another novel approach was introduced in [7], using an electromagnetic linear actuator. In this paper, the authors proposed a micro robotic system actuated by the electromagnetic actuator. The system consists of the main body, head, and fins or legs. The novel approach of this system relies on its unique tubular linear permanent magnet actuator. In this actuator, the permanent magnets are in the center, moving in one axis, back and forth, according to the applied voltage that is being supplied into the coil, which is static and is around the permanent magnets. The coil does not move and is positioned on the stator. The principle of the movement is based on the contraction of the bellows seal, which is contracting as the actuator is moving, thereby moving the robot in the desired direction. In [8], a device without external moving parts was introduced. The proposed robot was based on a capsule robot with the actuator placed inside of a capsule. The actuator was based on a controlled motion of mass inside the capsule. The main feature of this actuator lies in the collision of the inner mass and the capsule of the robot. As the mass moves, it collides with the capsule and creates the energy necessary for overcoming the friction force between the robot and the ground or pipeline. A similar type of robot was also analyzed in [9]. In this paper, the authors focused on a small bristled robot, where they analyzed its movement. The actuation of this robot was based on vibration and its frequency. The authors created a model where they analyzed the frequencies of these vibrations that actuated the robot and were able to determine at which frequencies the robot switches the direction of the movement. The bristle type of robot is also analyzed in [10]. In this paper, a miniature bristle robot is introduced. The actuation of this robot is based on the inertial stepping principle. The actuator consists of a permanent magnet, electromagnet, and a spring. The locomotion is divided into three phases, where the permanent magnet is repelled from the electromagnet and compresses the bottom spring. After this phase, the electromagnet is turned off and the permanent magnet travels in the opposite direction, as before. In the last phase, the permanent magnet collides with the drive body and creates an energy that is greater than the frictional force between the bristles and walls of the pipeline, thus enabling the movement of the robot. An in-pipe inspection robot was also presented in paper [11]. Here, the authors presented the robot for visual and non-destructive testing of the pipeline networks, based on a pneumatic actuator. The robot structure consisted of a pneumatic actuator, elastic elements to hold the robot in the pipeline, and a damper. Since the pneumatic actuator was being actuated with the compressed air, it caused the vibration of the robot as it was being stopped by the damper. These vibrations caused the robot's movement in the pipeline. Another approach of an in-pipe robot, that is, using compressed air, was introduced in the [12]. As described in [12], there are several general types of in-pipe robots. In that paper, the introduced robot used the mechanism of both inchworm and wall press types of robot. The robot consisted of two clamping devices, one on top and one on the bottom, and a pneumatic cylinder between them. When the clamping devices were actuated, they were pushed against the walls of the pipeline, holding the robot in place. The translation movement of the robot inside the pipeline was arranged by the pneumatic valve using the inchworm principle. A similar application of compressed air was also presented in [13]. In paper [13], an application of bristle-based robots was presented. The proposed motion principle is based on the pneumatic cylinder that was connected to two brushes, each on one end of the cylinder. As the pneumatic actuator was actuated, the brush moved with the cylinder, causing the robot to move in the desired direction based on differential friction. In this paper, a bristle traction model was extensively studied and tested. At the end, the proposed principle of the movement was tested on several robots. Three of the robots were tested in sewer pipelines with different diameters of brushes, and one modified robot with guidance wheels was tested in the gas pipeline. The aim of the proposed robots was to demonstrate the flexibility and adaptability of the bristle robots in ill-constraint pipelines and to help with the inspection process. The adaptability of the robot was also considered in [14]. In the presented paper, the MIRINSPECT V robot was introduced. The aim of this robot was to move in pipelines and adapt its control to the shapes of the pipelines. It was designed for pipelines with an 8-inch inner diameter, and it consisted of three driving units responsible for locomotion and one center unit connecting the driving units. The driving unit was comprised of four wheels, clutch, gears, and an electric motor. The focus in this paper was on the selective drive mechanism, which allowed the robot to use as many electric motors as it needed, depending on the situation. This algorithm therefore increased power efficiency and improved the ability to pass difficult spots in certain situations. Another approach to an in-pipe robot was presented in [15]. The proposed system consisted of three tracks that were distributed equally around the main body of the robot. By pressing the tracks against the walls, the robot finds enough traction to move forward. In order to determine the correct pressing force, an algorithm was created along with the PD controller, which helped in adapting the force to the variable pipe diameter. The experimentation was carried out in a horizontal, 30-degree-slope pipeline. In paper [16], a similar approach to the design of an in-pipe inspection robot was presented. In this paper, a robot for the inspection of oil pipelines was created. The presented robot therefore moved in large-diameter pipelines. It consisted of six sets of wheels that pressed against the walls and the center part carrying batteries and the control box with microcontroller as well as other circuitry required for the control of the robot and the inspection of the pipeline. The inspection was carried out via a camera unit and laser detection device recording the geometrical dimensions. The proposed robot was tested in the actual oil pipeline on various slopes. The importance of the inspection of oil and gas lines has led to the creation of devices or robots called 'pigs'. Pigs are robots or devices specifically developed for various applications such as cleaning and inspecting large-diameter oil and gas pipelines, but also other specific applications. The construction of each robot can be different; however, they have some common traits. The general construction of such robots consists of a main body and sealing discs [17–22]. The main body houses the equipment necessary for the inspection and the discs are important for the movement of the robot. In general, these robots do not use any kind of actuators for locomotion; they use the fluids that are inside of the pipelines to move them [20]. The purpose of these robots is cleaning and inspection of the pipeline. Because the number of the sensors used can be too much for one unit to handle, they can be composed for several units [22].

The above-mentioned studies led to the idea behind this paper. In this paper, the focus is on essentially in-pipe robots, whose main objective is to clean chimneys. Since the cross-section of the chimney is circular, the problems that in-pipe robots face are fairly similar. Due to the existence of soot in the chimney, the idea of robots with wheels or tracks [6,14–16] was abandoned. Seeing successful prototypes, using fins or bristles in [1,7–11], led to the idea of using chimney brushes instead. The brushes allow the robot to move and clean at the same time. By simply changing the size of the brushes, the robot can adapt to a variety of chimney diameters and be more modular. More in-depth analysis of the robot is presented in the following section.

The contributions of this paper are as follows. First is the development of a robot with a new concept, based on a single-acting pneumatic actuator, for chimney-cleaning and monitoring purposes. The principle of motion is based on pneumatic piston crash. Besides the utilization of compressed air for actuation, it can be also used for soot removal by high-pressure air. The next contribution of this paper is a numerical simulation model with an analysis of individual parameters for an optimal robot design. The experimental analysis proves the numerical simulation and verifies the functionality of the developed robot. To the best of our knowledge, the presented contributions have not been yet published.

3. Development of Chimney Robot

The following section introduces the principle of robot motion and its mechanical design.

3.1. Principle of Motion

The main principle of the motion of this robot relies on the inertial stepping principle. The main characteristic of this principle is based on the mass that is connected to a mechanical capacitor that pulls/pushes the mass. The mass subsequently collides with another mass or element that is connected to the body of the robot. Collision in the desired direction of the robot's motion is the key moment in the motion of the robot. At this moment, the required energy for the motion is translated into the robot, enabling the translation motion of the whole robot.

For the proposed robot, there were several configurations of the actuator, and using a pneumatic actuator was one of them. Based on the inertial stepping principle, there are several possible configurations of the actuator that were tested. This can be seen in Figure 1, where the pneumatic actuator hits the surface in front of it. The question raised during the initial testing of this idea was where to position the collision insert in order to achieve sufficient energy for the movement of the robot, at the contact of the collision insert and the actuator end mass. The number of different distances between the collision insert and end piece were tested and, as the result, the travel distance of 23.5 mm was chosen. This distance was chosen based on a number of criteria such as the travel time of the piston, the frequency of the supplied air, and the motion of the robot with the chosen travel distance.

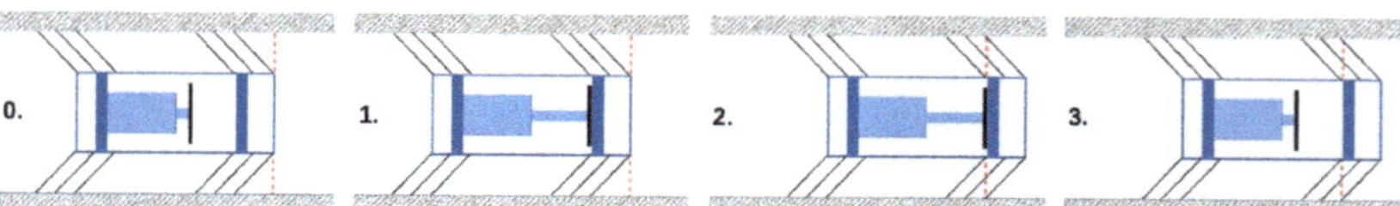

Figure 1. Motion phases of the proposed robot.

The principle of motion is shown in Figure 1. The motion consists of three phases. During the first phase, the cylinder extends until it hits the collision insert. The collision of these two objects represents the second phase. During this phase, the robot moves forward due to the impact force caused by collision. The red dashed line represents the initial position of the robot. During the third phase, the cylinder moves to its initial position. By repeating these three phases at a certain frequency, the robot moves forward.

3.2. Mechanical Design

In the previous section, the main principle of the robot motion was introduced. Here, the overall design of the proposed robot will be introduced. The basic structure of the robot consists of the actuator unit and the robot's body with brushes. The overall scheme of the robot can be seen in Figure 2.

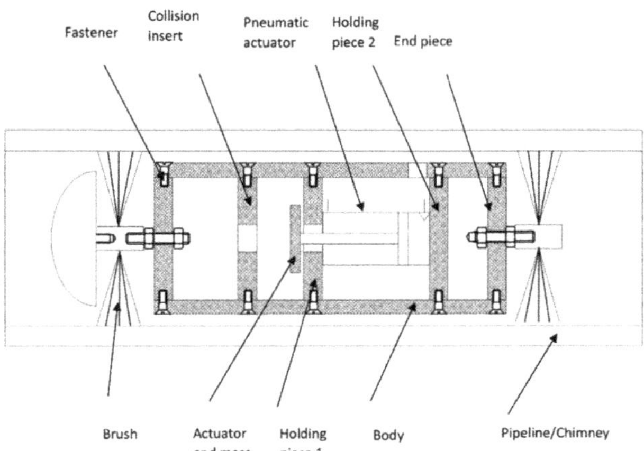

Figure 2. The robot's configuration.

As seen in Figure 2, at the top of the robot is a cap that could be made of plastic or aluminum, for the purpose of weight reduction, whose main job is to keep the robot straight during the motion. Since the service entrance of the chimney is often under the area where other appliances might be connected to the main chimney, the robot could derail and get stuck. Hence, the cap's job is to slide the brushes or the robot back to the mainline and, at the same time, to reject any excessive weight to the robot's body. The cap could be printed by a 3D printer. Immediately under the cap is a chimney brush, which is a regular chimney brush with a nut welded to the centre part of the brush so that the brush could be attached to the robot. It is then screwed to the end piece, which is connected to the body of the robot with fasteners. The purpose of this plug is to hold the brush and encapsulate the actuator so that it is shielded from any debris that is released from the walls of the chimney during the movement of the robot. Below that is a collision insert. The main purpose of this collision insert is in the actuation of the robot. This collision insert collides with the actuator end mass, and, during this process, the energy necessary for the motion is created. It is connected to the robot body with the fasteners in order to ensure stability. The actuator end mass is screwed on the end of the pneumatic actuator shaft that is moving according to the supplied pressure. As mentioned before, it collides with the collision insert and actuates the robot. Another part, as can be seen in Figure 2, is the holding piece 1 and 2. These pieces hold the pneumatic actuator in place so that it does not move during its actuation. Both of these pieces are again connected to the robot body with fasteners.

When it comes to brushes, it is assumed that the regular chimney brushes that are commercially available could be used, with some slight alterations such as welding the nut to them or, if required, cutting the length of the bristle. The assumption of using these brushes was made in regard to the overall price of the device because brushes are widely available in all different sizes. After the nut is welded onto the bottom part of the brush, the brush can be screwed onto the end piece that seals the main body. Another advantage of using regular, widely available chimney brushes is that they come in different sizes; therefore, by swapping the brushes, the robot could be used in various diameters of the pipelines or chimneys.

In the design of the robot, there are two end pieces, one at each end of the robot's body, so that the actuator inside is shielded away from the debris in the chimney. Both of the end pieces are 8 mm tall and 44 mm in diameter. Their other function is to hold the brushes attached to the robot's body. For that, there are four holes opposite each other on the sides of the end pieces in order for the fasteners to connect them with the main body and to keep them in place. Both of the end pieces were printed on a 3D printer with a 60%

infill of the material to make them more rigid. The material used for printing was ST-PLA, which is a strengthened PLA. The body of the robot is printed with a 3D printer using the same ST-PLA, as it is the case with the other parts. The body is 50 mm in diameter, with 3 mm thick walls and 45% infill of the material. The holes for connecting all the inner parts were drilled afterwards and were not made during 3D printing. The only other hole that was also drilled was the service hole for connecting the pneumatic line with the pneumatic actuator.

The whole motion of this robot relies on the pneumatic actuator, which is inside the robot's body. It is held in place with two holding pieces, printed on a 3D printer. The pneumatic actuator used in this robot is an SMC CD85E10-50s-B. This pneumatic actuator is a single-acting actuator, where the piston is pushed out by the supplied air and pushed back by the spring inside of the actuator. At the end of the moving piston of this actuator, there is a 3D printed part that is pushed repetitively against the collision inset, which is connected to the main body of the robot, hence conducting the motion of the robot.

The pneumatic scheme of the proposed chimney robot is shown in Figure 3. The robot uses polyurethane pipelines with an inner diameter of 4 mm and an electromagnetic valve controlled by PLC.

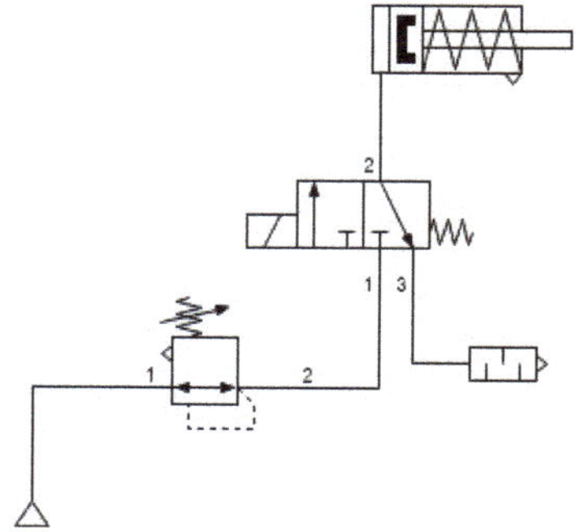

Figure 3. Pneumatic scheme of the robot.

3.3. Concept of the Proposed System

The concept of the proposed robot system is based on the premise of improving the chimney-sweeping process that has been stagnant for quite some time. By implementing a robotic system into this process, new data could be gained. The proposed robot is intended to carry a camera and make a recording that could help evaluate the condition of the chimney. A comprehensive concept of the developed system is shown in Figure 4.

The robot can be connected to a remote computer, where the operator can monitor the operation of the robot system and the camera feed from the robot. By using suitable sensors, the robot can become diagnosable, which is an important point of mechatronic systems [23]. This inspection process can also be important for insurance purposes because the recording can prove the condition of the chimney. Another important part of this concept is the cleaning of the chimney. Cleaning is an intentional collateral process of the movement of the robot. Another collateral process that is also helping to clean the chimney is the exhausting air from the pneumatic actuator. The use of compressed air can be also implemented on the top or bottom of the robot, where it can blow air and facilitate the cleaning process. The design of the robot system allows the interchangeability of the brushes, which can affect the cleaning performance but also the speed of the robot system and several parameters. This modularity is also an advantage when it comes to adaptability to different diameters of the chimneys.

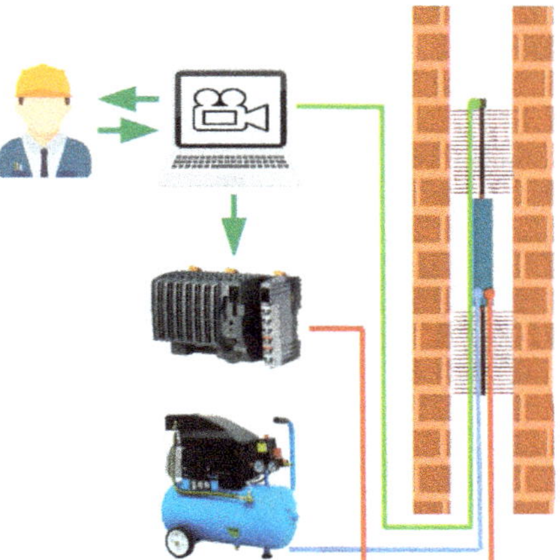

Figure 4. Concept of the chimney robot system.

4. Numerical Simulations and Experiments

4.1. Mathematical Model of the Proposed Solution

In the previous sections, the principle of the movement was introduced, along with the design of the proposed robot. To confirm the proposed principle of motion, a mathematical model is developed in this section, which will be further used for numerical simulation of the robot motion.

As mentioned, the robot motion is based on the collision of two masses. For this reason, the moment of collision can be modelled as a spring-damper mechanism [24–27].

Considering Figure 1, each phase of motion can be mathematically described as follows. During the first phase of motion, the actuator end mass moves forward according to Newton's law

$$m_1 \frac{dv_1}{dt} = F_C - G_1 \tag{1}$$

where F_C is the force of pneumatic cylinder and G_1 is the gravity of mass m_1. The mass m_1 represents the actuator end mass, which collides with the collision insert. The velocity of the mass m_1 is limited by the velocity of the pneumatic cylinder, which is 1.5 ms^{-1}. At the moment, when the pneumatic cylinder reaches its maximum set distance, in our case 23.5 mm, the model is changed to the spring-damper system, which imitates collision of the two masses. This can be expressed as

$$m_2 \frac{dv_2}{dt} = (y_1 - y_{1max})c + dy_1 b + F_f - G_2 \tag{2}$$

where F_f represents Coulomb friction force between brushes of the robot and the wall of the chimney, c is a coefficient of imaginary spring stiffness, b is the imaginary coefficient of damper, y_1 is the position of pneumatic piston relating to its initial position, y_{1max} is the distance between the initial position of the pneumatic piston and collision insert, and dy_1 is the velocity of the actuator end mass connected to the pneumatic piston. The value of F_f was determined experimentally. It should be noted that this phase lasts a very short time, and the time of this phase was determined by a set of simulations. The mass m_1 provides the energy for the mass of the robot. After the moment of collision, as described by Equation (2), which takes a very short time, the next phase follows. During this phase, the pneumatic piston comes to its initial position, while the robot moves forward or not, depending on the frequency of extension/insertion of the pneumatic piston.

4.2. Numerical Simulation

In this part, the focus is on the implementation of the mathematical background of the robot (mentioned in the previous section) into the numerical simulation and comparing the simulation with the real robot motion. For the simulation, MATLAB R2019b software was used.

The robot motion can be described by the following Algorithm 1.

Algorithm 1 Numerical simulation

1: Initialization of variables and constants
2: **FOR** cycle = 1 **TO** Number of required cycles **BY** 1
3: **FOR** i = 0 **TO** T **BY** dt // Forward motion of pneumatic piston
4:   **IF** (i > valveDelay) **THEN**
5:    $F_C = F_{max}$
6:   **END**
7:   Computation of mass 1 acceleration, velocity, position
8:   **IF** (contact < ρ) **AND** $(y_1 > y_{1max})$ **THEN**
9:    contact++
10:    $F_R = (y_1 - y_{1max})c + dy_1 b$
11:   **ELSEIF** (contact== ρ) **AND** $(y_1 > y_{1max})$ **THEN**
12:    $F_C = F_R = dv_1 = dy_1 = 0$
13:   **ELSEIF** (contact==0) **AND** $(y_1 < y_{1max})$ **THEN**
14:    $F_R = 0$
15:   **END**
16:    $m_2 \frac{dv_2}{dt} = F_R + F_f - G_2$
17:   Computation of mass 2 acceleration, velocity, position
18: **END**
19: **FOR** i = 0 **TO** T **BY** dt // Backward motion of pneumatic piston
20:   **IF** (i > valveDelay) **THEN**
21:    $F_C = F_{max}$
22:   **END**
23:    Computation of mass 1 acceleration, velocity, position
24:   **IF** $(y_1 \leq 0)$ **THEN**
25:    $dv_1 = dy_1 = y_1 = 0$
26:   **END**
27:   Computation of mass 2 acceleration, velocity, position
28: **END**
29: **END**

The algorithm is divided into two parts. The first part deals with the extension of the pneumatic piston, and the second part deals with the insertion of the piston. The algorithm uses mathematical background, which is described in the previous section. The constant *valveDelay* is the value from the datasheet of the pneumatic valve. It is the time delay of the used pneumatic valve. The force F_{max} represents the maximal force of the used pneumatic cylinder, which is given by the datasheet. As can be seen from the algorithm, there is variable *contact*, which stands for the simulation of the time period in which two masses are in contact. The value of this variable was empirically determined as constant ρ. The simulations use the constants given in the Table 1.

Choosing the correct damping and stiffness coefficients is very important for a successful simulation. In our case, the spring and damper do not represent the actual springs and dampers; hence, the theoretical approach of determining the values was chosen. Following the approach in [25], the stiffness of the contact can be calculated with the following equation:

$$c = 4\pi^2 f^2 m \tag{3}$$

The damping coefficient b_1 is also selected by a similar approach [26]. In this equation, the recommended value of β lies between 0.1 and 0.4.

$$b = 2\beta\sqrt{km} \tag{4}$$

Table 1. Simulation constants.

Parameter	Value (Units)
Mass 1	0.05 (kg)
Mass 2	1 (kg)
Maximum piston speed	1.5 (ms^{-1})
Maximum piston extension	0.0235 (m)
Maximum piston force	62.8 (N)
Time delay of valve	0.005 (s)
Time step	0.0001 (s)
Spring stiffness	800 (Nm^{-1})
Damping coefficient	240 (Nm^{-1}s)
Time period of impact	0.0006 (s)
Forward friction coefficient	0.8×50 (N)
Backward friction coefficient	50 (N)

The calculated stiffness and damping coefficient have to be checked with the simulation and, if necessary, adjusted. Note that the higher the stiffness of the contact is, the slower the calculations in the simulations will be. This means that the goal here is to choose the proper constant as low as possible while keeping the plausibility of the simulation.

In the first cycle, the algorithm starts with the calculations of the acceleration and velocity of the moving mass at the end of the pneumatic piston, which is changes as the piston extends toward the point of contact between the moving mass and the collision insert that is connected to the robot's body. The next step in the first cycle is the simulation of the contact between the moving mass and the collision insert. This contact is represented in the mathematical model with the damper and spring. The collision that happens is simulated by creating an impulse contact force that is the sum of the forces created by the spring and the damper, as described in Section 4.1. Initially, before the robot starts the motion, the gravitational force that pulls the robot down equals the friction force of the brushes and walls of the chimney/pipeline, which is why the robot stands still and does not fall toward the ground. The impulse force created at the contact triggers the start of the robot motion, as the equilibrium between the forces is disrupted by this impulse. In the last step of this cycle, the acceleration of the robot is calculated using Equation (2). From here, the velocity and the displacement of the robot can be calculated numerically based on the Euler integration method as well. The next part of the last step of this cycle is the calculation of the friction force that is acting between the brushes and the walls of the chimney/pipeline. This force is calculated using the following equation:

$$F_f = -\mu F_{fmax} sign(dy_2) \tag{5}$$

Of course, the Coulomb friction model is dependent on the velocity direction. This is taken into consideration by the algorithm, and the force is switched accordingly. In the forward direction of the motion, toward the top of the chimney, the actual friction force is slightly lower than in the opposite direction, due to the angle of the brushes that is naturally caused by the motion. The friction force in the opposite direction uses the same Equation (5), although with the coefficient $\mu = 1$.

In the second cycle, the algorithm looks very similar. However, there is no contact between the moving mass and the collision insert. The moving mass travels in a reversed direction compared to the previous cycle because the pneumatic piston is being retracted. After this, the whole process starts again and is repeated until it reaches the overall predefined time of the simulation.

4.3. Experiments

To evaluate the simulation model of the proposed robotic system and its plausibility, a set of experiments had to be carried out. The robot was built as described in Section 3 and can be seen in Figure 5. The main difference between a laboratory pipeline made from plexiglass and a real chimney is the absence of filth and dirt. Of course, the choice of suitable brushes with sufficient hardness is related to this point.

Figure 5. The developed chimney robot.

This section describes the objectives of the experiments, the experimental setup, and the results. The main objectives of the experiments are:

- Verification of the functionality and the principle of the motion of the designed robot moving in the pipeline similar to a chimney. The focus is on the functionality of the robot not on the quality of the motion. In other words, this objective is not focused on the maximum speed of the robot.
- Comparison of the experimental results with the simulation data in order to verify the quality and accuracy of the simulation model. Several measurements will be carried out, with different frequency of extension/insertion of the pneumatic piston. The travelled distance of the robot with changing frequency will be then compared with the simulated travelled distance.

Positive results of the mentioned experimental objectives have the potential to offer interesting contributions in the area of designing and modeling of in-pipe robots and mechanisms.

The testing setup can be seen in Figure 5. For the measurements and a better ability to track the movement of the robot, a transparent plexiglass pipeline was used, which was positioned vertically to imitate the movement of the robot inside of the chimney. This can be seen in Figure 5. For the actual measurement of the movement of the robot, a linear resistive sensor, Megatron MBH K r7.5k, was used. The linear resistive sensor acts as a potentiometer with a slider. The output of this sensor was fed right into the MATLAB/Simulink software through the measuring card Humusoft MF634; by creating the calibration characteristics of this sensor, it was possible to gain the travelled distance of the robot, in mm, as the result of the measurement.

Figure 6 shows the number of frames from the actual measurement of the proposed robot system. As can be seen from these frames, the increase of the travelled distance is gradual. The measurements were carried out in a controlled environment of 21 °C and atmospheric pressure. In order to gain reasonable data, it was opted to measure the travelled distance of the robot that was actuated with 20 different frequencies of actuation of the pneumatic actuator, which is responsible for the actuation of the robot. The supplied air was 0.8 MPa, and the duration of each measurement was 60 s, which allowed us to obtain a wide range of data. The overall mass of the robot with the brushes was measured at 1 kg, the mass of the actuator end mass was 0.05 kg, the delay of the electronic valve was 0.005 s, the force of the pneumatic cylinder was 62.5 N, the travel distance of the actuator end mass was 0.0235 m, and maximal velocity of the pneumatic cylinder was 1.5 m/s.

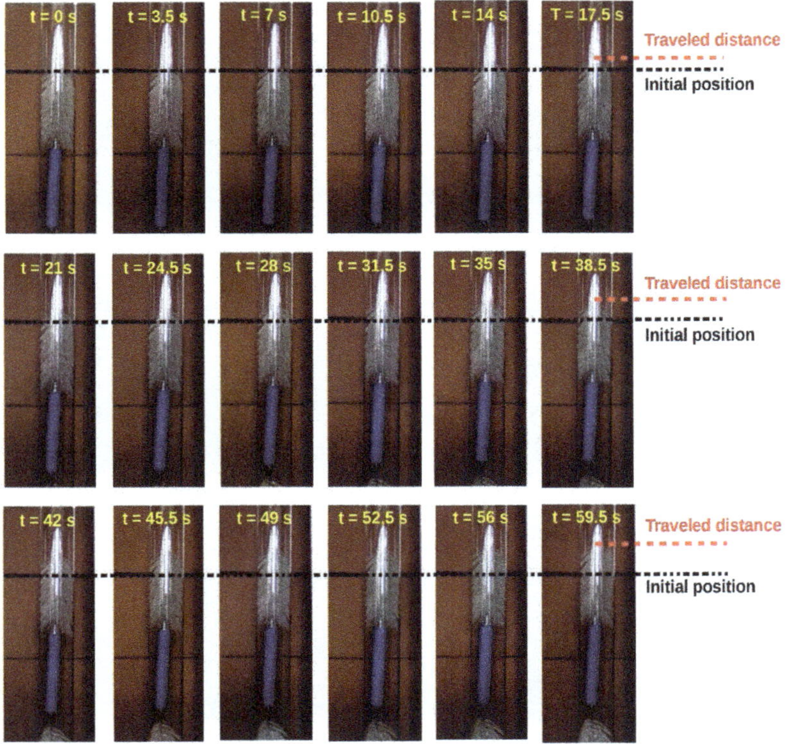

Figure 6. Video sequence of robot motion.

The example of travelled distance measurement during 60 s is shown in Figure 7. As mentioned earlier, both the simulation and measurement were tested on 20 different frequencies, starting with the frequency of 1 Hz and incrementing it by 1; therefore, a total of 20 measurements were carried out in order to evaluate the proposed concept of the robot. This frequency represents the frequency of switching between the extension and insertion of the pneumatic actuator, which is responsible for how often the pneumatic valve is turned on and off.

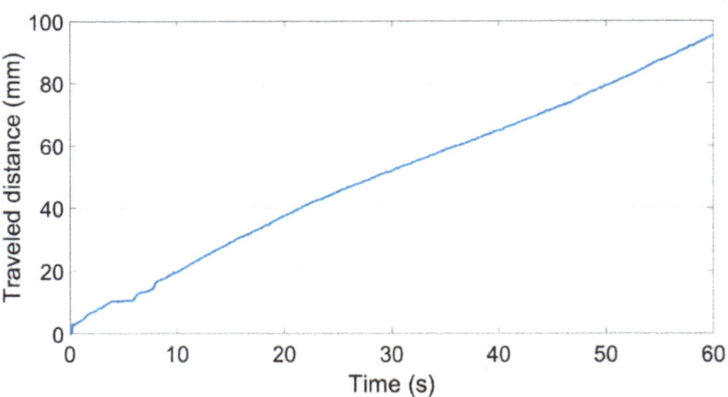

Figure 7. Example of measurement: the traveled distance of the robot, with an actuation frequency of 10 Hz.

In Figure 8, a comparison of the travelled distance, considering the experimental data and the simulation data, is shown. In Figure 8, the black line represents the measured data and the red line shows the results of the simulation. Further, it can be observed that the simulation follows the real measurement up until the frequency of 10 Hz, where the measured data start to deviate from the data obtained by the simulation. At this point, with frequencies above 10 Hz, the lack of supplied air starts to show, and the curve of the measured data starts to decrease. As frequency rises from this point, the air has not enough time to be supplied to the pneumatic piston, extend it, and escape from the pneumatic actuator. This effect results in a substantial decrease in the efficiency of the robot motion from this point on. This could be solved by increasing the diameter of the hose with the supplied air and the valve, which controls the air and shortens the length of the hose. However, this would cause difficulties if the device is used in a real setting, make the device usable only for laboratories. Another option is to put a smaller valve into the device; however, the flow would be again small, which would not solve the situation. Another possible option would be to add some small reservoirs in the device itself in order to compensate for the lack of flow; however, this solution would increase the mass, which would affect the travelled distance of the robot. Therefore, if the valve is bigger and the diameter of the hose is larger, then its length must be shorter due to the flow issue; and if the valve is smaller and can be placed on the robot, its flow is again smaller, which does not solve the problem. Because of these reasons, all of the following analyses are limited to 10 Hz, where the measurements and simulation match up.

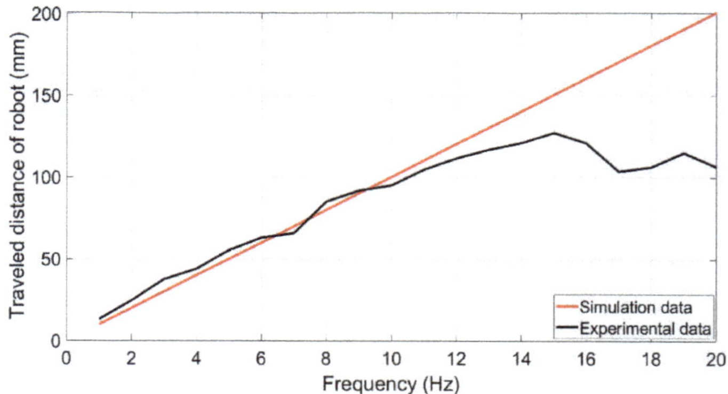

Figure 8. Comparison of experimental and simulation data.

4.4. Results of the Simulations and Analyses

The following section analyzes the influence of individual robot parameters to its motion. The first analysis focuses on the alteration of the robot mass and the mass of the actuator end mass or striking element mass and their impact on the travelled distance of the robot. For this analysis, the frequencies range from 1 to 10 Hz so that the simulation is still valid for these frequencies. The other parameters of the simulation remain the same, as described earlier in Table 1. The evaluation of the impact of the mass alterations are shown in Figure 9. On the y axis, the actuator end mass ranged from 0 to 1 kg. This mass was incremented by 0.01 kg on each iteration. The mass of the robot ranged from 0.5 to 5 kg because the mass of the robot is inevitably larger. This mass was incremented by 0.5 kg on each iteration. As can be seen in Figure 9, the changes in the mass of both bodies had quite some impact on the performance of the robot. In both cases, the impact was notable; however, the variation in the mass of the robot's body had a greater impact. This leads to the notion that by shaving off some mass from the robot's body, the travel distance of the robot can be further improved, which shows the significance of the robot's mass in regard to the travelled distance.

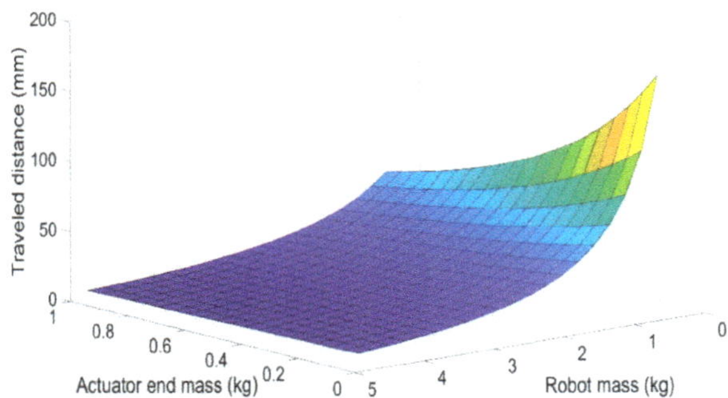

Figure 9. Correlation between the mass of the robot, actuator end mass, and the travelled distance.

Further, the influence of the friction force is shown in Figure 10. On the z axis, the travelled distance of the robot can be seen. On the y axis, the friction force is in the backward direction, meaning the robot would go down. On the x axis, the friction force is in forward direction, meaning the robot would go up. The force on the y axis is displayed in N and the force on the x axis is displayed in the percentage of the friction force in the backward direction.

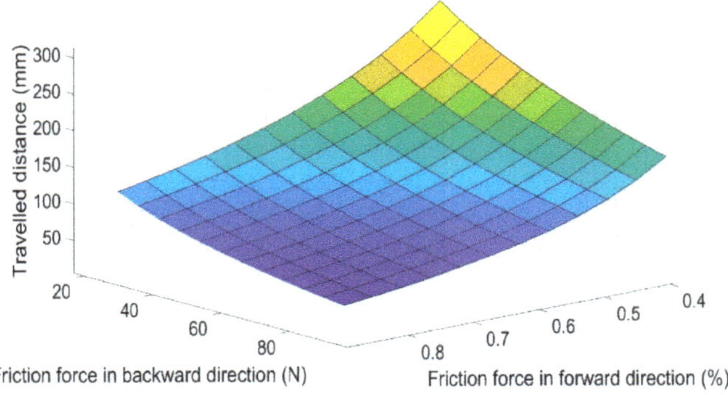

Figure 10. Correlation between the magnitude of the friction forces, the direction of the friction forces, and the distance travelled.

This was caried out due to the fact that the forward friction force is smaller due to the slope of the bristles, which are forced to bend into an angle due to the direction of the robot motion. The parameters of the simulation remain the same as in the previous analyses. In this figure, it can be observed that, in the given time, the smaller the forward friction is, the further the robot travels. This observation can be also confirmed in the mathematical model because the friction force is acting against the direction of the robot motion. Therefore, by decreasing it, the driving force has more influence over the travelled distance of the robot. The friction in the forward direction is highly influenced by the aforementioned slope of the bristles and by the material of the bristles and the wall. Of course, friction has also another purpose, which is to hold the robot in place and to stop it from falling when the robot is not being actuated. For these reasons, it cannot be smaller than the gravitational force pulling the robot down.

The last analysis was carried out to analyze the correlation between the velocity of the piston, its force, and their influence on the travelled distance of the robot. As can be seen in Figure 11, the y axis represents the velocity of the piston, x represents the piston's force, and the z axis represents the travelled distance of the robot. It can be observed that the piston's velocity has the most impact on the travelled distance. The distance gradually increases together with the increasing speed. The speed of the piston can be influenced, for example, by the change of airflow to the pneumatic cylinder. This cannot be said about the force. After a certain point of increasing the force, the travelled distance is not affected as much as in the case of increasing velocity. By increasing the velocity of the pneumatic piston, the frequency could be potentially increased. This is consistent with the simulation, where it could be seen that with the increasing frequency, the travelled distance increased, and so more hits per second could be executed with the actuator end mass. However, there are some hardware limitations of the pneumatic actuator as well as the pressurized air delivery system.

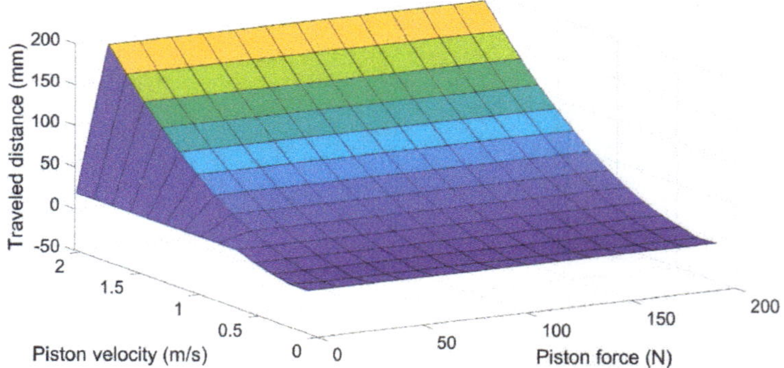

Figure 11. Correlation between the piston's velocity, the piston's force, and the travelled distance of the robot.

5. Results and Discussion

The proposed actuation of the robot is based on the inertial stepping principle, which is based on the collision of two masses. This collision causes the robot's forward motion. Since the possibility of using wheels or tracks for the movement was limited, we opted for using brushes instead. This way, the brushes acted instead of wheels and kept the robot inside the chimney, while cleaning the chimney as the robot moved. To validate the proposed design, a number of measurements with a real robot and a number of computer simulations were executed. The simulations and measurements were tested on 20 different frequencies, ranging from 1 to 20 Hz. The simulations followed the measurement up to 10 Hz. At this point, the measurements started to deviate from the simulation quite significantly, which was caused by the air supply chain. After the frequency of 10 Hz, the air was not supplied quick enough, and the flow of the air was limited. This resulted in the difference between the values that were measured and the ones that were simulated. To further validate the robot and the simulations, analyses of certain attributes were carried out, and their impact on the robot's travelled distance was investigated. One of the analyses was focused on the general mass of the robot, and the mass of the striking end mass. The analysis resulted in the conclusion that the variation in the robot's mass had a greater impact on the robot's performance compared with the variation in the mass of the striking end mass. This means that if the mass was further reduced, the performance could be improved. The following analysis investigates the influence of the friction force in both directions and their influence on the travelled distance of the robot. By decreasing the friction force in the forward direction, the travelled distance increases. This force cannot be decreased infinitely, because too small a friction force would result in the robot's falling. This force has to be at least at the level of the gravitational force that is acting on the robot; these two forces would cancel each other out, and the robot would stand still when it is not being actuated. The last analysis investigates the correlation of the piston force, its velocity, and the travelled distance. It shows that the importance of maximizing the piston's velocity is much higher than the importance of increasing the piston's force.

This paper contributed to the improvement of chimney sweeping using in-pipe robots. The novelty and contributions of this paper are as follows. As presented in the paper, the robot system based on a new concept, by using compressed air, has the ability to move vertically while at the same time cleaning the pipeline or a chimney. Having considered the previous works, as described in Section 2, one of the novelties of this paper is the principle of motion. Other robots' motion, as we have mentioned, is based on electromagnetic actuating with body impact; smooth piston extrusion without body impact and friction difference; wheel-based actuating; SMA-based actuating; and snake robot locomotion. However, our robot uses the impact of the bodies caused by a single-acting pneumatic cylinder. It uses the friction difference of the brushes to keep the robot in the chimney. Of course, the brushes also serve as cleaners of the chimney during the robot motion.

The next contribution of the paper is a model of numerical simulation. Some of the previous works introduced a mathematical model for the robot motion. However, a detailed view on the simulation model was absent. Our paper offers a detailed algorithm of numerical simulation, which was also experimentally verified. Thanks to the numerical simulation model, many of the analyses can be carried out without the need to of having an experimental model.

The next contribution of the paper is the provided analyses, which are based on the numerical simulation model. The analyses offer a deeper insight into the designing of chimney sweeping robots, showing the influence of individual parameters of the robot on its final motion properties.

The last contribution of the paper is an experimentally verified prototype of a proposed robot that is able to move vertically inside of the chimney/pipeline.

6. Conclusions

The paper investigates a chimney-robot's motion in the pipeline. We introduced a novel design of the robot, whose motion is ensured by single-acting pneumatic cylinder. Due to the impacting of the robot's collision insert by pneumatic piston, the robot performs a forward motion while it simultaneously cleans the chimney/pipeline. The paper also offers a detailed algorithm of the numerical simulation model, by which several analyses of the robot motion have been presented. Moreover, the influence of the robot mass and actuator end mass on the speed of the robot has been shown. The analyses also offer a view on the influence of friction and piston speed/force on the robot's final travelled distance per time. The experimental analysis proved, on the one hand, the functionality of the developed robot. On the other hand, it showed the accuracy of the developed numerical simulation model. The weakest element of the robot is between the pneumatic valve and the pneumatic cylinder. If the valve is placed in the robot, it needs to be small, which also causes a small air flow through it. Due to this fact, the pneumatic cylinder cannot work with the high frequency of piston extension/insertion, resulting in a slow robot speed. On the other hand, if the pneumatic valve has a high air flow, it needs to be outside of the robot (out of the chimney). For this reason, the pneumatic pipeline from the valve to the cylinder needs to be sufficiently long (especially in the case of a high chimney). This, again, results in the low frequency of piston extension/insertion.

The future work of this robot lies in the improvement of the air supply system, which greatly influences performance. By improving this part of the system, greater frequencies could be achieved, which should result in better performance. Another interesting area of research could be the ability to influence the friction force between the brushes and the walls of the chimney/pipeline. Being able to adjust the slope of the brushes could be another important area of improvement. Our next aim is also to secure the reverse motion of the robot.

Author Contributions: Conceptualization, I.V., P.J.S., and M.K.; methodology, I.V. and P.J.S.; software, I.V.; validation, P.J.S.; formal analysis, P.J.S. and T.K.; investigation, E.P., Z.B., and M.K.; resources, M.K., I.V., and E.P.; writing—original draft preparation, P.J.S. and I.V.; supervision, I.V.; project administration, I.V. All authors have read and agreed to the published version of the manuscript.

Funding: This research was funded by Slovak Grant Agency VEGA 1/0389/18 "Research on kinematically redundant mechanisms". This work was also supported by the European Regional Development Fund in the Research Centre of Advanced Mechatronic Systems project, project number CZ.02.1.01/0.0/0.0/16_019/0000867 within the Operational Programme Research, Development and Education.

Data Availability Statement: Data sharing is not applicable to this article.

Acknowledgments: The authors would like to thank the Slovak Grant Agency VEGA 1/0389/18 "Research on kinematically redundant mechanisms" and KEGA 030TUKE-4/2020 "Transfer of knowledge from the field of industrial automation and robotics to the education process in the teaching program mechatronics" They would also like to thank to the European Regional Development Fund in the Research Centre of Advanced Mechatronic Systems project, project number CZ.02.1.01/0.0/0.0/16_019/0000867 within the Operational Programme Research, Development and Education.

Conflicts of Interest: The authors declare no conflict of interest. The funders had no role in the design of the study; in the collection, analyses, or interpretation of data; in the writing of the manuscript; or in the decision to publish the results.

References

1. Ostertag, O.; Ostertagová, E.; Kelemen, M.; Kelemenová, T.; Buša, J.; Virgala, I. Miniature Mobile Bristled In-Pipe Machine. *Int. J. Adv. Robot. Syst.* **2014**, *11*, 189. [CrossRef]
2. Virgala, I.; Kelemen, M.; Prada, E.; Sukop, M.; Kot, T.; Bobovský, Z.; Varga, M.; Ferenčík, P. A snake robot for locomotion in a pipe using trapezium-like travelling wave. *Mech Mach. Theory* **2020**, *158*, 1–21.
3. Virgala, I.; Kelemen, M.; Božek, P.; Bobovský, Z.; Hagara, M.; Prada, E.; Oščádal, P.; Varga, M. Investigation of Snake Robot Locomotion Possibilities in a Pipe. *Symmetry* **2020**, *12*, 939. [CrossRef]
4. Kelemen, M.; Virgala, I.; Lipták, T.; Miková, Ľ.; Filakovský, F.; Bulej, V. A Novel Approach for a Inverse Kinematics Solution of a Redundant Manipulator. *Appl. Sci.* **2018**, *8*, 2229. [CrossRef]
5. Virgala, I.; Kelemen, M.; Varga, M.; Kuryło, P. Analyzing, Modeling and Simulation of Humanoid Robot Hand Motion. *Procedia Eng.* **2014**, *96*, 489–499. [CrossRef]
6. Kakogawa, A.; Ma, S. Design of a multilink-articulated wheeled inspection robot for winding pipelines: AIRo-II. In Proceedings of the 2016 IEEE/RSJ International Conference on Intelligent Robots and Systems (IROS), Daejeon, Korea, 9–14 October 2016; pp. 2115–2121. [CrossRef]
7. Lu, H.; Zhu, J.; Lin, Z.; Guo, Y. An inchworm mobile robot using electromagnetic linear actuator. *Mechatronics* **2009**, *19*, 1116–1125. [CrossRef]
8. Ivanov, A.P. Analysis of an impact-driven capsule robot. *Int. J. Non. Linear. Mech.* **2020**, *119*, 103257. [CrossRef]
9. Cicconofri, G.; DeSimone, A. Motility of a model bristle-bot: A theoretical analysis. *Int. J. Non. Linear. Mech.* **2015**, *76*, 233–239. [CrossRef]
10. Yum, Y.-J.; Hwang, H.; Kelemen, M.; Maxim, V.; Frankovský, P. In-pipe micromachine locomotion via the inertial stepping principle. *J. Mech. Sci. Technol.* **2014**, *28*, 3237–3247. [CrossRef]

11. Ragulskis, K.; Bogdevicius, M.; Mištinas, V. Behavior of Dynamic Processes in Self-Exciting Vibration of an Inpipe Robot. *J. Vibroengineering* **2008**, *10*. [CrossRef]
12. Kim, Y.; Yoon, K.; Park, Y. Development of the inpipe robot for various sizes. In Proceedings of the 2009 IEEE/ASME International Conference on Advanced Intelligent Mechatronics, Singapore, 14–17 July 2009; pp. 1745–1749. [CrossRef]
13. Wang, Z.; Gu, H. A Bristle-Based Pipeline Robot for Ill-Constraint Pipes. *IEEE/ASME Trans. Mechatron.* **2008**, *13*, 383–392. [CrossRef]
14. Roh, S.; Kim, D.W.; Lee, J.S.; Moon, H.; Choi, H.R. In-pipe robot based on selective drive mechanism. *Int. J. Control. Autom. Syst.* **2009**, *7*, 105–112. [CrossRef]
15. Kim, D.; Park, C.; Kim, H.; Kim, S. Force adjustment of an active pipe inspection robot. In Proceedings of the 2009 ICCAS-SICE, Fukuoka, Japan, 18–21 August 2009; IEEE: Piscataway, NJ, USA, 2009; pp. 3792–3797.
16. Li, H.; Li, R.; Zhang, J.; Zhang, P. Development of a Pipeline Inspection Robot for the Standard Oil Pipeline of China National Petroleum Corporation. *Appl. Sci.* **2020**, *10*, 2853. [CrossRef]
17. Zhu, X.; Zhang, S.; Wang, D.; Wang, W.; Liu, S. The impact of pipeline bend on bi-directional pig and the theories for the optimal pig design. In Proceedings of the 2013 IEEE International Conference on Mechatronics and Automation, Takamatsu, Japan, 4–7 August 2013; pp. 87–92. [CrossRef]
18. Guibin, T.; Shimin, Z.; Xiaoxiao, Z.; Liyun, S.; Qingbao, Z. Research on Bypass-valve and its Resistance Characteristic of Speed Regulating PIG in Gas Pipeline. In Proceedings of the 2011 Third International Conference on Measuring Technology and Mechatronics Automation, Shanghai, China, 6–7 January 2011; pp. 1114–1117. [CrossRef]
19. Alnaimi, F.B.I.; Mazraeh, A.A.; Sahari, K.S.M.; Weria, K.; Moslem, Y. Design of a multi-diameter in-line cleaning and fault detection pipe pigging device. In Proceedings of the 2015 IEEE International Symposium on Robotics and Intelligent Sensors (IRIS), Langkawi, Malaysia, 18–20 October 2015; pp. 258–265. [CrossRef]
20. Xinyu, Z.; Bo, Z.; Yuanzhang, J.; Bo, L.; Rui, L.; Lin, Z.; Qiao, S.; Xiaoli, Z. Analysis of rotation of pigs during pigging in gas pipeline. In Proceedings of the 2015 International Conference on Fluid Power and Mechatronics (FPM), Harbin, China, 5–7 August 2015; pp. 892–895. [CrossRef]
21. Xingping, X.; Hai, W.; Fuzheng, X.; Bin, W. Study of the Drive and Speed Governing Control for a Pipeline Pig. In Proceedings of the 2018 10th International Conference on Modelling, Identification and Control (ICMIC), Guiyang, China, 2–4 July 2018; pp. 1–6. [CrossRef]
22. Wang, Z.; Cao, Q.; Luan, N.; Zhang, L. Development of new pipeline maintenance system for repairing early-built offshore oil pipelines. In Proceedings of the 2008 IEEE International Conference on Industrial Technology, Chengdu, China, 21–24 April 2008; pp. 1–6. [CrossRef]
23. Božek, P.; Nikitin, Y.; Krenický, T. *Diagnostics of Mechatronic Systems (Studies in Systems, Decision and Control, 345)*; Springer: Berlin/Heidelberg, Germany, 2021; ISBN 978-3030670542.
24. Sentyakov, K.; Peterka, J.; Smirnov, V.; Bozek, P.; Sviatskii, V. Modeling of Boring Mandrel Working Process with Vibration Damper. *Materials* **2020**, *13*, 1931. [CrossRef] [PubMed]
25. Son, L.; Bur, M.; Rusli, M.; Matsuhisa, H.; Yamada, K.; Utsuno, H. Fundamental Study of Momentum Exchange Impact Damper Using Pre-straining Spring Mechanism. *Int. J. Acoust. Vib.* **2017**, *2*, 422–430. [CrossRef]
26. Jönsson, A.; Bathelt, J.; Broman, G. Implications of Modelling One-Dimensional Impact by Using a Spring and Damper Element. *Proc. Inst. Mech. Eng. Part. K J. Multi-Body Dyn.* **2005**, *219*, 299–305. [CrossRef]
27. Kuinian, L. Antony Darby. Modelling a buffered impact damper system using a spring-damper model of impact. In *Structural Control and Health Monitoring*; Wiley-Blackwell: Hoboken, NJ, USA, 2009; Volume 16, pp. 287–302. [CrossRef]

applied
sciences

MDPI

Article

Determination of the Function of the Course of the Static Property of PAMs as Actuators in Industrial Robotics

Darina Matisková [1], Tomáš Čakurda [1], Daniela Marasová [2,*] and Alexander Balara [1]

[1] Department of Industrial Engineering and Informatics, Faculty of Manufacturing Technologies with a Seat in Prešov, Technical University of Košice, Bayerova 1, 08001 Prešov, Slovakia; darina.matiskova@tuke.sk (D.M.); tomas.cakurda@tuke.sk (T.Č.); alek.balara@gmail.com (A.B.)

[2] Faculty of Mining, Ecology, Process Control and Geotechnologies, Technical University of Košice, Letná 9, 04200 Košice, Slovakia

* Correspondence: daniela.marasova@tuke.sk; Tel.: +421–55–602–3125

Abstract: Current efforts are focused on assembling new constructions while applying non-conventional actuators, for example, artificial pneumatic muscles, in engineering manufacturing processes. The reason is to eliminate stiffness and inflexibility of equipment structures that make sharing the working space of the technological equipment complicated. This article presents the results of experimental measurements of pressures in artificial muscles and rotations of the actuator with artificial muscles at various loads, using a testing device of an antagonistic actuator. The measurement results were used to create the function of the course of the static property of the antagonistic actuator with artificial muscles of the Festo type and to determine a mathematical model of the actuator dynamics, while applying the method of least squares.

Keywords: antagonistic system; pneumatic artificial muscles; function; model

Citation: Matisková, D.; Čakurda, T.; Marasová, D.; Balara, A. Determination of the Function of the Course of the Static Property of PAMs as Actuators in Industrial Robotics. *Appl. Sci.* **2021**, *11*, 7288. https://doi.org/10.3390/app11167288

Academic Editors: Pavol Božek, Yuri Nikitin and Tibor Krenicky

Received: 29 June 2021
Accepted: 6 August 2021
Published: 8 August 2021

Publisher's Note: MDPI stays neutral with regard to jurisdictional claims in published maps and institutional affiliations.

1. Introduction

In the recent past, a rather fast growth of automated and robotised technological workplaces has been observed in various industries, including the logistics processes which have been, until now, avoided by automation and robotisation [1]. Paper [2] deals with the results of the application of a speed signal averaging device and the acceleration loop. This device is applied in electric servo systems as well as electric servo systems of industrial robots. By applying this circuit, the robot positioning uncertainty was reduced by as much as 47% [2]. Manipulation equipment used in automated technological processes is rather accurate and efficient. Its disadvantage is a heavy weight and stiff and inflexible structure. Such drawbacks may be partially solved by applying pneumatic artificial muscles (PAMs) [3]. Pneumatic artificial muscles are referred to in the literature by different terms, such as Air Muscle, Fluidic Muscle, Pneumatic Muscle Actuator, Fluid Actuator, Fluid-Driven Tension Actuator, Axially Contractible Actuator, Tension Actuator, and Braided Pneumatic Muscle Actuator [4–6], whereas, from a structural point of view, in most cases it is in principle the McKibben's pneumatic artificial muscle. As for its structure, it is a rather simple device consisting, in its basic form, of the internal layer, external layer, and end pieces [7]. Generally, it is a braided artificial muscle. Its behaviour depends on its shape, contraction, and force when pressurised. These muscles are usually of a cylindrical shape [8]. In addition to significant hysteresis of properties and insensitivity, the conventional McKibben's pneumatic artificial muscle has also a short service life.

In an effort to eliminate these drawbacks, several improvements have been suggested to substantially increase the applicability of PAMs as actuators in the field of industrial robotics. J. M. Winters [9] published a muscle with external braiding. It differs from the McKibben's muscle in the internal layer structure. The main advantage of this muscle is the simplicity of assembly. J. M. Yarlott has patented the mesh artificial muscle [10]. It consists of an elastomer tube and a series of fibres stretched axially to both muscle ends that transform the pressure energy into the tensile energy. Unlike braided muscles, meshed artificial muscles are of lower density of the mesh surrounding the membrane. The mesh has larger openings and is more adjoined to the membrane. Meshed muscles can thus only stand lower pressures. M. Kukolj [11] has patented the artificial muscle. As compared to the McKibben's muscle, it has a different external layer. McKibben's muscle has a thick membrane braid, while Kukolj uses open openings, whereas the fibres are at certain spots firmly connected together and form a mesh. In the non-pressurised state, the openings are visible, but after the pressurisation, the openings disappear because all the layers will adjoin to each other. The problem of friction between the elastic layer and the braid is solved by lubrication. The reason for creating such muscle was the finding that a thick mesh tends to constrict faster than the membrane. G. Immeg and M. Kukolj [12] have patented a robotically active component based on an artificial muscle (Robotic Muscle Actuator). It is not reminiscent of any of the previous muscles. It is made of non-elastic material consisting of plenty of juts. The muscle coating is characterised with high tensile stiffness and flexibility. Friction is minimal and, therefore, it can develop a greater force. Such force approaches to the maximum theoretical value; it excels in high relative shortening and minimum hysteresis.

A disadvantage of the conventional structure of the McKibben's pneumatic artificial muscle is the formation of rather high friction between the fibres in the external layer and the rubber wall of the tube in the internal layer. Such friction reduces the force developed by the muscle and is also manifested as hysteresis that complicates the muscle control. This type of muscle is typical for threshold insensitivity when the pressure increases. Muscle contraction occurs only after a certain level of pressure increase is achieved [13,14]. The elimination of such friction and thus also of hysteresis will facilitate the muscle control by reducing its threshold insensitivity. It was achieved by constructing a membrane with longitudinal grooves. The grooves are free to broaden due to radial pressure when the muscle is being filled. Tensile forces are transmitted through extremely strong polymer fibres located in every groove of the membrane [15–17]. Such pneumatic artificial muscle was named the Pleated Pneumatic Artificial Muscles—PPAM. Increasing the service life by eliminating the effect of dry friction between the tube and the fibres, as well as between the fibres themselves, represents the connection of the elastic tube and glass fibres into a single unit. This facilitates a much smaller width of hysteresis and a narrower insensitivity zone [18]. This principle is applied in MAS–20 pneumatic artificial muscles that appear to be very appropriate for applications in industry due to their robust structure. Properties, shapes, and behaviour of pneumatic artificial muscles are comparable to human muscles; this makes them easy to combine into more complex manipulation mechanisms. Other important properties of pneumatic artificial muscles are listed in papers [19,20] and their basic drawbacks are described by Mižáková [21] and Wickramatunge et al. [22] The basic principle of their action is presented by Trojanova and Čakurda [23], Hošovský et al. [24], Tóthová and Piteľ [25], and Sárosi et al. [26].

Behind the effective movement of the robot or manipulator are many aspects that have to be mastered in motion control. Modelling of kinematics and dynamics of manipulators plays an important role in the control accuracy of these devices. Trojanova et al. in [27] describe the dynamics of a manipulator with two degrees of freedom, while the dynamic model of the manipulator's arm is derived using Lagrangian formalism, which considers the difference between the kinetic and potential energy of the system. In [28], Nezhad and Korayem described the dynamic and kinematic model of industrial manipulator Hyundai HS165. The kinematic model is obtained using the D-H parameters. The obtained model is compared to the simulations and the experimental test. The difference between the computed and real load is slight, which indicates that the dynamic model is quite accurate. Ref. [29] describes the kinematic model of a five-DOF spatial manipulator used in robot-assisted surgery. In [30], the dynamic model of a robotic manipulator is described using the Euler–Lagrange formulation. Sung et al. [31] derived the exact solution of kinematics, and they provide geometry intuition and meaning of kinematics equations.

The manuscript deals with improvement of the layer coating technology of painting robots, resulting in the formation of a new painting robot concept. Requirements for painting robots with low energy demands lead ultimately to a solution of a light robotic arm driven by pneumatic artificial muscles in an antagonistic configuration (pneumatic actuator). Its application in robotics is directly dependent on the perfection of a computerized control system. Designed original control was carried out on the real system through a PC using the software Matlab/Simulink. The concept of the control system ensures maximum stiffness of actuator mechanism.

2. Materials and Methods

2.1. Problem Formulation

Engaging pneumatic artificial muscles in antagonistic connection facilitates the regulation of their own stiffness/compliance, which is not usual in conventional types of actuators. The primary problem is complicated modelling and control while using pneumatic artificial muscles. Complicated modelling and control are caused by non-linear muscle properties, medium compressibility, and friction of the internal structure. Modelling and simulation of actuators based on pneumatic artificial muscles require the knowledge of the mathematical relationship between the muscle strength and its contraction at various pressures in the muscle. For this purpose, we carried out the measurements of static properties of the MAS–20 pneumatic artificial muscle used in experiments [32].

2.2. Experimental Material

Experimental research was focused on MAS–20 pneumatic artificial muscles (PAM) with a robust structure The internal layer of this type of PAM consists of a rubber tube and the external layer is formed by fibres. The layers are interconnected to minimise the friction between the fibres and the tube that causes hysteresis (Figure 1). Materials typically used for the manufacture thereof are chloroprene and aramid.

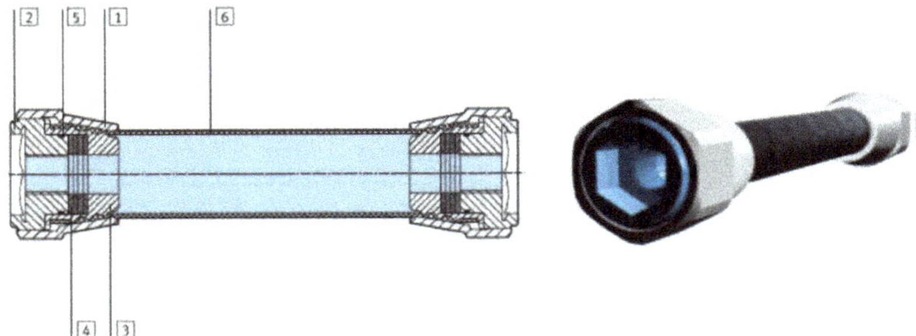

Figure 1. Pneumatic artificial muscle of MAS–20 type (**1**) connecting matrices (wrought aluminium alloy); (**2**) flange (wrought aluminium alloy); (**3**) internal cone (wrought aluminium alloy); (**4**) disc springs (steel); (**5**) sealing ring (nitrile rubber); (**6**) tube (chloroprene and aramid) [33].

The elastic tube and the glass fibres of MAS–20 are connected into a single unit. This facilitates a much smaller width of hysteresis and a narrower insensitivity zone. MAS–20 artificial muscles were chosen as the research object because in terms of their application in industry, they appear to be highly applicable due to their long service life, as compared to other types of PAM.

2.3. Testing Device

The testing device is the experimental actuator with pneumatic artificial muscles in antagonistic connection. The actuator with PAMs in antagonistic connection (Figure 2) consists of two artificial muscles interconnected with a roller chain. The chain transmits the forces from the muscles to the gear wheel that is firmly attached to the output shaft performing a rotary motion (not depicted). An experimental actuator based on PAM with one distance level (Figure 2) was designed at the Department of Industrial Engineering and Informatics, Faculty of Manufacturing Technologies with a Seat in Prešov, Technical University of Košice.

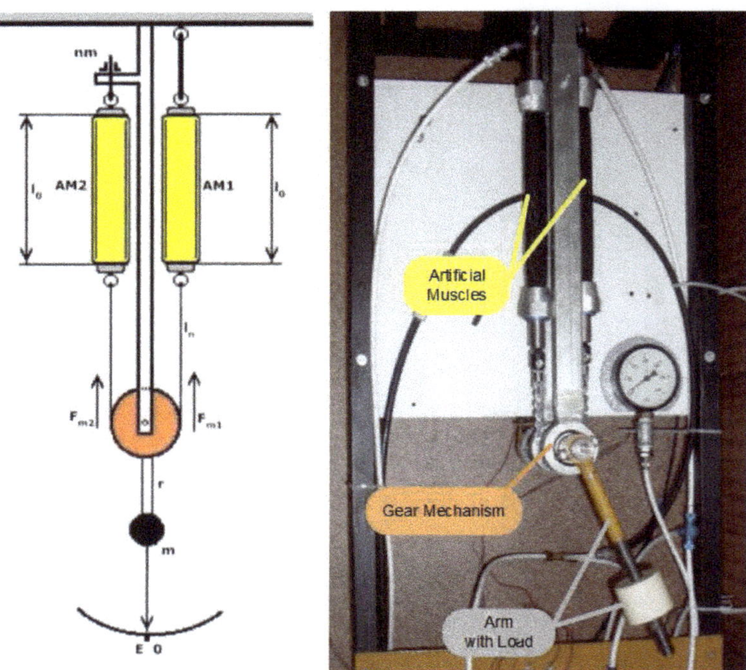

Figure 2. Experimental actuator based on PAM—An antagonistic connection of pneumatic artificial muscle.

The actuator is a kinematic structure consisting of two antagonistic pneumatic artificial muscles (Figure 2), interconnected with a pull chain that carries the gear wheel mounted on and attached to the output revolving shaft. Rotation (position) of the shaft is proportional to the difference in pressures in individual muscles. The stiffness of the shaft's position in the respective direction is determined by the magnitude of the pressure in the respective pneumatic muscle that is subjected to tension.

The experiments were carried out with the actuator comprising a pair of pneumatic artificial muscles of the MAS–20 type; their movement is transmitted through a pulley with a chain gear. An arm with a load is attached to the shaft. The active component consists of two pairs of directly controlled two-way electromagnetic valves, for each muscle 1 inlet and 1 outlet valve, MATRIX EMX 821.104C224. The arm's position is measured by the IRC 120 incremental encoder with 2500 pulses/rev. The pressure in artificial muscles is sensed by pressure transducers PMD 60G with an analogue output signal of 0–10 V DC.

2.4. Experiment Methodology

The basic component of the control unit was a PC with an Intel Core 2 Quad processor with a clock frequency of 2.33 GHz and a 4 GB operating memory. The main component for interconnecting the real system with the computer was the input–output measuring card of the MF 624 type by Humusoft. It contains a 14-bit A/D converter with a simple shaper, 4 software-adjustable ranges, and an 8-channel multiplexer at the inlet, 8 independent

14-bit D/A converters with a double buffer, 8-bit numerical input and 8-bit numerical output, and 4 inputs for an incremental sensor (simple or differential). One of them is used in one mode. It is supplied with the Real Time Toolbox for Matlab. The card is intended for a PCI slot. Four bits of numerical output, compatible with TTL logic circuits, were used for the control. Analogue outputs use 4 voltage ranges (±10 V, ±5 V, 0–10 V, 0–5 V) selected by the software. A signal from the analogue output is normed into the range of ±1. The maximum sampling frequency determined by the used hardware was 500 Hz. Although this hardware facilitated increasing the sampling frequency, it would require using input blocks with the buffer. These are only applicable in the control because it is possible to use delayed processing of the obtained specimens by the software. With regard to the requirement of continuous control of the actuator, such blocks could not be used; nevertheless, the above-mentioned sampling frequency was sufficient. The control block also included a module with the EPAM (Electro-Pneumatic Action Module) working designation, facilitating adjustments of the control parameters from the measuring card output for the valve inputs (the voltage of the signal corresponding to the logic unit was 5 V, the control voltage of valves was 24 V). Within the implementation of automatic control using a computer, the interface with manual control was maintained; it proved to be very useful for adjusting the basic status of the actuator as well as fast verification of functionality. Schematic diagram of the experimental platform is depicted in Figure 3.

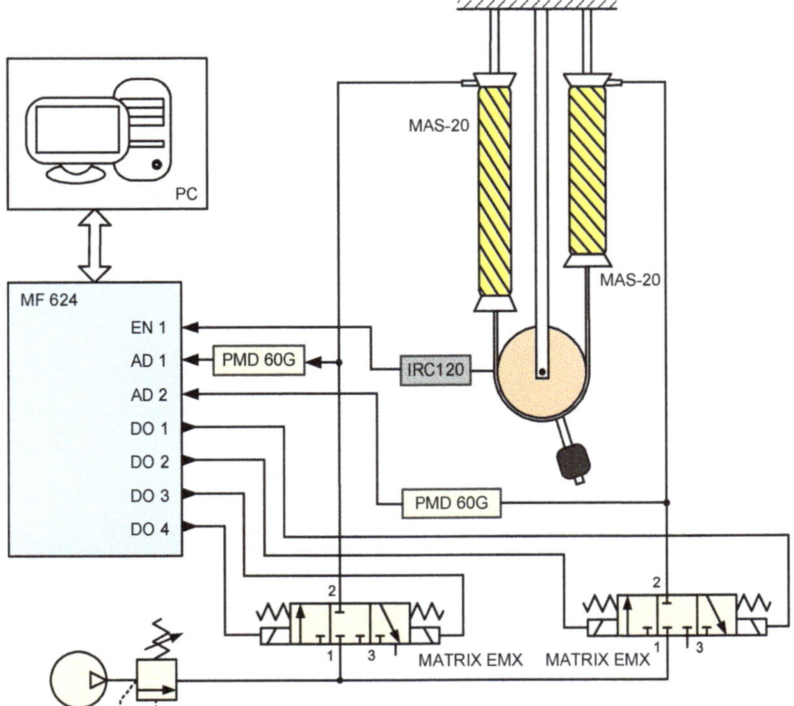

Figure 3. Schematic diagram of the experimental platform.

3. Theory/Calculation

3.1. Physical Fundamentals

The principle of the action of such an actuator with artificial muscles (AMs) in antagonistic connection can be explained by Figure 4.

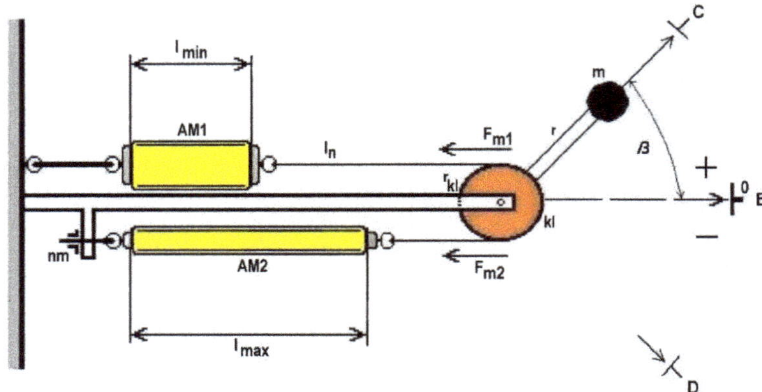

Figure 4. Pneumatic artificial muscles in antagonistic connection.

If we have two identical muscles, AM1 and AM2, in antagonistic connection (Figure 2), and these muscles are filled with an identically large amount of compressed air, they shorten to the length l_0, and the following applies [33]:

$$l_0 = l_{max} \frac{\Delta l_{max}}{2} \tag{1}$$

where:

l_{max}—maximum AM length,

Δl—change in AM length.

AM1 has the tensile force F_{m1} which is transmitted through a pulley to the AM2 muscle that develops the tensile force F_{m2}. With identical filling pressures in AM1 and AM2, tensile forces become identical at identical values of their contractions and the actuator's arm stabilises in the initial position. In such an initial position, the actuator stiffness is the highest if air pressure in both muscles is at its maximum.

At non-identical filling pressures in the muscles (Figure 4), the actuator's arm stabilises in the position corresponding to the equality of tensile forces in both muscles. The angular displacement β for the arm exposed to the external load with the mass m depends on the pulley radius r_{kl} for identical changes in the length of pneumatic artificial muscles. The rotation of the arm is subject to the equation:

$$\beta = \frac{\Delta l}{r_{kl}} \tag{2}$$

If we designate the derivation of the muscle volume according to the length dV/dl generally as a non-linear function f, for the two muscles we can write the following:

$$F_{m1} = -P_1 f_1(l_{10} - \Delta l), F_{m2} = -P_2 f_2(l_{20} + \Delta l) \tag{3}$$

where:

P_1, P_2—pressure in the respective muscle,

l_{10}, l_{20}—initial lengths of both muscles.

The above-mentioned equation (3) indicates that the length of each muscle was changed in the same value Δl, while the shortening of one muscle became larger and of the other muscle became smaller (Figure 4). It is interesting that the stiffness of such a mechanism can be changed according to the determined requirement, as the arm's position is proportionate to the difference in pressures in the muscles, whereas the stiffness is proportionate to the sum of pressures in the muscles. Hence, the same pressure difference may be achieved at different values of their sum, i.e., the same arm displacement can be achieved at various resulting stiffnesses. The stiffness of the mechanism may be derived on the basis of the expression of the total force affecting the load, as follows:

$$F = F_{m1} - F_{m2} = -P_1 \frac{dV_1}{dl_1} + P_2 \frac{dV_2}{dl_2} = P_1 \frac{dV_1}{d(\Delta l)} + P_2 \frac{dV_2}{d(\Delta l)}. \tag{4}$$

In which case, the following applies to the stiffness:

$$K = \frac{dF}{d(\Delta l)} = -\frac{dP_1}{dV_1} \left(\frac{dV_1}{d(\Delta l)} \right)^2 - \frac{dP_2}{dV_2} \left(\frac{dV_2}{d(\Delta l)} \right)^2 - P_1 \frac{d^2 V_1}{d(\Delta l)^2} + P_2 \frac{d^2 V_2}{d(\Delta l)^2}. \tag{5}$$

According to the new concept of actuator action prepared by the authors, if we change the pressure only in one muscle, we distinguish between active and passive AM. The active muscle is always the muscle with a variable air pressure. The passive muscle serves as a non-linear spring at constant air pressure, thus ensuring the stiffness of the actuator's mechanism and equality of forces for every actuator position. When the air pressure changes (decreases), for example, in the AM2 muscle, the muscle contraction changes (decreases) as well (active muscle). As a result, the mass of the load on the arm attached to the pulley axis makes a rotary motion. Such motion direction is regarded as positive (+) with regard to the initial point. Negative position values (−) are achieved by the actuator in the same manner. Only the roles of the muscles are switched. The air pressure of AM1 is variable; AM2 serves as a non-linear pneumatic spring.

Dependencies of the end position of the actuator represent a non-linear function of the filling air pressure in AMs and their courses are different at different values of actuator load. This follows from non-linear muscle properties. It is also necessary to respect a non-identical tensile force (or the torque) of the actuator, the magnitude of which changes with the value of angular displacement of the actuator pulley shaft (or AM contraction, AM displacement). This property causes also non-identical mechanism stiffness at different position values, whereas the highest and two-side symmetrical stiffness of the actuator is reached in the reference point, when the pressures in AMs are identical and reach the maximum. Besides overcoming the forces from the load, the active muscle must also overcome the variable directive force of the passive muscle. That is why the requirements put on nominal parameters (tensile force) of artificial muscles are higher than those that would only be based on the load size. With regard to the need for appropriate mechanism stiffness, the forces of the actuator muscles are expected to be higher than the (maximum) load force.

3.2. Model Formation

The pneumatic artificial muscle AM2 extends its length while the compressed air is released, and at the same time it exerts the tensile force F_{m2}, the value of which gradually decreases. The relationship between the tensile force F_m, the air pressure p, and the length l (contraction k magnitude) of this artificial muscle is expressed by the properties presented in Figure 5 that are depicted together with the property of the AM1 artificial muscle.

In the case of this artificial muscle, the air pressure does not change, so only one of its properties is depicted, the one corresponding to the initial filling pressure p_m. The property is drawn so that it describes antagonistic forces of AM1 against AM2. The AM1 muscle serves as a non-linear pneumatic spring. This ensures equality of forces for every position of the positive position value. It also provides stiffness of the actuator's mechanism. Property intersection points in the EC section correspond to the course of increasing concentration of the passive AM of AM1 during a gradual pressure decrease in the active AM of AM2. The above-described procedure facilitates achieving an arbitrary positive value l_d (up to l_{dmax}). Actuator's negative position values are reached through the same procedure as described in the previous paragraph. Only the roles of individual muscles are switched. AM1 has a variable air pressure; AM2 serves as a non-linear pneumatic spring.

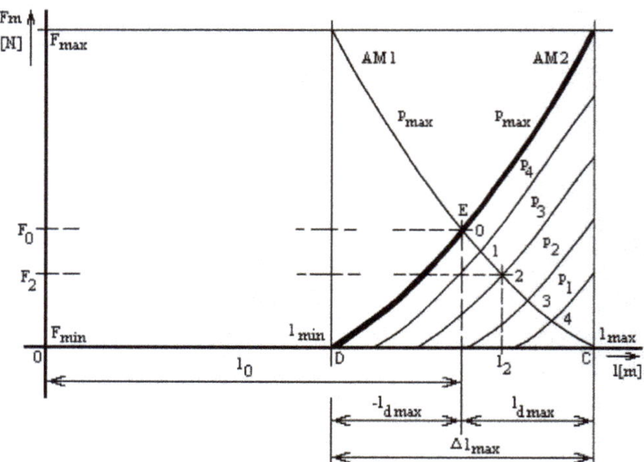

Figure 5. Static properties of PAMs in antagonistic connection at variable filling pressure.

In the configuration of the antagonistic actuator, as shown in Figure 5 (position in the reference point E), while identical pneumatic artificial muscles are used, their non-linear functions g_N will be identical too, as well as their tensile forces:

$$F_{m1} = g_N(l_1) = F_{m2} = g_N(l_2) = F_0, \tag{6}$$

At the same time, the following applies: $l_1 = l_2 = l_0$ and $p_1 = p_2 = p_m$.

In the configuration of the antagonistic actuator, as shown in Figure 5, while identical pneumatic artificial muscles are used, their grids of non-linear functions g_M will be identical, as well as their tensile forces (position in the reference point E):

$$F_{m1} = g_M(l_1, p_1) = F_{m2} = g_M(l_2, p_2), \tag{7}$$

while the following applies: $l_1 <> l_2$ and $p_1 <> p_2$ and $k_{AM1} + k_{AM2} = k_{max}$ (k_{AM1} is contraction of the actuator AM1, k_{AM2} is contraction of the actuator AM2 and k_{max} is maximum contraction of the AMs antagonistic connection).

If in Equation (7), we know the non-linear function g_M (l, p), then on the basis of the given equality of forces we can express the relationship between the output from the actuator (length l or l_d) and its input (air pressure p) in the form of a non-linear function f_N:

$$abs(l_d) = f_N(p), \tag{8}$$

$$abs(\beta) = \frac{l_d}{r_{kl}} = \frac{f_N(p)}{r_{kl}}. \tag{9}$$

Equations (8) and (9) clearly indicate that the values of respective parameters β and l_d will be achieved with a positive and a negative polarity. Therefore, the value of the input pressure p must also have assigned both polarities, despite the fact that the pressures into individual muscles have positive values ($p_{AM1} > 0$ and $p_{AM2} > 0$). This procedure is carried out depending on the sign of the required displacement of AM l_{dD} or the required angular displacement β_D as follows:

if $sign\ \beta_D = +$, then $p = p_{AM2}.sign\ \beta_D > 0$ whereas $sign\ \beta = +$ and $p_{AM1} = const > 0$,

if $sign\ \beta_D = -$, then $p = p_{AM1}.sign\ \beta_D < 0$ whereas $sign\ \beta = -$ and $p_{AM2} = const > 0$.

Such requirements for the angular displacement β apply also to the displacement of AM l_d, instead of $sign\ \beta_D$ the requirements contain $sign\ l_{dD}$:

if sign $l_{dD} = +$, then $p = p_{AM2}.sign\ l_{dD} > 0$ whereas $sign\ l_d = +$ and $p_{AM1} = const > 0$,

if sign $l_{dD} = -$, then $p = p_{AM2}.sign\ l_{dD} < 0$ whereas $sign\ l_d = -$ and $p_{AM2} = const > 0$.

The following shall apply to the total angular displacement of the actuator pulley shaft:

$$\beta = \frac{l_d}{r_{kl}} = \frac{abs(l_d)\cdot sign(l_d)}{r_{kl}},\tag{10}$$

$$\beta = abs(\beta)\cdot sign(\beta),\tag{11}$$

whereas the following applies to the displacement and the total length of the respective AM of the actuator:

$$l_d = abs(l_d)\cdot sign(l_d),\tag{12}$$

$$l = l_0 + l_d = l_0 + abs(l_d).sign(l_d)\tag{13}$$

additionally, the following applies to the magnitude of the actuator AM contraction k:

$$k = l_{max} - l = l_{max} - l_0 - abs(l_d)\cdot sign(l_d),\tag{14}$$

Displacement l_d, as compared to the reference point E, will correspond to the intersection of relevant AM properties. For example, in Figure 5 they are points 1–4. The figure depicts a situation in which AM1 has a constant air pressure and the pressure of AM2 changes from the maximum to the zero value. At that moment, the actuator arm reaches the position of point C, i.e., the maximum positive displacement from the reference point E. The same procedure can also be applied in the case of opposite distribution of air pressures in the AM. Then, the actuator arm reaches the position of point D, i.e., the maximum negative displacement. With changing pressure in individual muscles, the working point of the actuator position will move from the initial reference point E along the relevant property with pressure pm either to point C or to point D.

4. Results and Discussion

4.1. Creation of the Model of Static Property of PAMs in Antagonistic Connection

The calculation of the function of the course of the measured static property of the actuator was carried out using the course of one half of the static property quadrant 1 (Figure 6) [33]. A general form of the equation for the solution of the linear regression task (by the method of least squares) in the matrix form for the basic form (without expressing the error) is:

$$y = a_0 + a_1 x_1 + a_2 x_2 + \cdots + a_k x_k,\tag{15}$$

and in the matrix form

$$X = XA,\tag{16}$$

with the dimensions $Y = Nx_1$, $X = Nx(1 + k)$, $A = (1 + k)x_1$, where N is the number of measured values.

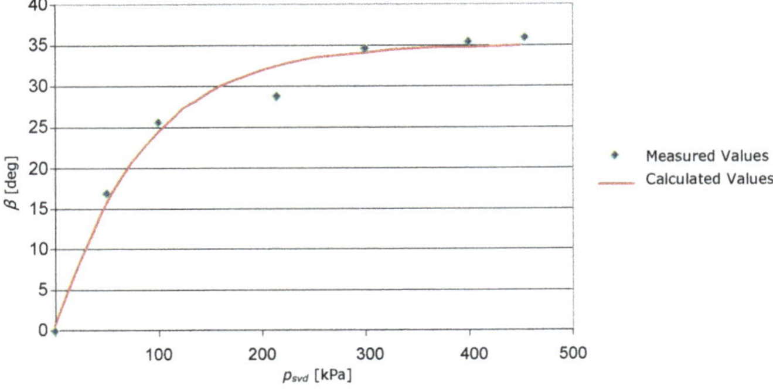

Figure 6. Approximation of the actuator static characteristic by exponential function.

The selected function form was

$$y = a_0 + a_1 e^{-x} + a_2 x e^{-x},\tag{17}$$

If in the following expression of the relationship between the actuator arm displacement at the input air pressure into the AM, according to Function (17), we introduce for the given physical parameters by the substitution:

$$y = \beta,\tag{18}$$

$$x = p_{svd} = p_1 - p_2\tag{19}$$

whereas the given parameters may have the following values:

$$\alpha = (\beta_1, \beta_2, \dots \beta_3)\tag{20}$$

$$p_{svd} = (p_1, p_2, \dots p_3)\tag{21}$$

The set of equations in the matrix form then has the following form

$$\begin{bmatrix} \beta_1 \\ \beta_2 \\ \dots \\ \beta_3 \end{bmatrix} = \begin{bmatrix} 1 & e^{-p_1} & p_1 e^{-p_1} \\ 1 & e^{-p_2} & p_2 e^{-p_2} \\ \dots & \dots & \dots \\ 1 & e^{-p_3} & p_3 e^{-p_3} \end{bmatrix} \cdot \begin{bmatrix} a_0 \\ a_1 \\ a_2 \end{bmatrix}\tag{22}$$

By solving this set of equations, the following values of coefficients were obtained for the used AMs

$$a_0 = 35.135$$
$$a_1 = -34.4443$$
$$a_2 = 5.470$$

and therefore

$$\beta(p_{svd}) = 35.135 - 34.4443 e^{-p_{svd}} + 5.470 p_{svd} e^{-p_{svd}}\tag{23}$$

As the antagonistic actuator with two AM moves, depending on the sign of pressure p_{svd} in two directions from the initial (reference) point, corresponding to the equality of maximum pressures in both muscles, the static property of such an actuator will be four-quadrant and symmetric. It is assumed that the properties of both AMs are approximately the same.

4.2. Creation of the Model of the Dynamic Model Section

The model creation was based on experimental measurements of pressures in AMs and actuator rotation at various loads and at large step changes in the input pressure. Results of measurements are listed in Table 1 and graphical responses are in Figure 7. Results of measurements at various loads and at small step changes in the input pressure are listed in Table 2 and graphical responses are in Figure 8.

Table 1. Values of pressures in MAS 20–250 and actuator rotation at large step changes of the input pressure.

Measurement No.	p_1 [kPa]	p_2 [kPa]	p_{svd} [kPa]	Load [kg]	β [Increment]	β [deg]
1	487	0	487	0	253	36.43
2	0	489	−489	0	−254.4	−36.63
3	490	0	490	1.2	255.5	36.80
4	0	464	−464	1.2	252	−36.29
5	454	0	454	2.14	250	36.00
6	0	455	−455	2.14	−245	−35.28

Table 1. *Cont.*

Measurement No.	p_1 [kPa]	p_2 [kPa]	p_{svd} [kPa]	Load [kg]	β [Increment]	β [deg]
7	216	0	216	2.14	183.3	26.40
8	0	226	−226	2.14	−192	−27.65
9	220	0	220	1.2	183.3	26.39
10	0	229	−229	1.2	−190	−27.36
11	227	0	227	0	183.3	26.39
12	0	232	−232	0	−197	−28.37

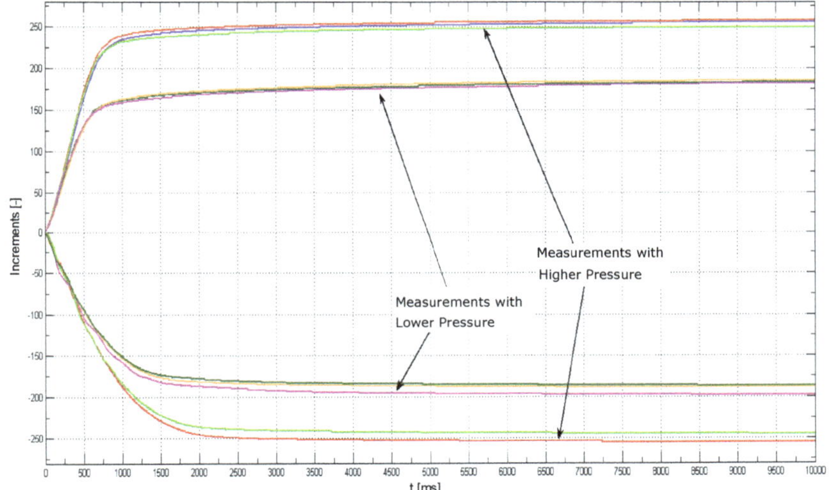

Figure 7. The actuator responses measured at large step changes of the input pressure.

Table 2. Values of pressures in MAS 20–250 and actuator rotation at small step changes of the input pressure.

Measurement No.	p_1 [kPa]	p_2 [kPa]	p_{svd} [kPa]	Load [kg]	β [Increment]	β [deg]
13	247	160	87	2.14	121.43	17.48
14	155	198	−43	2.14	−48.21	−6.94
15	190	123	67	1.2	71.69	10.32
16	121	178	−57	1.2	−56.43	−8.12
17	171	108	63	0	67.41	9.70
18	105	167	−62	0	−65.00	−9.36
19	446	339	107	0	124.99	17.99
20	332	375	−43	0	−53.57	−7.71
21	459	371	88	1.2	121.14	17.44
22	247	160	87	2.14	121.43	17.48
23	155	198	−43	2.14	−48.21	−6.94
24	190	123	67	1.2	71.69	10.32

The comparison of actuator responses at low, moderate, and high input pressure values clearly indicates that in the case of first two types of input, the responses are monotonous, whereas at low inputs and the concurrent impact of a failure, these responses contain oscillations prior to the stabilisation. They are probably caused by the effects of the failure gravitation with a constant effect acting in the same direction causing, together with the air compressibility in the AM, the formation of a short-term damped marginal cycle prior to the stabilisation of the moment equilibrium on the actuator axis. Therefore, the determination of the mathematical model of the actuator dynamics was carried out while considering mainly the responses at great input changes, without the effects of failures; however, a general expression of the model must contain also the properties manifested at small inputs and failures.

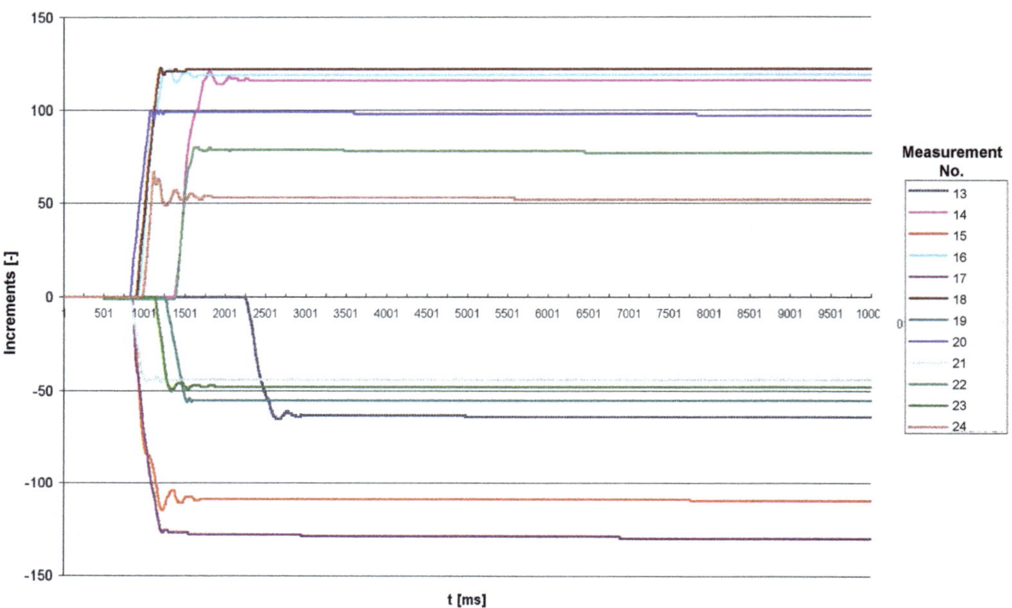

Figure 8. The actuator responses measured at small step changes of the input pressure.

The resulting transfer function would have the following form:

$$G_1(s) = \frac{K_p}{Ts + 1} \text{ or } G_2(s) = \frac{K_p}{(T_1 s + 1)(T_2 s + 1)} \tag{24}$$

The model might be identified while applying the method of least squares (in a discrete form), provided that the differential equation is as follows:

$$y(t) = -a_1 y(t-1) - \cdots - a_n y(t-n) + b_1 u(t-1) + \cdots + b_m u(t-m) \tag{25}$$

where $y(t)$ is the output from the system, $a_1, \ldots, a_n$ are the output coefficients, $u(t)$ is the input to the system, and $b_1, \ldots, b_m$ are input coefficients. The vector form is then as follows

$$\theta = \begin{bmatrix} a_1 & \cdots & a_n & b_1 & \cdots & b_m \end{bmatrix}^T$$
$$\phi(t) = \begin{bmatrix} -y(t-1) & \cdots & -y(t-n) & u(t-1) & \cdots & u(t-m) \end{bmatrix}^T \tag{26}$$

where θ represents a *parametric vector* and $\phi(t)$ represents a *regression vector*. The task is to minimise the argument of the function $V_N(\theta, Z^N)$ by identifying relevant values of coefficients if $Z^N = \{u(1), y(1), \cdots, u(N), y(N)\}$ (i.e., N input–output pairs), while applying the known procedure.

Then, we obtain

$$V_N(\theta, Z^N) = \frac{1}{N} \sum_{t=1}^{N} (y(t) - \hat{y}(t|\theta))^2 = \frac{1}{N} \sum_{t=1}^{N} (y(t) - \phi^T(t)\theta)^2 \tag{27}$$

$$\hat{\theta}_N = \underset{\theta}{\mathrm{argmin}}\, V_N\left(\theta, Z^N\right) \tag{28}$$

If we set the derivation $V_N\left(\theta, Z^N\right)$ as equal to zero and after editing, we obtain an estimation of the parametric vector

$$\hat{\theta} = \left[\sum_{t=1}^{N} \phi(t)\phi^T(t)\right]^{-1} \sum_{t=1}^{N} \phi(t)y(t). \tag{29}$$

Then, the actuator can be described by the first order transfer function

$$G_1(s) = \frac{1}{0.64595s + 1}. \tag{30}$$

or by the second order transfer function

$$G_2(s) = \frac{1}{(0.090314s + 1)(0.57873s + 1)} = \frac{1}{0.052267s^2 + 0.669s + 1} \tag{31}$$

Responses of the linear model of the first and second orders after the identification are presented in Figure 9.

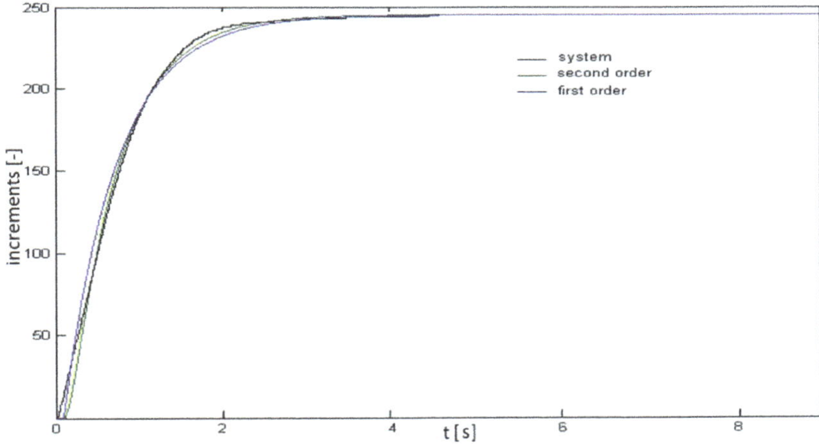

Figure 9. Responses of the linear model of the first and second orders after the identification.

The actuator consisting of pneumatic artificial muscles in an antagonistic arrangement has, in each half of the angular displacement of the pulley shaft, regulated air pressure only in one artificial muscle. The other artificial muscle has a constant pressure and serves as a non-linear and pneumatic spring. Such an actuator is a non-linear system with the end position being a non-linear centrally symmetrical function of the filling air pressure in artificial muscles.

With decreasing pressure in the active AM, the angular displacement of the actuator pulley shaft increases along the curve with a decreasing inclination. The static property of such a system proves by its course that actuator strengthening depends on the arm's position which depends on pressures and forces in individual artificial muscles. Actuator configuration, together with a simple control system, facilitates the construction of a rather simple positional servo system with adequate control requirements and execution costs.

Dependencies of the actuator end position represent a non-linear function of the filling air pressure in the artificial muscle and their courses are different at different values of actuator load. This follows from non-linear properties of artificial muscles. It is also necessary to respect a non-identical tensile force (or torque) of the actuator, the magnitude of which is changing with the value of angular displacement of the actuator pulley shaft (or artificial muscle contraction or displacement). This property also causes non-identical mechanism stiffness at different position values, whereas the highest and two-side symmetrical stiffness is reached in the reference point, when the pressures in the AMs are identical and reach the maximum.

5. Conclusions

The ability of artificial muscles to work in antagonistic connection facilitates the regulation of their own stiffness, thus providing a number of benefits in individual applications of actuators based on pneumatic artificial muscles. It offers the possibility to create various concepts for the control of systems comprising pneumatic artificial muscles and the use thereof in industry, under an assumption of minimum contact between the manipulation equipment and staff.

Mathematical modelling of PAM is important for the simulation of the dynamics of the movements of pneumatic actuators with artificial muscles, applied in the designing and execution stages, as well as for the creation of algorithms for the control of this type of actuators. In the creation of a mathematical model, it is important to know the geometrical properties of PAM and physical phenomena running inside the muscle. A significant role in PAM modelling is also played by the flexibility of materials used within the muscle assembly. Additionally, in this type of muscle, there is dry friction between the rubber tube and the braid; this indicates that PAM is regarded as a non-linear component with the insensitivity zone and hysteresis. Due to incomplete knowledge of all physical phenomena running inside the PAM, it is not possible to construct a perfect mathematical model thereof. It is therefore necessary to design deterministically a mathematical model of the lowest possible sensitivity to uncertainties, able to also provide the required responses in the event of changed parameters and failures.

Besides overcoming the forces from the load, the active muscle must also overcome the variable directive force of the passive muscle. That is why the requirements put on nominal parameters (tensile force) of artificial muscles are higher than those that would only be based on the load size. With regard to the need of appropriate mechanism stiffness, the forces of the actuator muscles are expected to be higher than the (maximum) load force.

Author Contributions: Conceptualization, T.Č.; methodology, D.M. (Darina Matisková), A.B. and D.M. (Daniela Marasová); validation, T.Č., A.B. and D.M. (Darina Matisková), formal analysis, D.M. (Daniela Marasová); investigation, A.B., T.Č. and D.M. (Darina Matisková); resources, D.M. (Daniela Marasová); data curation, T.Č. and A.B.; writing—original draft preparation, D.M. (Daniela Marasová); writing—review and editing, D.M. (Daniela Marasová) and A.B.; visualization, D.M. (Daniela Marasová) and A.B.; supervision, D.M. (Darina Matisková); project administration, D.M. (Darina Matisková) and D.M. (Daniela Marasová); funding acquisition, D.M. (Darina Matisková). All authors have read and agreed to the published version of the manuscript.

Funding: This research was funded by NATIONAL RESEARCH AGENCY OF SLOVAKIA, grant number VEGA 1/0393/18Research of Methods for Modeling and Compensation of Hysteresis in Pneumatic Artificial Muscles and PAM-actuated Mechanisms to Improve the Control Performance Using Computational Intelligence.

Institutional Review Board Statement: Not applicable.

Informed Consent Statement: Not applicable.

Data Availability Statement: Not applicable.

Acknowledgments: The article was written as a result of the successful solving of the Project of the Structural Funds of the EU, ITMS code: 26220220103.

Conflicts of Interest: The authors declare no conflict of interest.

References

1. Čujan, Z.; Marasová, D. Evaluation of the Logistic process robotisation using the multiple-criteria decision-making. *TEM J.* **2018**, *7*, 501–506.
2. Balara, M.; Dupláková, D.; Matisková, D. Application of a signal averaging device in robotics. *Measurement* **2018**, *115*, 125–132. [CrossRef]
3. Dorf, C.R. *Modern Control Systems*, 3rd ed.; Addison-Wesley Longman: Boston, MA, USA, 1980; p. 425.
4. Plettenburg, D.H. Pneumatic actuators: Comparison of energy-to-mass ratio's. In Proceedings of the 9th International Conference on Rehabilitation Robotics, Chicago, IL, USA, 28 June–1 July 2005; pp. 545–549.
5. Ramasamy, R.; Juhari, M.R.; Mamat, M.R.; Yaacob, S.; Mohd Nasir, N.F.; Sugisaka, M. An application of finite element modelling to pneumatic artificial muscle. *Am. J. Appl. Sci.* **2005**, *2*, 1504–1508. [CrossRef]
6. Kerscher, T.; Albiez, J.; Zollner, J.M.; Dillman, R.F. Dynamic modeling of fluidic muscles using quick-release. In Proceedings of the 3rd International Symposium on Adaptive Motion in Animals and Machines, Ilmenau, Germany, 25–30 September 2005; pp. 1–6.
7. Šitum, Ž. Pneumatic muscle as an actuator. *Znan. Pop. Časopis Sustavi.* **2009**, *3*, 54–60. (In Slovenian)
8. Hosovsky, A.; Trojanova, M.; Piteľ, J.; Svetlik, J. Identification of DMSP-5 fluidic muscle dynamics using hammerstein model. In Proceedings of the IEEE 12th International Symposium on Applied Computational Intelligence and Informatics, Timisoara, Romania, 17–19 May 2018; pp. 319–324.
9. Winters, J.M.; Savio, L.-Y.S. *Multiple Muscle Systems: Biomechanics and Movement Organization*; Springer: New York, NY, USA, 1990; ISBN 978-1-4613-9030-5.
10. Yarlott, J.M. Fluid Actuator. US Patent No. 3645 173, 1972.
11. Kukolj, M. Axially Contractable Actuator. US Patent US4733603A, 29 March 1988.
12. Immega, G.; Kukolj, M. Axially Contractable Actuator. US Patent No. US4939982A, 10 July 1990.
13. Sarosi, J.; Pitel, J.; Tothova, M.; Hosovsky, A.; Biro, I. Comparative survey of various static and dynamic models of pneumatic artificial muscles. *Trans. Can. Soc. Mech. Eng.* **2017**, *41*, 825–844. [CrossRef]
14. Verrelst, B.; Daerden, F.; Lefeber, D.; Van Damme, M.; Vanderborght, B.; Van Ham, R.; Vermeulen, J. Pleated pneumatic artificial muscles for robotic application. In Proceedings of the Industry-Ready Innovative Research, 1st Flanders Engineering PhD Symposium, Brussels, Belgium, 11 December 2003; p. Mech11.
15. Verrelst, B.; Daerden, F.; Lefeber, D.; Van Ham, R.; Fabr, T. Introducing Pleated pneumatic artificial muscles for the actuation of legged robots: A one-dimensional set-up. In Proceedings of the 3rd International Conference on Climbing and Walking Robots, Madrid, Spain, 2–4 October 2000; pp. 583–590.
16. Ambrisko, L.; Cehlar, M.; Marasova, D. The rate of stable crack growth (scg) in automotive steels sheets. *Metalurgija* **2017**, *56*, 396–398.
17. Ambrisko, Ľ.; Pesek, L. The stretch zone of automotive steel sheets. *Sadhana* **2014**, *39*, 525–530. [CrossRef]
18. Villegas, D.; Michaël Van Damme, M.; Vanderborght, B.; Beyl, P.; Lefeber, D. Third–generation pleated pneumatic artificial muscles for robotic applications: Development and comparison with McKibben muscle. *Adv. Rob.* **2012**, *26*, 1–23. [CrossRef]
19. Davis, S.; Caldwell, D.G. Braid effects on contractile range and friction modeling in pneumatic muscle actuators. *Int. J. Rob. Res.* **2006**, *25*, 359–369. [CrossRef]
20. Serres, J.L. Dynamic Characterization of a Pneumatic Muscle Actuator and Its Application to a Resistive Training Device. Ph.D. Thesis, Wright State University, Dayton, OH, USA, 2008; p. 201.
21. Mizakova, J.; Piteľ, J.; Tothova, M. Pneumatic artificial muscle as actuator in mechatronic system. *Appl. Mech. Mater.* **2014**, *460*, 81–90. [CrossRef]
22. Wickramatunge, K.C.; Leephakpreeda, T. Study on mechanical behaviors of pneumatic artificial muscle. *Int. J. Eng. Sci.* **2009**, *48*, 188–198. [CrossRef]
23. Trojanová, M.; Čakurda, T. Validation of the dynamic model of the planar robotic arm with using gravity test. *MM Sci. J.* **2020**, *5*, 4210–4215. [CrossRef]
24. Hošovský, A.; Piteľ, J.; Židek, K.; Tóthová, M.; Sárosi, J.; Cveticanin, L. Dynamic characterization and simulation of two-link soft robot arm with pneumatic muscles. *Mech. Mach. Theory* **2016**, *103*, 98–116. [CrossRef]
25. Tóthová, M.; Piteľ, J. Testing of two types of membership functions in fuzzy adaptive controller of pneumatic muscle actuator. In Proceedings of the 11th IEEE International Symposium on Applied Computational Intelligence and Informatics SACI 2016, Timişoara, Romania, 12–14 May 2016; pp. 497–502.
26. Sárosi, J.; Piteľ, J.; Šeminský, J. Static force model-based stiffness model for pneumatic muscle actuators. *Int. J. Eng. Res. Af.* **2015**, *18*, 207–214. [CrossRef]
27. Trojanová, M.; Čakurda, T.; Hošovský, A.; Krenický, T. Estimationof grey-box dynamic model of2-DOF pneumatic actuator robotic arm using gravity tests. *Appl. Sci.* **2021**, *11*, 4490. [CrossRef]
28. Nezhad, M.N.; Korayem, A.H. Dynamic modeling of industrial manipulator Hyundai HS165 in order to determine the dynamic load-carrying capacity for a specified trajectory. In Proceedings of the 7th International Conference on Robotics and Mechatronics (ICRoM), Teheran, Iran, 20–21 November 2019; pp. 538–543.
29. Al-Qahtani, H.M.; Mohammed, A.A.; Sunar, M. Dynamics and control of a robotic arm having four links. *Arab. J. Sci. Eng.* **2017**, *42*, 1841–1852. [CrossRef]

30. Singh, S.; Singla, A.; Singh, A.; Soni, S.; Verma, S. Kinematic modelling of a five-DOFs spatial manipulator used in robot-assisted surgery. *Perspect. Sci.* **2016**, *8*, 550–553. [CrossRef]
31. Sung, H.; Lee, J.H.; Lee, C.; Seo, T.; Lee, J.W. Geometrical kinematic solution of serial spatial manipulators using screw theory. *Mech. Mach. Theory* **2017**, *116*, 404–418.
32. Zidek, K.; Vasek, V.; Pitel, J.; Hošovsky, A. Auxiliary device for accurate measurement by the smartvision system. *MM Sci. J.* **2018**, *1*, 2136–2139. [CrossRef]
33. Balara, A. The contribution to the enhancement of the computer aided methods of technology plant actuators movement. *Diss. Work* **2011**, 149. (In Slovak)

Review

Automation and Robotization of Underground Mining in Poland

Łukasz Bołoz [1,*] **and Witold Biały** [2]

[1] Department of Machinery Engineering and Transport, Faculty of Mechanical Engineering and Robotics, AGH University of Science and Technology, A. Mickiewicza Av. 30, 30-059 Krakow, Poland
[2] Department of Production Engineering, Faculty of Management and Organization, Silesian University of Technology, Roosevelta 26, 44-800 Zabrze, Poland; witold.bialy@polsl.pl
* Correspondence: boloz@agh.edu.pl; Tel.: +48-12-617-30-81

Received: 11 August 2020; Accepted: 21 September 2020; Published: 16 October 2020

Abstract: The article concerns the condition of automation and robotization of underground mining in Poland. Attention has been focused on the specific character of the mining industry. This limits the possibility of using robotization, and sometimes even the mechanization of certain processes. In recent years, robotic and automated machines and machine system solutions have been developed and applied in Poland. They are autonomous to a various degree, depending on the branch. The type of automation and artificial intelligence depends on the specific use. Some examples presently being used include the MIKRUS automated longwall system and autonomous device(s) for breaking rocks or mining rescue work. In Poland, fully automated plow systems produced by foreign companies are also used. Companies in Poland and international research centers are also actively engaged in the development of underwater and space mining. where robotization is of key importance. Research is also being undertaken by Robotics in Mining, euRobotics and PERASPERA as well as Space Mining Conference.

Keywords: underground robots; automation in mines; remote control; autonomous underground machinery

---

1. Introduction

Underground mining exploits various useful minerals. The exploitation and transport of minerals are mostly mechanized. Some machines are automated and, to varying degrees, autonomous [1,2]. In Poland, the largest market for underground mining machinery is the mining of hard coal. This is followed by mining for copper ores and other raw materials, such as rock salt or zinc and lead ores. The share of mining for these other raw materials is small. Difficult conditions of machine operation have resulted in noticeable changes in the approach to safety and comfort at work. This is the reason for seeking a method to eliminate or reduce human participation in mining operations. The progress in mobile control and navigation techniques has led to an interest in automation and robotization. Measurable economic benefits have been obtained by using such solutions. Another important consideration is increasing the safety and comfort of underground workers. Robotization can contribute to enhanced safety when withdrawing people from the most threatened zones. These zones are where mining usually takes place. A separate and increasing need is using robots in underground rescue operations [3].

Coal mines are difficult workplaces, not only for humans, but also for machines. Natural hazards, limited space, a lack of natural light, dustiness, humidity, high temperature and mine atmosphere hinder mechanization. Despite these adversities, Poland boasts many solutions developed in national research centers in cooperation with machine manufacturers. The state of underground mining automation and robotization in Poland will be presented through several selected examples.

2. Solutions Applied to Mining in Poland

In Poland, various solutions for mining machines and machine systems with varying degrees of autonomy are designed and manufactured. Several such examples will be briefly described.

2.1. MIKRUS Automated Longwall System

Raw materials found in coal seams are exploited by means of mechanized longwall systems [4]. Longwall systems are systems of compatible and cooperating machines that simultaneously carry out the process of mining (loading and hauling of mined rock) as well as securing the roof. The extraction from ever-thinner seams causes working conditions which make manual control difficult [5,6]. Therefore, automated longwall systems need to be developed. The American company Caterpillar Inc. is the leader in the field of automatic plowing systems. Their most advanced product, GH1600, is used by two coal companies in Poland. A competitive longwall technique is the shearer technology, which has been fully automated for use in thin seams. The KOPEX S.A. group (Katowice, Poland) has developed and constructed a longwall system for thin seams. Currently, the solution belongs to the FAMUR S.A group. The system, called MIKRUS ("midget" in Polish), is equipped with a GUŁ-500 (stands for Polish words głowica urabiająco-ładująca) cutting and loading head with two cutting drums. GUL-500 is moved by a longwall conveyor along the coal side wall by means of a cable system under the powered roof support units (Figure 1).

Figure 1. MIKRUS system produced by FAMUR S.A.

The whole complex is powered and controlled by an integrated system located on the surface of the mine (Figure 2). In addition, in the event of a failure, the complex is equipped with a central desktop at the operator's station, which is located in the haulage heading. Trouble-free, automatic and optimal operation is ensured by the EH-WallControl (Elgór + Hansen WallControl) automation system. Information on the work of all longwall face machines is supplied to the automation system. Based on this information, the system generates signals to the control systems of machines (shearer, conveyors, support units, pumps). In the event that information about hazards or pre-emergency conditions appears in individual devices, the system signals it at the operator's workstation. If the permissible operating parameters are exceeded or pre-emergency conditions occur, the system will turn off individual devices or will stop the entire system. For security reasons, the operator can start up individual devices even if there is a risk of failure. Such a status is separately and clearly recorded.

All the data about the operation of longwall equipment are available in particular menus called on the operator's command and are visualized at his workstation. In the automatic work cycle, the operator only controls the speed of the loading cutterhead feed, while the work of other devices is controlled by the master system of the longwall automation complex. The operator can switch over to manual control and change the operating parameters of the devices at any time. The combination of the features of plow and longwall shearer systems in the MIKRUS complex translates into increased operational efficiency in low faces. Moreover, the use of advanced control and diagnostic systems has enabled the construction of a fully automated longwall complex. The MIKRUS longwall system for mining thin seams is a unique solution on a global scale. The MIKRUS longwall system allows for profitable and safe extraction in thin coal seams. Innovative solutions based on automation contribute to employees' comfort [7,8].

 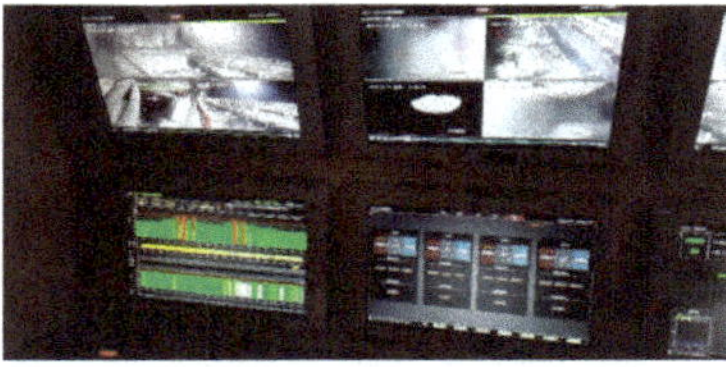

Figure 2. Cabin for monitoring the automatic operation of the MIKRUS system from the surface [8].

2.2. Mine Master Self-Propelled Mining Machines

In ore mines, such as KGHM Polska Miedź S.A. (Lubin, Poland), the useful mineral is usually mined by means of explosives. Self-propelled drilling machines are used for harsh operating conditions, characterized by an ambient temperature exceeding 35° Celsius and air humidity up to 98%. The machines are operated at depths ranging from 600 m to 1200 m, where there are strong saline watercourses, which force the use of subassemblies with high IP 67 (International Protection). In such extreme conditions, the human factor is increasingly often an unreliable element, so a chance to improve work efficiency is seen in the automation and monitoring of mining processes. Mine Master, in cooperation with the AGH University of Science and Technology, based on joint experiences in drilling process automation, designed two modular drilling machine monitoring systems. The system was implemented jointly with the machine user—KGHM Polska Miedź—which works on selected Face Master 1.7 machines [9–11]. Monitoring of drilling machines was first introduced for systems that have a decisive impact on achieving the assumed productivity by machines. In the case of Face Master 1.7 drill rigs, it was a system for monitoring the drilling parameters and drilling frame settings in accordance with the assumed blasting pattern (Figure 3). The natural consequence of monitoring was the implementation of systems for diagnosing the drive system and the hydraulic system of machines. Monitoring has control over the machine, not only during the process of drilling, but also when traveling to and from the workplace on roads with a slope of up to 15°.

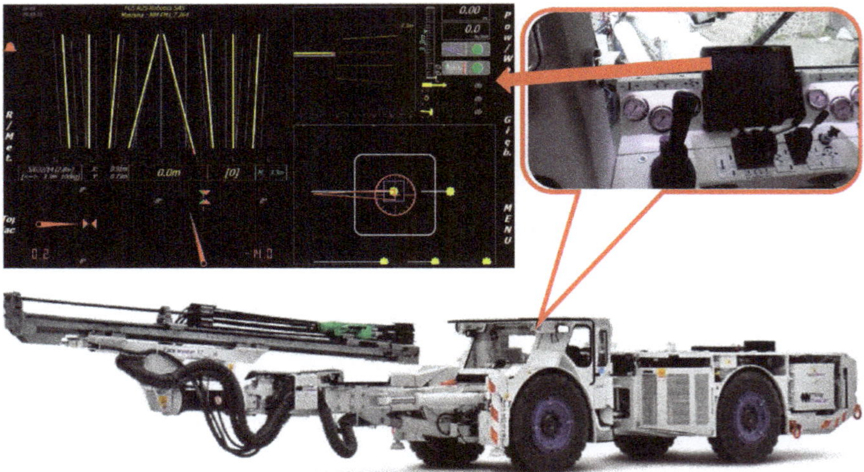

Figure 3. Face Master 1.7 machine with a control panel and a view of the Feeder Guide System (FGS) drill rig positioning system [9].

Monitoring systems for drilling machines used in underground conditions of KGHM Polska Miedź, which successfully completed their tasks, include the following scope:

- Mapping of the blast pattern on the heading face and its visualization on the operator's panel;
- Repeatability of the drilling process in the face geometry;
- Monitoring and control of drilling parameters;
- Quick diagnosis of the drilling system;
- Increased work efficiency;
- Recording of work done;
- USB or WiFi data transfer;
- Generating reports on the work performed (Figure 4);
- Work "in the background" of systems so that their possible failure will not limit the basic functions of the machines;
- Increasing the availability of machines through more precise and faster damage diagnostics.

The FGS system (Feeder Guide System) is the successor of the DMS system (Drilling Monitoring System). DMS monitored only drilling parameters, such as speed, depth or pressure in the hydraulic system.

The basic element of the system is the operator's panel, which contains software that controls the drilling support process. The operator's panel is resistant to operating conditions and has a high-contrast display with built-in function buttons or a touch screen facilitating correct readings. The operator's panel is additionally equipped with an integrated USB port, which enables data transfer. Another element necessary for the proper operation of the FGS system is sensors, the basic task of which is to determine the drill rig's position between the positioning and measuring elements. Thus, it is possible to reflect the position of the drilling frame in the system and the control panel. The software allows employees to create 3D blasting patterns, including the preparation of previous patterns and their uploading via a USB flash drive. The FGS system allows obtaining one plane of the face regardless of the initial state. The length of individual holes is calculated by the system in such a way that they end in one plane. Obtaining this effect with manual control is not possible. The program for creating blasting patterns enables changing the direction of the excavation. The result is the program choosing the appropriate length of the holes. The FGS system in drilling machines has contributed to the improvement of drill rigs performance in the area where tests were carried out at Polkowice-Sieroszowice mine, KGHM Polska Miedź S.A. The FGS system has also enhanced work safety for persons staying in the faces. This repeatability in the drilling process results in the proper geometry of the face.

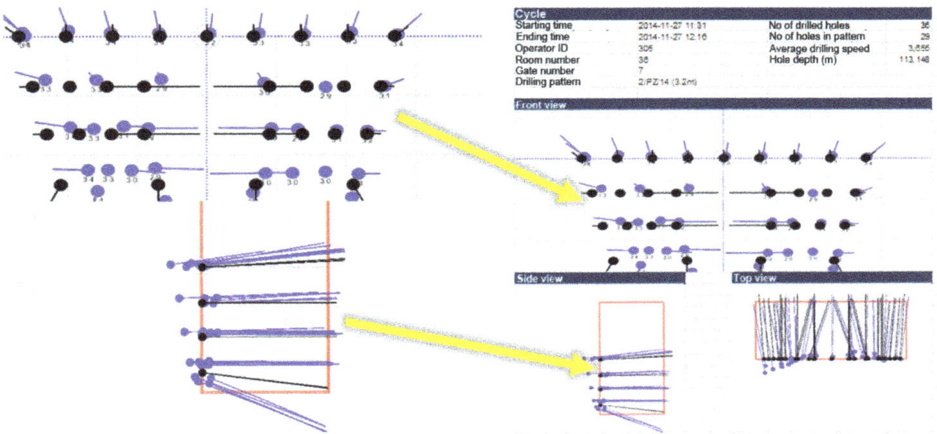

Figure 4. FGS report on drilling with a comparison of the assumed (**black**) with actually completed blasting pattern (**blue**) [9].

2.3. Famur Roadheader

Making drifts in underground mines is one of the most important processes. In most cases, headings are made with roadheaders controlled locally by the operator. This work is dangerous due to falling rocks, dust, noise and temperature. It was, therefore, decided to withdraw the staff from the area of shearer operation and introduce systems for supervising and automating its work. Elgór-Hanses S.A., which is part of the FAMUR S.A. group(Katowice, Poland), developed and implemented a shearer remote control system. EH-RemoteHeadControl v2 is a system enabling the remote control of roadheaders designed for operation in particularly dangerous zones due to the risk of gas and rock outburst as well as rock burst hazards. The system allows the service staff to work in safe work conditions, as during the shearer operation, they stay in a non-hazardous area. The safety of underground miners is the most important issue of all matters related to the operation of underground mining plants [8].

The system consists of a few elements (Figure 5).

- Up to ca 50 m–100 m behind the shearer, a box with a laser shining parallel to the axis of the excavation is mounted on the roof. The laser position is corrected using a wireless remote control.
- The shearer has a set of devices for automatic positioning of the shearer in the heading. The devices include a set of cameras, sensors and a signal processing controller. A radar is used to communicate with the laser.
- In the safe zone outside the directly threatened area is an operator's workstation in the form of a cage with monitors with a remote-control panel. The operator's workstation is in the form of a cage with monitors and a remote-control panel located in a safe zone outside the directly threatened area.
- Diagnostic of head tools wear [12,13].

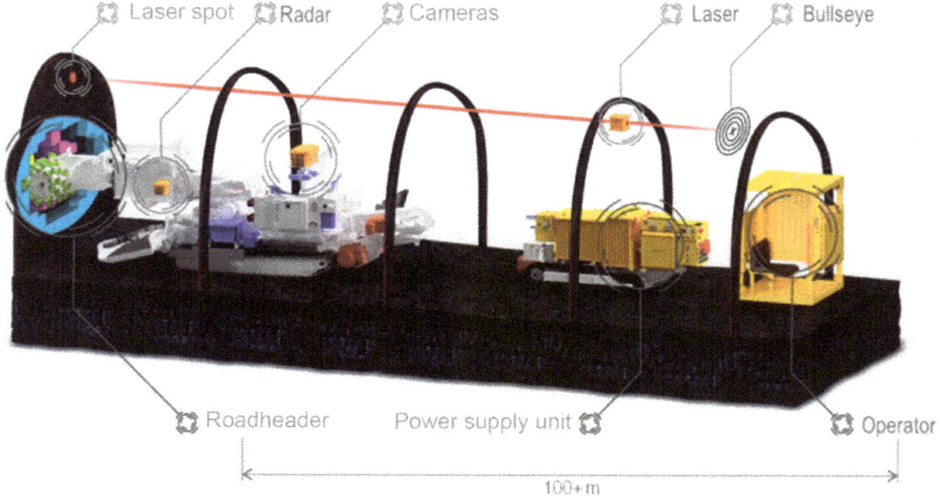

Figure 5. Structure of the EH-RemoteHeadControl v2 system by Elgór + Hansen [8].

The operator's station is equipped with computers presenting diagnostic information and the image from the cameras installed on the shearer as well as visualizing the operation of the machine. The real-time graphic representation of the cutting head makes it easier for the operator to drill the right profile. Audio and video are transmitted from three cameras located in the shearer's surroundings. This allows the operator to better understand the current conditions of the machine. The laser is installed in the excavation every 100 m and gives a beam of light in two directions. The laser is set using a wireless remote control and/or PC application.

The data recorded by the laser sensor, radar sensor and cameras allow the visualization station to determine the position of the shearer and the head. It also checks the possibility of their movement on an ongoing basis and determining a possible collision. The 3D application additionally maps the state of cutting in the face. It uses various indicators and lights and provides information on the shearer's condition, including the signaling of warnings and emergencies (Figure 6). Data transmission to the surface of the mine enables generating reports on the operation of the system.

Figure 6. 3D application of the EH-RemoteHeadControl v2 system by Elgór + Hansen.

3. Research and Development in the Field of Robotization of Mining in Poland

Research centers in Poland are engaged in works in the field of mining robotization, which concern not only traditional mining, but also space and underwater mining. The effects of these works have, in most cases, been tested in underground mine conditions, and some of them have already been implemented.

3.1. Autonomous Machine for Breaking Rocks

The most important elements of the ore mining process have been presented earlier. In this process, the mined rock from the face is transported to the surface. The transport system includes non-continuous haulage to handling points, also referred to as transfer points, where the mined rock is loaded onto conveyor belts. Large lumps of mined rock can damage the conveyor, so the handling points are fitted with grates mounted by means of manipulators with hydraulic hammers. The operator manipulates the boom to clear the grate of mined rock in the quickest possible way. The operator is exposed to many adverse factors. First, a hammer remote control followed by an autonomous machine were developed and installed. The project was implemented by the consortium composed of KGHM ZANAM S.A., AGH University of Science and Technology and KGHM CUPRUM Sp. z o. o. (Wroclaw, Poland). Research and Development Centre [14–16]. The main goal of the project was to develop an automatic grate cleaning system.

Constructing an autonomous machine required equipping it with an automatic control system, which, based on information from a number of sensors, would react to changes in the environment and generate control signals in such a way that the main goal, i.e., the cleaning of the grate, could be achieved. The main element was a hydraulic boom with a hydraulic hammer attached at the end. The boom has 4 degrees of freedom. An analysis of the functioning of the existing manual system was carried out and consultations with future users of the transfer points were held. The conducted analyses enabled specifying the tasks to be performed by the machine. the most important restrictions and requirements formulated in the form of the following assumptions.

The control system should operate in an autonomous manner.

- The operator can take over control at any time and turn off the autonomous system.
- At least 80% of activities related to rock crushing will be performed by the machine automatically.
- The operator interfering with the situation on the grate 20% of the time.
- The process of clearing the grate of mined rock will last up to several minutes.
- Constructional changes must be as small as possible in relation to the existing equipment.

The system is, among other things, equipped with hydraulic cylinder sensors and a mined rock heap scanner. The system was installed at the test stand at KGHM ZANAM S.A. (Figure 7).

The stand was designed and constructed in a manner enabling its use in the underground mines of KGHM Polska Miedź S.A. The device should be adapted to work in the following conditions:

- Relative humidity up to 95% at a temperature of up to + 40 °C
- Degree of corrosive aggressiveness—C according to PN-71/H-04651
- Maximum relative humidity at a temperature of + 25 °C or at lower temperatures with 100% steam condensation.

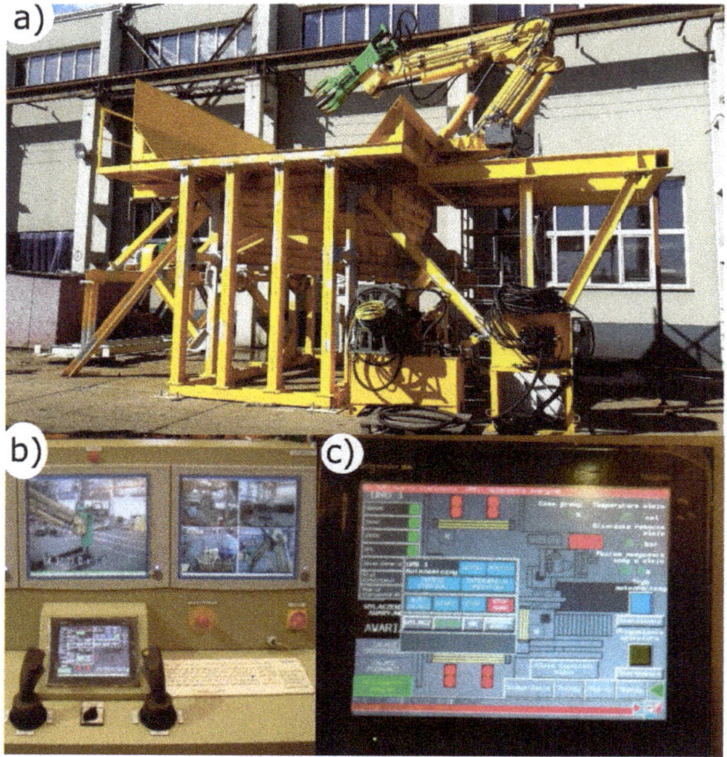

Figure 7. Station for testing; (**a**) autonomous equipment; (**b**) operator's panel; (**c**) GUI (Graphical User Interface) view.

The remote-controlled stand is monitored by cameras. It has been equipped with optical barriers and signaling devices to supervise and ensure traffic safety in the vicinity of the grate. The automation system controls the boom by means of signals determined by the current state of the system and the setpoint. This system consists of a master control layer and a direct control layer. In the superior layer, after identifying the shape and dimensions of the mined rock heap is identified (Figure 8). The control system generates a trajectory that will be implemented by the direct control system. The trajectory was developed using heuristic algorithms and observations of the grate cleaning process. The automatic control system consists of four main modules: mined rock identification, determination of hammer motion trajectory, inverse kinematics and direct control. After being equipped with a shape identification assembly, the device becomes a robot, which autonomously detects the position of mined rock on the grate and, then, starts removing it.

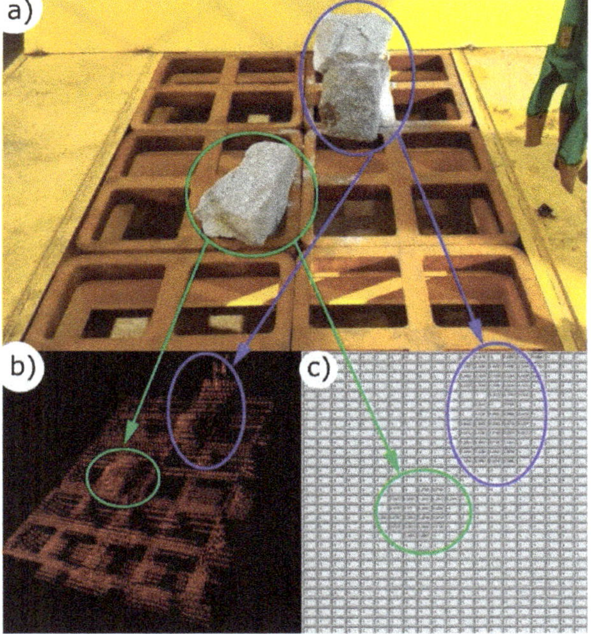

Figure 8. Analysis of the situation on the grate by the system: (**a**) view of mined rock on the grate; (**b**) view of the scan; (**c**) table of mined rock height on the grate [15].

This autonomous robot system for breaking rocks performs tasks in accordance with the assumptions. The systems of mined rock identification, trajectory determination and implementation work well. Research has shown that the system can clear the grate of large chunks of mined rock in a satisfactory manner [15]. The next step was to mount the system on a real grate in a mine. The works were completed successfully.

3.2. Research Projects and Prototype Solutions

Below we present some examples of projects and solutions that significantly contribute to increasing the level of mining robotization and pave the way for the future mine, which is to be fully robotic.

An interesting project has been implemented by the Space Research Centre of the Polish Academy of Sciences and the AGH University of Science and Technology—"Development of a model of an automatic core drilling rig for work in extreme conditions, in particular in the space environment." The aim of the project is to design, construct and test a model of an automatic core drilling rig, the task of which is the unmanned sampling of material from a depth of several meters. The expected operating environments of the machine are both hard-to-reach places on Earth as well as the surface of planets and asteroids. Specific requirements are related to available power and mass. In addition, work under vacuum causes difficulty. However, it can fill the market niche, especially in the context of Poland's accession to the European Space Agency [17].

Another interesting example of a project is the automation of the shaft support control and monitoring system. This solution was developed by the Central Mining Institute. The laser system of automatic geometry measurement has been tested in terms of the recording, visualization and signaling of emergencies. This system increases the operational safety of shaft equipment and allows for continuous supervision at lower costs than periodically performed classic geodetic measurements [18].

In the introduction, attention is drawn to threats occurring in underground mines, which are the main reason for interest in the robotics of mining works. However, the problem does not only concern regular, typical mining works, but also rescue operations. Robots equipped with sensors for measuring concentrations of dangerous gases and climatic conditions should also participate in rescue operations. Robots provide reconnaissance for rescue teams giving them advance information on the conditions prevailing in the excavation. Therefore, robots should ensure the greater safety of people. This necessity is recognized all over the world, as evidenced by the multitude of construction solutions of mining robots from various countries. Groundhog, Wolvarine V-2, Gemini-Scout robots developed in the US, Numbat and the Water Corporation robots constructed in Australia, or Telerescuer, implemented by an international consortium, are examples of some robots. Tangshan Kaicheng Electronic from China offers robots for the hard coal mining industry. As part of the project "Research and feasibility study of a model of the M1 category mobile inspection platform with electric drives for potentially explosive areas", implemented by the consortium of the Institute of Innovative Techniques EMAG and the Industrial Institute of Automation and Measurements PIAP, the Mobile Inspection Platform (MPI) was developed. The most important functionality of MPI is to measure the concentration and climate parameters of the mine atmosphere on a continuous basis or at the request of the operator. MPI also sends the measurement results to the control and measurement console. The measurement results are archived together with images from cameras operating in the visible and infrared band.

The MPI's place of operation is a potentially explosive zone. So, since the very beginning, the robot has been intended to work as a machine that meets the requirements of Directive 94/9/EC (ATEX), as well as the Machinery Directive 2006/42/EC (MD) and Directive 2004/108/EC(EMC). The robot has been designed to overcome various obstacles, such as debris, water, mud and mining floor railways [19]. The above-mentioned units have also developed the Mining Mobile Inspection Robot (GMRI) [20]. Figure 9 shows both MPI and GMRI mining inspection robots. These solutions are the first step for robots that will not only be able to reach casualties, but will also carry out basic activities related to the protection and removal of people from endangered areas.

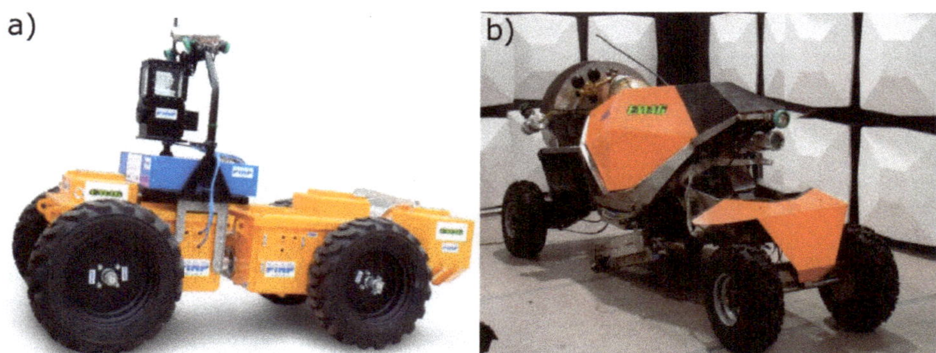

Figure 9. Mining inspection robots: (**a**) MPI [19], (**b**) GMRI [20].

4. Activities Promoting Robotization in Mining

Poland is actively engaged in the development of robotization in underground and surface mining as well as in solutions for underwater and space mining. In addition to the three selected initiatives described in Sections 4.1–4.3, meetings, symposia and conferences are organized in Poland. An important direction of the development of robotization is also the Polish consortium, EX-PL, as well as the Centre for Space Studies of the Kozminski University.

4.1. euRobotics Topic Group on Mining

The well-known European association euRobotics has over 250 members. The aim of euRobotics is to increase European research, development and innovations in the field of robotics. euRobotics also supports its positive perception. The AGH University of Science and Technology leads the "Mining" group, one of 30 thematic groups in euRobotics. Thematic groups develop the content of the Strategic Research Agenda (SRA) and Multi-Annual Roadmap (MAR). Both SRA and MAR distributed by SPARC connects euRobotics with the European Commission in a public-private partnership. The SPARC partnership is the world's largest publicly funded robotics innovation program. Thematic groups (TG) identify the current challenges of their domain and describe the required progress in the capabilities of robots needed to meet these challenges. By connecting research, industry and end-users, TG can provide knowledge about the potential effects of robotics progress and enable knowledge transfer among the shareholders.

In 2014, AGH University of Science and Technology in Cracow took the initiative to create a new working group in the structures of the euRobotics association, in the field of mining (Robotics in Mining). The idea of this project was conceived during the First International MARG Conference. (MARG stands for Mechanizacja, Automatyzacja i Robotyzacja w Górnictwie). In March 2015, during the European Robotics Forum, TG Mining was officially appointed. The AGH University of Science and Technology in Cracow became its coordinator. A series of meetings for potential new members of the association and new members of TG Mining were organized in the first year. As a result of this activity, AGH in Cracow, as the coordinator, defined the Multi Annual Roadmap for mining for the coming years, which was submitted to the euRobotics association and, next, to the European Commission. In subsequent years, TG Mining members took part in conferences and forums organized in various places in Europe. Several European Robotics Forums (ERF) have taken place—ERF 2016 in Ljubljana, ERF 2017 in Edinburgh, ERF 2018 in Tamper and ERF 2019 in Bucharest. During ERF 2017, in some of these meetings, participants discussed the possibility of initiating cooperation between TG Mining and Construction Robotics, Nuclear Inspection, Inspection and Maintenance or Oil and Gas groups. However, this year's ERF 2020 meeting took place in March, in Malaga [21–23].

4.2. Space Mining Conference

In the middle of 2016, a team consisting of students and academic teachers enthusiastic about space exploration focused on the space industry. All its aspects related to mining were established at the AGH University of Science and Technology. Ideas for Space Mining Engineering (ISME) was created to prepare and conduct the Space Mining Conference organized by the AGH University of Science and Technology in Cracow. The first Student Space Mining Conference 2018 was a significant success and an important event. ISME organized a second conference in 2019. A third conference is being planned for 2020. The ISME Group (Krakow, Poland) covers all aspects (technological, mechanical, economic, legal and ethical) related to space exploration, including extra-terrestrial resource use. This initiative has attracted an increasing number of participants and listeners. Analyzing the speeches of the participants of these conferences, one can notice a great emphasis on the problems of automation and robotization of mining works as the only possibility of space exploitation. Although the topic is relatively new, it is rapidly developing. Current information can be found on the Conference website [24].

4.3. PERASPER Research Cluster

The PERASPERA project (in Latin "Per aspera ad astra" means "Through hardships to the stars"), created in 2014, is financed by the European Union under the Horizon 2020 program. The goal of the PERASPERA initiative is to support industry competence in Europe. The focus is planetary and orbital robotics, and to demonstrate key technologies related to these fields in space. In 2019, Poland joined the PERASPER Research Cluster. Poland's membership in the consortium may facilitate the participation of domestic space sector entities in innovative space projects and cooperation with large European entities, as well as the testing of technologies in the field of orbital and planetary robotics in space. Polish companies, institutes and universities have extensive experience in the field of ground robotics and implemented projects related to space robotics systems. The field of space robotics in which Poles have the widest competence is underground exploration. This exploration includes devices for sampling, mechanisms working in vacuums, devices for underground testing as well as systems and control sensors. Experience and potential in the production of subsystems and components for orbit robotics systems are also evident in Poland. Servicing objects in orbit can be accomplished because of expertise in Poland with gripping and holding mechanisms, control systems, connectors, motion sensors and antenna systems [25].

5. Conclusions

The underground mining industry has difficult conditions and high costs of implementation. The development of new technical measures is exposed to high risk in terms of both technical and financial capabilities of the contractor. Research and development (R&D) entail the necessity to carry out basic tests in the first phase, including design works. The second phase of R&D involves carrying out field tests and implementing new machinery and equipment solutions. These tests need to be carried out under specific mining, geological, technical and organizational conditions. Tests must also comply with the regulations of the State Mining Authority (Poland) [26]. The existing solutions that have been used for many years in various industries cannot be easily transferred to underground mines. Robotic vehicles are currently able to respond to typical road situations, including pedestrians, traffic lights and traffic control by a police officer. However, underground loaders or dump trucks still do not have such autonomy. Moreover, typical anthropomorphic robotic solutions work well in the production process. These solutions include the implementation of assembly or welding processes, while the process of installing support arches in underground excavations has only been mechanized.

In mining, especially underground, the simplest solutions are the best. Difficult conditions and high costs sometimes limit the use of technical means to the necessary minimum. However, automation and robotization in mining are developing slowly and encounter many obstacles. Some solutions have already been implemented in mining practice. Solutions include systems supporting the work of self-propelled drilling carts and roadheaders, an autonomous machine for breaking rocks, automated longwall systems, and rescue works. The effects of implementing these solutions are tangible. The presented solutions make it possible to work in severe environment and conditions. People are not able to operate in these environments and conditions. For more specific effects, more research and analyses need to be conducted.

Poland has many underground mining plants and producers of mining machinery and equipment. Poland is good training ground for the development and testing of automated as well as robotic machines and machine systems. The establishment of TG Mining (part of euRobotics, a rapidly growing ISME group) organize space mining conferences. Other initiatives in Poland, like consortium EX-PL and the Centre for Space Studies of the Kozminski University, PERASPERA are also important for robotization.

Author Contributions: Both Ł.B. and W.B. collated the topics within this review and contributed to the writing of the text. Both authors have read and agreed to the published version of the manuscript.

Funding: This research received no external funding.

Acknowledgments: Acknowledgements to FAMUR S.A., Mine Master Sp. z o. o. and KGHM ZANAM S.A. for figures and materials.

Conflicts of Interest: The authors declare no conflict of interest.

References

1. Semykina, I.; Grigoryev, A.; Gargayev, A.; Zavyalov, V. Unmanned Mine of the 21st Centuries. In Proceedings of the IInd International Innovative Mining Symposium (Devoted to Russian Federation Year of Environment), Kemerovo, Russia, 20–22 November 2017; EDP SCIENCES: Les Ulis, France, 2017; Volume 21.
2. Noort, D.; McCarthy, P. The Critical Path to Automated Underground Mining. In Proceedings of the First International Future Mining Conference and Exhibition, Sydney, Australia, 19–21 November 2008; Australasian Inst Mining & Metallurgy: Carlton, Australia, 2008.
3. Jonak, J.; Gajewski, J. Mining machines robotization on the example of the heading machine. *Transport. Przemysłowy i Maszyny Robocze* **2011**, *4*, 66–69.
4. Biały, W. Determination of workloads in cutting head of longwall tumble heading machine. *Manag. Syst. Prod. Eng.* **2016**, *1*, 45–54. [CrossRef]
5. Bołoz, Ł. Unique project of single-cutting head longwall shearer used for thin coal seams exploitation. *Arch. Min. Sci.* **2013**, *4*, 1057–1070.
6. Bołoz, Ł. Longwall shearers for exploiting thin coal seams as well as thin and highly inclined coal seams. *MIAG* **2018**, *2*, 59–72.
7. Dziura, J. Mikrus system—New technology of low seams exploitation. *Maszyny Górnicze* **2012**, *3*, 3–11.
8. Famur, S.A. Available online: https://famur.com (accessed on 30 July 2020).
9. Mendyka, P.; Nawrocki, M.; Ostapow, L. Calibration of mining drilling rig boom equipped with feeder guiding system FGS. In Proceedings of the 14th International Conference Mechatronics Systems and Materials, Zakopane, Poland, 4–6 June 2018; Robak, G., Ed.; AIP Publishing LLC: College Park, MD, USA, 2018.
10. Mine Master. Available online: https://www.minemaster.eu/pl/pl (accessed on 30 July 2020).
11. Mendyka, P.; Kotwica, K.; Stopka, G.; Czajkowski, A.; Ostapów, L.; Karliński, J. Possibilities and Barriers in Implementation of Drilling Rig Remote Control. In *Modern Technology and Safety in Mining*; Kotwica, K., Ed.; Department of Mining, Dressing and Transporting Machines: Krakow, Poland, 2019; pp. 183–190.
12. Hasilová, K.; Gajewski, J. The use of kernel density estimates for classification of ripping tool wear. *Tunn. Undergr. Space Technol.* **2019**, *88*, 29–34. [CrossRef]
13. Jedliński, Ł.; Gajewski, J. Optimal selection of signal features in the diagnostics of mining head tools condition. *Tunn. Undergr. Space Technol.* **2019**, *84*, 451–460. [CrossRef]
14. Krauze, K.; Rączka, W.; Sibielak, M.; Konieczny, J.; Kubiak Culer, H.; Bajus, D. Automated transfer point URB/ZS-3. *MIAG* **2017**, *2*, 80–85. [CrossRef]
15. Krauze, K.; Rączka, W.; Konieczny, J.; Sibielak, M. Automatic ore loading point URB/ZS-3. *CUPRUM* **2016**, *4*, 5–11.
16. Sibielak, M.; Rączka, W.; Konieczny, J.; Kowal, J. Optimal control based on a modified quadratic performance index for systems disturbed by sinusoidal signals. *Mech. Syst. Signal. Process.* **2015**, *64–65*, 498–519. [CrossRef]
17. AGH w Krakowie. Available online: https://www.agh.edu.pl/nauka/info/article/kosmiczna-wiertnica (accessed on 30 July 2020).
18. Szade, A.; Szot, M.; Królak, M.; Urbańczyk, T. Multi-level intrinsically safe monitoring of the shafts' lining. *Wiadomości Górnicze* **2016**, *7–8*, 416–422.
19. Kasprzyczak, L.; Szwejkowski, P.; Cader, M. Robotics in mining exemplified by Mobile Inspection Platform. *MIAG* **2016**, *2*, 23–28.
20. Kasprzyczak, L.; Trenczek, S. Mining inspective mobile robot for monitoring hazardous explosive zones. *Pomiary Automatyka Robotyka* **2011**, *2*, 431–440.
21. Kasza, P. euRobotics—AGH koordynatorem Grupy Tematycznej: Górnictwo (TG Mining). In Proceedings of the MARG Conference Summary, Wisla, Poland, 17 June 2016.
22. Kasza, P. euRobotics—Way to the mining of the future. In Proceedings of the MARG Conference, Wisła, Poland, 26 June 2017.
23. Eurobotics. Available online: https://www.eu-robotics.net/eurobotics (accessed on 30 July 2020).
24. Konferencja Górnictwa Kosmicznego. Available online: www.kgk.agh.edu.pl (accessed on 30 July 2020).

25. Urania—Postępy Astronomii Centrum Astronomii UMK. Available online: www.urania.edu.pl (accessed on 30 July 2020).
26. AGH. Available online: www.agh.edu.pl/nauka/info/article/macgyver-z-agh-czyli-prototyp-robota-gorniczego (accessed on 30 July 2020).

Publisher's Note: MDPI stays neutral with regard to jurisdictional claims in published maps and institutional affiliations.

Article

Methods of Pre-Identification of TITO Systems

Milan Saga [1], Karel Perutka [2], Ivan Kuric [1], Ivan Zajačko [1,*], Vladimír Bulej [1], Vladimír Tlach [1] and Martin Bezák [3]

[1] Faculty of Mechanical Engineering, University of Zilina, Univerzitna 8215/1, 01026 Zilina, Slovakia; milan.saga@fstroj.uniza.sk (M.S.); ivan.kuric@fstroj.uniza.sk (I.K.); vladimir.bulej@fstroj.uniza.sk (V.B.); vladimir.tlach@fstroj.uniza.sk (V.T.)
[2] Faculty of Applied Informatics, Tomas Bata University in Zlin, Nad Stráněmi 4511, 760 05 Zlin, Czech Republic; kperutka@utb.cz
[3] VIPO, a. s., Gen. Svobodu 1069/4, 95801 Partizanske, Slovakia; martin.bezak@vipo.sk
* Correspondence: ivan.zajacko@fstroj.uniza.sk; Tel.: +421-910956861

Abstract: The content of this article is the presentation of methods used to identify systems before actual control, namely decentralized control of systems with Two Inputs, Two Outputs (TITO) and with two interactions. First, theoretical assumptions and reasons for using these methods are given. Subsequently, two methods for systems identification are described. At the end of this article, these specific methods are presented as the pre-identification of the chosen example. The Introduction part of the paper deals with the description of decentralized control, adaptive control, decentralized control in robotics and problem formulation (fixing the identification time at the existing decentralized self-tuning controller at the beginning of control and at the beginning of any set-point change) with the goal of a new method of identification. The Materials and methods section describes the used decentralized control method, recursive identification using approximation polynomials and least-squares with directional forgetting, recursive instrumental variable, self-tuning controller and suboptimal quadratic tracking controller, so all methods described in the section are those ones that already exist. Another section, named Assumptions, newly formulates the necessary background information, such as decentralized controllability and the system model, for the new identification method formulated in Pre-identification section. This section is followed by a section showing the results obtained by simulations and in real-time on a Coupled Drives model in the laboratory.

Keywords: pre-identification; least squares method; instrumental variable method; robotics; sensor

Citation: Saga, M.; Perutka, K.; Kuric, I.; Zajačko, I.; Bulej, V.; Tlach, V.; Bezák, M. Methods of Pre-Identification of TITO Systems. *Appl. Sci.* **2021**, *11*, 6954. https://doi.org/10.3390/app11156954

Academic Editors: Yuri Nikitin and Anton Civit

Received: 21 April 2021
Accepted: 23 July 2021
Published: 28 July 2021

Publisher's Note: MDPI stays neutral with regard to jurisdictional claims in published maps and institutional affiliations.

1. Introduction

Most processes in practice are processes that have multiple inputs and multiple outputs, and these are influenced by interactions. These systems can be controlled by a centralized or decentralized controller. The main advantages of decentralization include simplifying the overall task by dividing it into a set of sub-tasks. Decentralized control is very often used in practice. Decentralized charge control of electric vehicles is a nice application. There was introduced a fully decentralized and participatory learning mechanism for privacy-preserving coordinated charging control of electric vehicles that regulates three Smart Grid socio-technical aspects: (i) reliability, (ii) discomfort and (iii) fairness [1]. Another good application of decentralized control is a quadrocopter, namely outdoor flocking of quadcopter drones with decentralized model predictive control [2]. From the theoretical point of view, there was proposed a decentralized explicit (closed-form) iterative formula that solves convex programming problems with linear equality constraints and interval bounds on the decision variables [3], or decentralized control problem for non-affine large-scale systems with nonaffine functions possibly being discontinuous [4], or decentralized adaptive tracking control strategy consisting of a steady-state controller and modified optimal feedback controller. Design parameters-dependent feasibility conditions were formulated by using Lyapunov theory to guarantee the existence of the proposed decentralized control scheme [5]. Decentralized voltage control is another example of a decentralized strategy. It includes network partitioning strategy for the optimal voltage control of Active Distribution Networks actuated by means of a limited number of Distributed Energy Storage Systems [6]. Another paper is concerned with the problem of decentralized event-triggered dynamic output feedback control for large-scale systems with unknown interconnections. By using a modified cyclic-small gain condition and introducing a free-matrix-based integral inequality, a sufficient condition was derived to ensure that the overall closed-loop system is asymptotically stable with a prescribed H∞ performance [7]. There was also implemented fuzzy decentralized control, for example an adaptive fuzzy decentralized control approach for a class of nonlinear systems with unknown control direction and different performance constraints. In the control design, the different performance constraints, that were the prescribed performance error constraints for some subsystems and the asymmetric time-varying output constraints for the others, could be unified as one form by selecting proper performance functions [8]. Adaptive control is another area that has expanded the use of decentralized control, and here are at least a few such examples. A minimal-neural-networks-based design approach was presented for the decentralized output feedback tracking of uncertain interconnected strict-feedback nonlinear systems with unknown time varying delayed interactions unmatched in control inputs [9]. The decentralized output-feedback adaptive backstepping control scheme was also proposed for a class of interconnected nonlinear systems with unknown actuator failures by introducing a kind of high gain K-filters [10], or decentralized output-feedback adaptive control scheme for a class of interconnected nonlinear systems with input quantization. Both logarithmic quantizers and improved hysteretic quantizers were studied, and a linear time-varying model was introduced to handle the difficulty caused by quantization [11]. A decentralized adaptive backstepping control scheme was also proposed for a class of interconnected systems with nonlinear multisource disturbances and actuator faults. The nonlinear multisource disturbances comprised two parts: one was the time-varying parameterized uncertainty; the other was the dynamic unexpected signal formulated by a nonlinear exogenous system [12]. Additionally, the problem of decentralized adaptive backstepping control for a class of large-scale stochastic nonlinear time-delay systems with asymmetric saturation actuators and output constraints was also solved [13]. It can be mentioned that a backstepping-based robust decentralized adaptive neural H ∞ tracking control method was addressed for a class of large-scale strict feedback nonlinear systems with uncertain disturbances [14]. Decentralized control was implemented in robotics. The example where a discrete-time decentralized neural identification and control for large-scale uncertain nonlinear systems at a two degree of freedom planar robot was implemented can be mentioned [15], or the work that investigated the use of a decentralized control system for suppressing the vibration of a multi-link flexible robotic manipulator using embedded smart piezoelectric transducers [16]. Decentralized motion coordination algorithms were proposed for the robots so that they collectively moved in a rectangular lattice pattern from any initial position [17]. Mobile robot formations differ in accordance to the mission, environment and robot abilities. In the case of decentralized control, the ability to achieve the shapes of these formations has to be built in the controllers of each autonomous robot [18]. A decentralized control algorithm for the robots to accomplish the sweep coverage was also proposed. The sweep coverage was achieved by coordinating the robots to move along a given path that was unknown to the vehicles a priori [19].

During the simulation experiments in the real-time laboratory in recent years, we reveal the fact that some time is needed to get the appropriate behaviour of control when the self-tuning controller is used. This time depends on the type of the controlled system. This is the known problem of self-tuning control because the controller needs the adequate model of the system. This is one problem that we solve by a new approach described in this paper, by the method named as pre-identification. Another problem comes from the fact that we used the decentralized controllers for the control of multivariable systems. If one set-point changes its value, it influences all other main subsystems by interactions and therefore the model of subsystems changes. This could be also fixed by a self-tuning controller but some time is needed to obtain the stable model. Therefore, by the new method described in this paper, named as pre-identification, we also solved this problem.

2. Materials and Methods

2.1. Decentralized Control

Using the decentralized approach, the control is divided into a set of sub-tasks that are matched by simple controllers. These partial tasks will then give us the overall course of control. The main advantages of decentralized control are primarily that a more complex system is divided into a set of simple tasks and the resulting controller is more flexible [20].

A special example of multidimensional systems is a system with two inputs and two outputs. This can be realized by the so-called P structure, see Figure 1. In this case, the inputs to the systems describing the interactions are the values of the action signals of the SISO controllers and their outputs are added to the opposite outputs of the main diagonal systems. From this figure, we get the transfer function equations of the model in the form

$$G_{S1} = \frac{Y_1}{U_1} = G_{S11} - \frac{G_{S21} G_{S12} G_{R22}}{1 + G_{S22} G_{R22}} \tag{1}$$

$$G_{S2} = \frac{Y_2}{U_2} = G_{S22} - \frac{G_{S21} G_{S12} G_{R11}}{1 + G_{S11} G_{R11}} \tag{2}$$

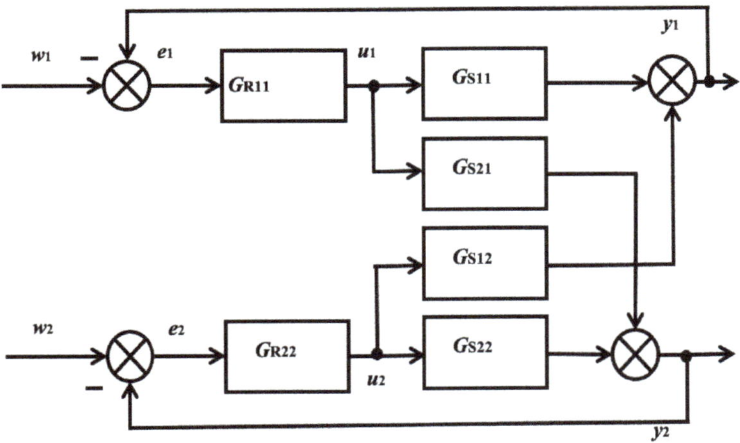

Figure 1. Decentralized system control with two inputs and two outputs, the so-called P structure.

2.2. Recursive Identification Using Approximation Polynomials

A prerequisite for good control is the most accurate description of the regulated system. Identification is the procedure by which the mathematical model of a system is obtained. The beginnings of identification based on continuous models date back to the middle of the 20th century. For continuous-time identification, the identified model is in the form of the differential equations. Differential equations contain expressions with derivatives over time that are not measurable. It is possible to replace the segment by an approximation polynomial whose derivatives can be calculated analytically in advance and then calculated numerically, see Figure 2. This approach was for example used by Perutka [20].

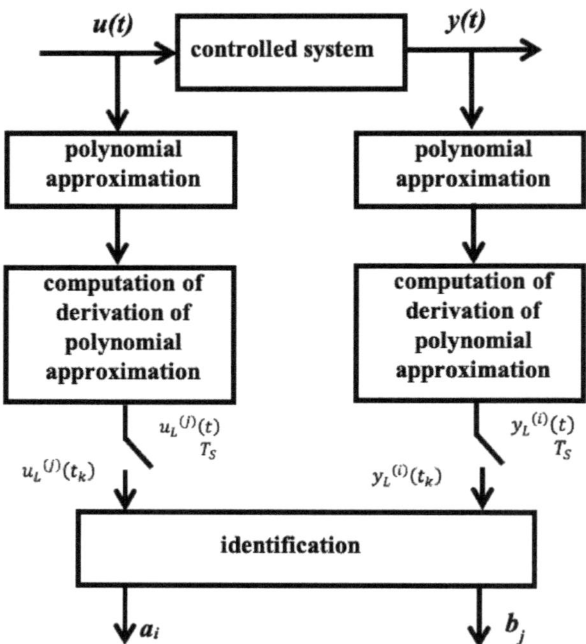

Figure 2. Identification scheme for continuous-time systems.

2.3. Least Squares Method with Exponential Forgetting

The estimation of model parameters is computed as

$$\hat{\theta}(k) = \hat{\theta}(k-1) + K(k)\hat{e}(k) \tag{3}$$

The gain vector is calculated as

$$K(k) = \frac{C(k-1)\phi(k)}{1 + \phi^T(k)C(k-1)\phi(k)} \tag{4}$$

and covariance matrix

$$C(k) = C(k-1) - \frac{C(k-1)\phi(k)\phi^T(k)C(k-1)}{1 + \phi^T(k)C(k-1)\phi(k)} \tag{5}$$

The following applies to the calculation of the prediction error

$$\hat{e}(k) = y(k) - \phi^T(k)\hat{\theta}(k-1) \tag{6}$$

In the case of exponential forgetting, the criterion of identification is

$$J = \sum_{i=k_0}^{k} \left(\varphi^{k-i}e(i)\right)^2 \tag{7}$$

where the exponential forgetting factor is chosen in the range of 0 to 1, the most common near 1.

If

$$\phi^T(k)C(k-1)\phi(k) > 0 \tag{8}$$

Then

$$C(k) = C(k-1) - \frac{C(k-1)\phi(k)\phi^T(k)C(k-1)}{\eta^{-1} + \phi^T(k)C(k-1)\phi(k)} \tag{9}$$

where

$$\eta(k) = \varphi(k) - \frac{1 - \varphi(k)}{\xi(k)} \tag{10}$$

If

$$\phi^T(k)C(k-1)\phi(k) = 0 \tag{11}$$

Then

$$C(k) = C(k-1) \tag{12}$$

Furthermore

$$\varphi(k) = \left\{ 1 + (1+\rho)[\ln(1 + \xi(k-1))] + \left[\frac{(v(k-1)+1)\eta(k-1)}{1 + \xi(k-1) + \eta(k-1)} - 1 \right] \frac{\xi(k-1)}{1 + \xi(k-1)} \right\}^{-1} \tag{13}$$

$$\eta(k) = \frac{e^2(k)}{\lambda(k)} \tag{14}$$

$$v(k) = \varphi(k)[v(k-1) + 1] \tag{15}$$

$$\lambda(k) = \varphi(k)\left[\lambda(k-1) + \frac{e^2(k)}{1 + \xi(k-1)} \right] \tag{16}$$

$$\xi(k) = \phi^T(k)C(k-1)\phi(k) \tag{17}$$

The parameters estimation vector is in the form

$$\hat{\theta}^T(k) = \left(\hat{a}_0, \hat{a}_1, \ldots, \hat{a}_{\deg(a)}, \hat{b}_0, \hat{b}_1, \ldots, \hat{b}_{\deg(b)}, d \right) \tag{18}$$

and regressor

$$\phi^T(k) = \left(-y(t_k), \ldots, -y_L^{(n-1)}(t_k), u(t_k), \ldots, u_L^{(m)}(t_k), 1 \right) \tag{19}$$

2.4. Self-Tuning Controller

The main reason for using adaptive control is that the systems change over time or the characteristics of the controlled system are unknown. The basic principle of adaptive systems is to change the characteristics of the controller based on the characteristics of the controlled process [21]. The general scheme of the self-tuning controller is shown in Figure 3.

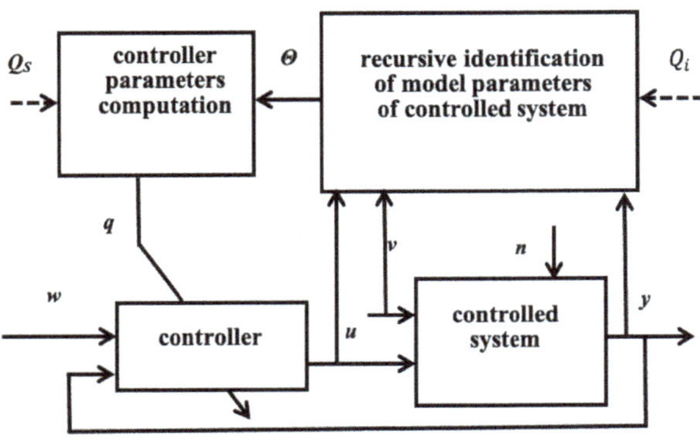

Figure 3. The general scheme of the self-tuning controller.

2.5. Suboptimal Linear Quadratic Tracking Controller

The method was introduced by Dostál [22]. If the system of Figure 4 is considered

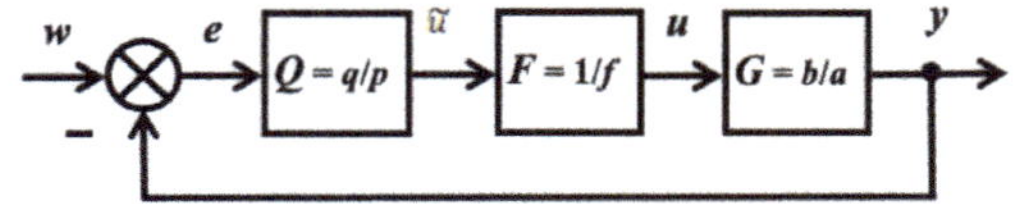

Figure 4. System with feedback controller.

Let us minimize a quadratic functional with two penalty constants

$$J = \int_0^\infty \left\{ \mu e^2(t) + \varphi \tilde{u}^2(t) \right\} dt \tag{20}$$

The Laplace image of the set point holds

$$w(s) = \frac{h_w(s)}{s f_w(s)} \tag{21}$$

It holds for degrees of polynomials

$$\deg(h_w) \leq \deg(f_w), \; f_w(0) \neq 0 \tag{22}$$

We calculate stable polynomials g and n as results of spectral factorizations

$$(as) * \varphi as + b * \mu b = g * g, n * n = a * a \tag{23}$$

We solve the following diophantine equation

$$asp + bq = gn \tag{24}$$

Considering the transfer function of the system

$$G(s) = \frac{b_0}{s^2 + a_1 s + a_0} \tag{25}$$

then the controller is

$$F(s)Q(s) = \frac{q_2 s^2 + q_1 s + q_0}{s(p_2 s^2 + p_1 s + p_0)} \tag{26}$$

In this case, the polynomials have the form

$$g(s) = g_3 s^3 + g_2 s^2 + g_1 s + g_0 \tag{27}$$

$$n(s) = s^2 + n_1 s + n_0 \tag{28}$$

and to calculate their coefficients obtained by spectral factorization

$$g_0 = \sqrt{\mu b_0^2} \tag{29}$$

$$g_1 = \sqrt{2 g_2 g_0 + \varphi a_0^2} \tag{30}$$

$$g_2 = \sqrt{2 g_3 g_1 + \varphi(a_1^2 - 2a_0)} \tag{31}$$

$$g_3 = \sqrt{\varphi} \tag{32}$$

$$n_0 = \sqrt{a_2^0} \tag{33}$$

$$n_1 = \sqrt{2n_0 - a_1^2 - 2a_0} \tag{34}$$

2.6. Calculation of Derivatives Using Approximation Functions

To calculate the derivatives, we approximate the closest neighborhood for a given time by the approximation function. For example, we will use the Lagrange polynomial in the form

$$P_2(x) = \frac{(x-b)(x-c)}{(a-b)(a-c)}f(a) + \frac{(x-a)(x-c)}{(b-a)(b-c)}f(b) + \frac{(x-b)(x-b)}{(c-a)(c-b)}f(ac) \tag{35}$$

whose first derivative is

$$\begin{aligned} f'(x) \cong P_2'(x) &= \frac{2x-(b+c)}{(a-b)(a-c)}f(a) + \frac{2x-(a+c)}{(b-a)(b-c)}f(b) \\ &+ \frac{2x-(a+b)}{(c-a)(c-b)}f(c) \end{aligned} \tag{36}$$

and second derivative is

$$f''(x) \cong P_2''(x) = \frac{2f(a)}{(a-b)(a-c)} + \frac{2f(b)}{(b-a)(b-c)} + \frac{2f(c)}{(c-a)(c-b)} \tag{37}$$

2.7. Recursive Instrumental Variable Method

The instrumental variable method is a modification of the least squares method. The least squares method uses the quadratic criterion and the existence of one global minimum. However, a prerequisite for successful least-squares modelling is that the fault is represented by white noise with zero mean value. The instrumental variable method does not make it possible to determine the noise properties, but is based on weaker assumptions than the least squares method. The identification proceeds according to number 5. As with the least squares method, the method of the instrumental variable can also be formulated recursively [22–25]. The parameter estimation vector has the form

$$\hat{\theta}^T(k) = \left(\hat{a}_0, \hat{a}_1, \ldots, \hat{a}_{\deg(a)}, \hat{b}_0, \hat{b}_1, \ldots, \hat{b}_{\deg(b)}, d \right) \tag{38}$$

and data vector

$$\hat{\theta}^T(k) = \left(-y(t_k), \ldots -y_L^{(n-1)}(t_k), u(t_k), \ldots, u_L^{(m)}(t_k), 1 \right) \tag{39}$$

The gain vector is calculated by relation

$$L(k) = \frac{C(k-1)z(k)}{1 + \phi^T(k)C(k-1)z(k-1)} \tag{40}$$

In addition to the data vector, it is necessary to know the covariance matrix

$$C(k) = C(k-1) - \frac{C(k-1)z(k)\phi^T(k)C(k-1)}{1 + \phi^T(k)C(k-1)z(k)} \tag{41}$$

and instrument vector

$$z(k) = \left(u(t_k), u(t_{k-1}), \ldots, u(t_{k-n-m}), \right. \tag{42}$$

which we choose as a set of delayed inputs. The prediction error is calculated by

$$\hat{e}(k) = y(k) - \phi^T(k)\hat{\theta}(k-1) \tag{43}$$

and estimating the parameters according to

$$\hat{\theta}(k) = \hat{\theta}(k-1) + L(k)\hat{e}(k) \tag{44}$$

3. Assumptions

3.1. Decentralized Controllability

Assume the existence of a stable minimum phase Linear Time Invariant (LTI) in time of a continuous multidimensional system of the dimension $N \times N$. Its Laplace transformation $S(s) : S(t) > S(s)$, which we call the transformed system is in the form

$$S(s) = \begin{pmatrix} S_{11}(s)S_{12}(s) & \cdots & S_{1N}(s) \\ \vdots & \ddots & \vdots \\ S_{N1}(s)S_{N2}(s) & \cdots & S_{NN}(s) \end{pmatrix} \tag{45}$$

where $S_{ij}(s)$, $\quad i = 1, 2, \ldots, N$, $\quad j = 1, 2, \ldots$ is Laplace transformation of the ij-th subset of $S_{ij}(t)$ of the transformed system $S(s)$. The transformed system $S(s)$ je is decentrally controllable only when its main diagonal is dominant.

3.2. System Model and Shape of Reference Signal

Suppose there exists a system $S(t)$ and a transformed system $S(s)$ as described above. Then we formulate a model created for the purpose of decentralized control, which we call $M(t)$, and its Laplace transformation $M(s)$. Consider $M(s)$ as a stable minimum phase linear time invariant multivariate diagonal matrix in the form

$$M(s) = \begin{pmatrix} M(s) & 0 & \cdots & 0 \\ 0 & M(s) & \cdots & 0 \\ \vdots & \vdots & \ddots & \vdots \\ 0 & 0 & \cdots & M_{N(s)} \end{pmatrix} \tag{46}$$

where $M_i(s)$, $\quad i = 1, 2, \ldots, N$ is the Laplace transformation of the i-th submodule $M_i(t)$ of the model $M(t)$ of the transformed system $S(s)$. This assumes minimal impact of extra-diagonal transmissions, which is important because of the deployment of a decentralized controller. Simplification of the N-dimensional system to N-dimensional systems is simplified.

Furthermore, suppose that the Laplace transformation of the reference signal vector $r(s)$ is always in the form

$$r(s) = (R_1(s), R_2(s), \ldots, R_N(s)) \tag{47}$$

where $R_i(s), i = 1, 2\ldots, N$ is the i-th Laplace reference signal of Laplace transformation of the reference signal vector $r(s)$ and has the form

$$R_i(s) = \frac{h_i}{s} \tag{48}$$

where $h_i \in R, i = 1, 2, \ldots, N$, is the i-th part by constant function, i.e., reference signal, which is only a combination of p-l step changes of the signal of its different constant values

$$h_i = \begin{cases} j_{i1} \; for \; t \in \langle 0, t_{i1} \rangle \\ j_{i2} \; for \; t \in \langle t_{i1}, t_{i2} \rangle \\ \quad \vdots \\ j_{ip} \; for \; t \in \langle t_{ip-1}, t_{ip} \rangle \end{cases} , \; 0 < t_{i1} < t_{i2} < \cdots < t_{ip-1} < t_{ip} \tag{49}$$

where $j_{il} \in R$, $\quad i = 1, 2, \ldots, N$, $\quad l = 1, 2, \ldots, p$, the l-th constant function i-th in parts by the constant function h_i, t is the time $t_{il} \in R^+$, $\quad i = 1, 2, \ldots, N$, $\quad l = 1, 2, \ldots, p$, the l-th moment of the i-th in portions of the constant function h_i. This means that each non-zero element of the matrix $M(s)$ has exactly one non-zero element of the vector $r(s)$, i.e., that each partial transmission of the overall system model has a reference signal defined for it. As for the form of the reference signal, it is a constant function in parts. This function is approximated from an arbitrary but predetermined number of p segments of a different but concise value, i.e., it varies over time.

4. Pre-Identification

Consider the validity of the assumption of decentralized controllability, system description and system model. Then, the transformed system S(s) can be viewed as a black box model, and let it be identified by direct and/or indirect time-continuous algorithms. In time, continuous model identification can be realized by following steps: The controller is not connected in the closed circuit. The values of the vector of difference of output quantities and reference signals E(t) are sent to the input of the system S(t). The values of the reference signals are the same and at the same time as those that will be used during regulation.

1. The controller is not connected in the closed circuit. The values of the vector of difference of output quantities and reference signals E(t) are sent to the input of the system S(t). The values of the reference signals are the same and at the same time as those that will be used during regulation.
2. If switching control is considered, each time interval of the control of the system S(s) at which all reference signals have a constant value is identified separately, in so-called Identification Elements (IE).
3. Each identification element is identified several times, each time by a different identification algorithm, and the obtained model can be verified by comparison with the measured data. The obtained model, which is most consistent with the measured data, is then used for control. Let us call this method of Identification More Than One Method (IMTOM).

The system S(s) is completely identified by the above procedure before the actual regulation begins, therefore identification during the regulation is not necessary. This procedure is suitable for processes where the same procedure is repeated many times.

5. Results

5.1. Simulation Results

The verified system is described as

$$G_S(s) = \begin{pmatrix} \frac{3.7}{s^2+5.2s+4.6} & \frac{0.4}{s^2+4.4s+3.8} \\ \frac{0.6}{s^2+10.6s+10.2} & \frac{8.7}{s^2+7.4s+8.2} \end{pmatrix} \tag{50}$$

Since it is the system with two inputs and two outputs, we obtained two suboptimal linear quadratic controllers in the form that was described in the previous section. These controllers had the following penalty constants

$$\mu_1 = 1, \; \varphi_1 = 30, \mu_2 = 1, \; \varphi_2 = 30 \tag{51}$$

We used our algorithm, pre-identification, at this system and we obtained the following results, see Figures 5–8. First, we performed system response on the given set-points depicted in Figure 5. Together with this system response, we obtained the system pre-identified parameters shown in Figure 6, for the first subsystem in the left one and for the second subsystem in the right one. According to these pre-identified parameters we performed the simulation of control which is shown in Figure 7. During the control, the controller parameters were changing, and they are recorded in Figure 8, for the first controller in left one and and for the second controller in the right one.

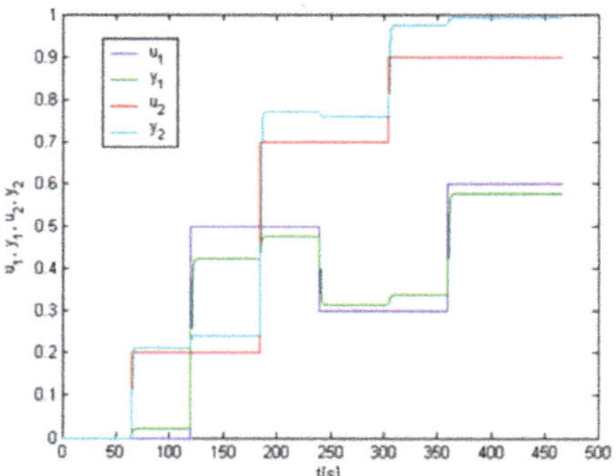

Figure 5. System response (green and cyan) on the given set-points (blue and red).

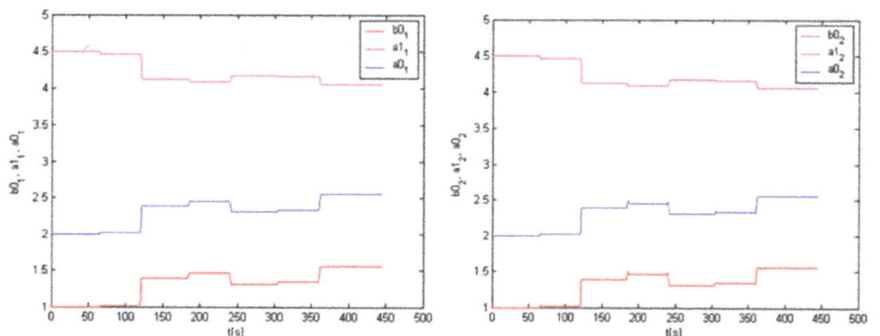

Figure 6. System pre-identification using recursive intrumental variable—1st subsystem (**left**), 2nd (**right**).

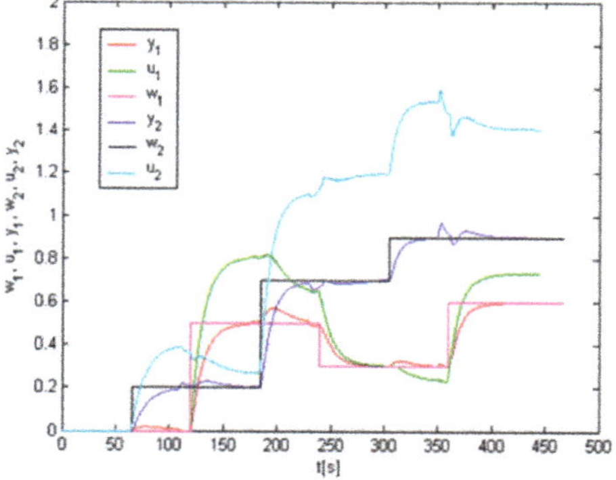

Figure 7. Output of control.

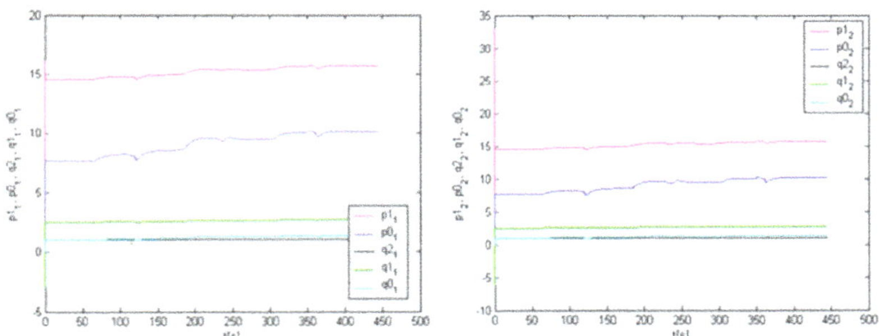

Figure 8. Controllers parameters during simulation—1st (**left**), 2nd (**right**).

5.2. Results in Real-Time at Laboratory Model

We verified the presented method in real-time control using MATLAB at CE108 Coupled Drives Apparatus Model [26] which is shown in the Figure 9. The laboratory model CE108 allows solving practical problems of tensioning and speed of material movement in continuous processes. An example is the speed and tension of the thread when rewinding from one spool to another, which must be controlled. This situation is adapted for laboratory measurements so that the elastic band is mounted on three wheels. The lower two wheels are fixed, their speed can be measured and regulated. The third wheel can move (located on a movable arm suspended on a spring) and simulates a workstation along with tension and speed measurements. Two servomotors control the speed of movement and tensioning of the belt. The speed is 0–3000 rpm, which corresponds to a voltage of 0–10 V. Tension measurement is indirect through the angle of the movable arm, from $-10\,^\circ$ to $10\,^\circ$, which corresponds to a voltage from -10 V to 10 V. Inputs and outputs are located on the front panel of the device; it is the control voltage to the servomotor amplifiers, which are bidirectional, and which are inputs. There are four outputs, the voltage corresponding to the speed of the two servomotors and the voltage corresponding to the tension and the speed of the belt.

Figure 9. Photo of CE108 Coupled Drives Apparatus model connected with PC with MATLAB.

Using the pre-identification method and fully implementing interactions in the main plants, we obtained the following mathematical model to be used at control of speed

$$G_S(s) = \begin{pmatrix} \frac{0.78}{s^2+2.66s+1.33} & 0 \\ 0 & \frac{4.16}{s^2+11.66s+16.66} \end{pmatrix} \tag{52}$$

Since it is the system with two inputs and two outputs, we obtained two suboptimal linear quadratic controllers in the form that was described in the previous section. These controllers had the following penalty constants

$$\mu_1 = 1,\ \varphi_1 = 1, \mu_2 = 1,\ \varphi_2 = 0.85 \tag{53}$$

We used our algorithm, pre-identification, at this system and we obtained the following results, see Figures 10–13. First, we performed system response on the given set-points depicted in Figure 10. Together with this system response, we obtained the system pre-identified parameters shown in Figure 11, for the first subsystem in the left one and for the second subsystem in the right one. According to these pre-identified parameters we performed the simulation of control which is shown in Figure 12. During the control, the controller parameters were changing, and they are recorded in Figure 13, for the first controller in left one and and for the second controller in the right one.

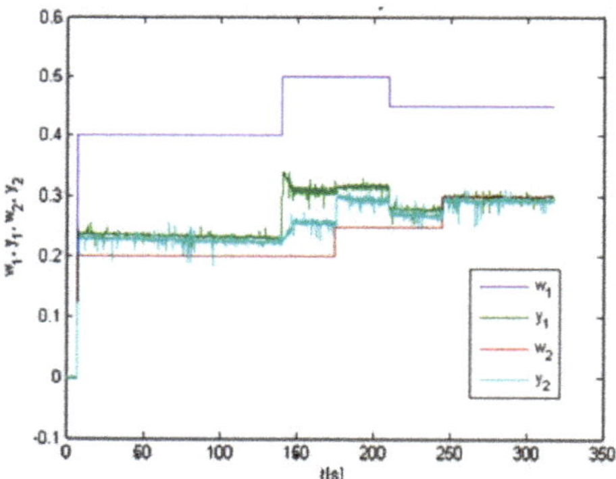

Figure 10. System response (green and cyan) on the given set-points (blue and red).

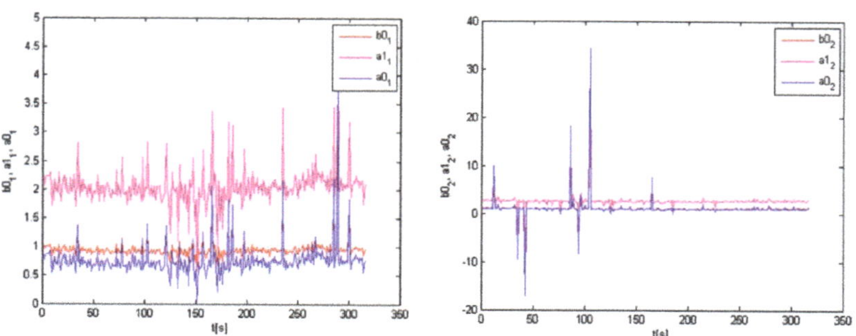

Figure 11. System pre-identification using recursive instrumental variable—1st subsystem (**left**), 2nd (**right**).

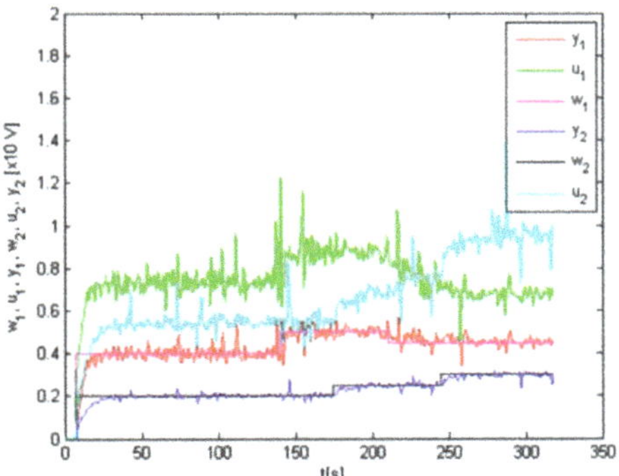

Figure 12. Output of control.

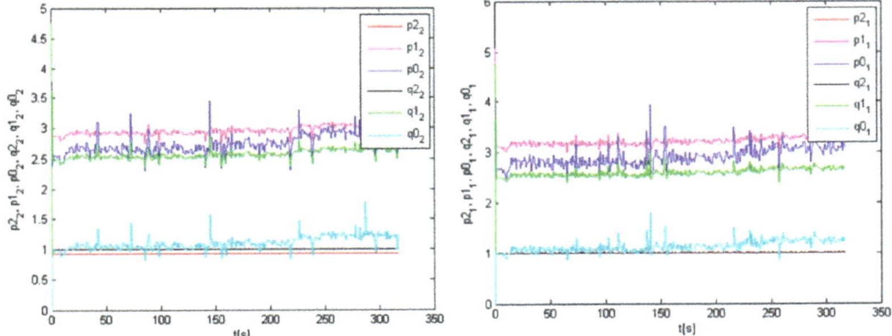

Figure 13. Controllers parameters during simulation—1st (**left**), 2nd (**right**).

6. Conclusions

This paper presents the new method of identification named as pre-identification on the theoretical level and subsequently verified it by simulations and in the real-time experiments at Coupled Drives model in the laboratory. The results confirm that the method can be successfully used and future work will focus on the verification of this method on more examples both in simulation and in laboratory conditions. This new method enhances the usage of a decentralized self-tuning controller in the way that it fixes the time the adaptive controller needs to adapt its model.

Author Contributions: Conceptualization, K.P. and I.K.; methodology, K.P. and I.K.; software, I.Z.; validation, V.B., K.P. and V.T.; formal analysis, M.B.; investigation, K.P. and V.T.; resources, K.P.; data curation, K.P. and M.B.; writing—original draft preparation, K.P.; writing—review and editing, I.Z.; visualization, K.P. and I.Z.; supervision, K.P. and M.S.; project administration, K.P. and I.K.; funding acquisition, M.S. All authors have read and agreed to the published version of the manuscript.

Funding: This research was funded by by the European Regional Development Fund under the project CEBIA-Tech No. CZ.1.05/2.1.00/03.0089. This research was funded by the Ministry of Education, Science, research and Sport of the Slovak Republic under the project STIMULY MATADOR 1247/2018. Project title: Research and development of modular reconfigurable production systems using Smart Industry principles for automotive with pilot application in MoBearing Line industry.

Institutional Review Board Statement: Not applicable.

Informed Consent Statement: Not applicable.

Data Availability Statement: Not applicable.

Acknowledgments: This work was supported by the European Regional Development Fund under the project CEBIA-Tech No. CZ.1.05/2.1.00/03.0089. This article was made under support of project: STIMULY MATADOR 1247/2018. Project title: Research and development of modular reconfigurable production systems using Smart Industry principles for automotive with pilot application in MoBearing Line industry.

Conflicts of Interest: The authors declare no conflict of interest.

References

1. Pournas, E.; Jung, S.; Yadhunathan, S.; Zhang, H.; Fang, X. Socio-technical smart grid optimization via decentralized charge control of electric vehicles. *Appl. Soft Comput. J.* **2019**, *82*, 10573.
2. Yuan, Q.; Zhan, J.; Li, X. Outdoor flocking of quadcopter drones with decentralized model predictive control. *ISA Trans.* **2017**, *71*, 84–92. [CrossRef] [PubMed]
3. Blanchini, F.; Casagrande, D.; Fabiani, F.; Giordano, G.; Pesenti, R. Network-decentralized optimization and control: An explicit saturated solution. *Automatica* **2019**, *103*, 379–389. [CrossRef]
4. Wang, Y.; Hu, J.; Zheng, Y. Improved decentralized prescribed performance control for non-affine large-scale systems with uncertain actuator nonlinearity. *J. Frankl. Inst.* **2019**, *356*, 7091–7111. [CrossRef]
5. Qu, Q.; Zhang, H.; Feng, T.; Jiang, H. Decentralized adaptive tracking control scheme for nonlinear large-scale interconnected systems via adaptive dynamic programming. *Neurocomputing* **2017**, *225*, 1–10. [CrossRef]
6. Bahramipanah, M.; Cherkaoui, R.; Paolone, M. Decentralized voltage control of clustered active distribution network by means of energy storage systems. *Electr. Power Syst. Res.* **2016**, *136*, 370–382. [CrossRef]
7. Liu, D.; Yang, G.H. Decentralized event-triggered output feedback control for a class of interconnected large-scale systems. *ISA Trans.* **2019**, *93*, 156–164. [CrossRef] [PubMed]
8. Shi, W.; Yan, F.; Li, B. Adaptive fuzzy decentralized control for a class of nonlinear systems with different performance constraints. *Fuzzy Sets Syst.* **2019**, *374*, 1–22. [CrossRef]
9. Choi, Y.H.; Yoo, S.J. Decentralized adaptive output-feedback control of interconnected nonlinear time-delay systems using minimal neural networks. *J. Frankl. Inst.* **2018**, *355*, 81–105. [CrossRef]
10. Wang, C.; Wen, C.; Guo, L. Decentralized output-feedback adaptive control for a class of interconnected nonlinear systems with unknown actuator failures. *Automatica* **2016**, *71*, 187–196. [CrossRef]
11. Wang, C.; Wen, C.; Lin, Y.; Wang, W. Decentralized adaptive tracking control for a class of interconnected nonlinear systems with input quantization. *Automatica* **2017**, *81*, 359–368. [CrossRef]
12. Wang, Z.; Zhang, B.; Yuan, J. Decentralized adaptive fault tolerant control for a class of interconnected systems with nonlinear multisource disturbances. *J. Frankl. Inst.* **2018**, *355*, 4493–4514. [CrossRef]
13. Si, W.; Dong, X.; Yang, F. Decentralized adaptive neural control for interconnected stochastic nonlinear delay-time systems with asymmetric saturation actuators and output constraints. *J. Frankl. Inst.* **2018**, *355*, 54–80. [CrossRef]
14. Li, X.; Liu, X. Backstepping-based decentralized adaptive neural H_∞ tracking control for a class of large-scale nonlinear interconnected systems. *J. Frankl. Inst.* **2018**, *355*, 4533–4552. [CrossRef]
15. Tellez, F.O.; Loukianov, A.G.; Sanchez, E.N.; Corrochano, J.B. Decentralized neural identification and control or uncertain nonlinear systems: Application to planar robot. *J. Frankl. Inst.* **2010**, *347*, 1015–1034. [CrossRef]
16. Halim, D.; Luo, X.; Trivailo, P.M. Decentralized vibration control of a multi-link flexible robotic manipulator using smart piezoelectric transducers. *Acta Astronaut.* **2014**, *104*, 186–196. [CrossRef]
17. Cheng, T.M.; Savkin, A.V. Decentralized Control of Multi-robot Systems for Rectangular Aggregation. *IFAC Proc. Vol.* **2011**, *44*, 11574–11579. [CrossRef]
18. Necsulescu, D.; Pruner, E.; Kim, B.; Sasiadek, J. Decentralized Control of Autonomous Mobile Robots Formations using Velocity Potentials. *IFAC Proc. Vol.* **2012**, *45*, 31–36. [CrossRef]
19. Cheng, T.M.; Savkin, A.; Javed, F. Decentralized Control of a group of mobile robots for deployment in sweep coverage. *Robot. Auton. Syst.* **2011**, *59*, 497–507. [CrossRef]
20. Perutka, K. Adaptive LQ control with pre-identification of two tanks laboratory model. In *Annals of DAAAM and Proceedings of the International DAAAM Symposium*; Danube Adria Association for Automation and Manufacturing: Wienna, Austria, 2009; pp. 439–440.
21. Bobal, V.; Böhm, J.; Fessl, J.; Machacek, J. *Digital Self-Tuning Controllers*; Springer: London, UK, 2009.
22. Dostal, P.; Bobal, V. The suboptimal tracking problem in linear systems. In Proceedings of the 7th Conference on Control and Automation, Haifa, Israel, 28–30 June 1999; pp. 667–673.
23. Perutka, K. Decentralized Adaptive Control. Ph.D. Thesis, Tomas Bata University iz Zlin, Zlín, Czech Republic, 2007.
24. Velíšek, K.; Holubek, R.; Sobrino, D.R.; Ružarovský, R.; Vetríková, N. Design of a robotized workstation making use of the integration of CAD models and Robotic Simulation software as way of pairing and comparing real and virtual environments. In *MATEC, 4th International Conference on Computing and Solutions in Manufacturing Engineering 2016—CoSME '16*; EDP Sciences: Youlis, France, 2017.

25. Holubek, R.; Ružarovský, R.; Delgado Sobrino, D.R.; Košťál, P.; Švorc, A.; Velíšek, K. Novel trends in the assembly process as the results of human—The industrial robot collaboration. In *MATEC Web of Conferences*; EDP Sciences: Youlis, France, 2017; Volume 137.
26. TQ Tecquioment Academia-CE108 Coupled Drives Apparatus. November 2018. Available online: https://www.tecquipment.com/assets/documents/datasheets/CE108-Coupled-Drives-Datasheet.pdf (accessed on 28 July 2021).

Article

The Influence of Automated Machining Strategy on Geometric Deviations of Machined Surfaces

Ján Varga [1], Teodor Tóth [2], Peter Frankovský [3,*], Ľudmila Dulebová [1], Emil Spišák [1], Ivan Zajačko [4] and Jozef Živčák [2]

1 Department of Engineering Technologies and Materials, Faculty of Mechanical Engineering, Technical University of Košice, Letná 9, 040 02 Košice, Slovakia; jan.varga@tuke.sk (J.V.); ludmila.dulebova@tuke.sk (Ľ.D.); emil.spisak@tuke.sk (E.S.)
2 Department of Biomedical Engineering and Measurement, Faculty of Mechanical Engineering, Technical University of Košice, Letná 9, 040 02 Košice, Slovakia; teodor.toth@tuke.sk (T.T.); jozef.zivcak@tuke.sk (J.Ž.)
3 Department of Applied Mechanics and Mechanical Engineering, Faculty of Mechanical Engineering, Technical University of Košice, Letná 9, 040 02 Košice, Slovakia
4 Department of Automation and Production Systems, Faculty of Mechanical Engineering, University of Žilina, Univerzitná 8215/1, 010 26 Žilina, Slovakia; ivan.zajacko@fstroj.uniza.sk
* Correspondence: peter.frankovsky@tuke.sk; Tel.: +421-55-602-2457

Citation: Varga, J.; Tóth, T.; Frankovský, P.; Dulebová, Ľ.; Spišák, E.; Zajačko, I.; Živčák, J. The Influence of Automated Machining Strategy on Geometric Deviations of Machined Surfaces. *Appl. Sci.* **2021**, *11*, 2353. https://doi.org/10.3390/app11052353

Academic Editor: Pavol Božek

Received: 15 February 2021
Accepted: 2 March 2021
Published: 6 March 2021

Abstract: This paper deals with various automated milling strategies and their influence on the accuracy of produced parts. Among the most important factors for surface quality is the automated milling strategy. Milling strategies were generated from two different programs, CAM system SolidCAM, with the help of workshop programming in the control system Heidenhain TNC 426. In the first step, simulations of different toolpaths were conducted. Using geometric tolerance is becoming increasingly important in robotized production, but its proper application requires a deeper understanding. This article presents the measurement of selected planes of robotized production to evaluate their flatness, parallelism and perpendicularity deviations after milling on the coordinate measuring machine Carl Zeiss Contura G2. Total average deviations, including all geometric tolerances, were 0.020 mm for SolidCAM and 0.016 mm for Heidenhain TNC 426. The result is significantly affected by the flatness of measured planes, where the overlap parameter of the tools has a significant impact on the flatness of the surface. With interchangeable cutter plate tools, it is better to use higher overlap to achieve better flatness. There is a significant difference in production time, with SolidCAM 25 min and 30 s, and Heidenhain 48 min and 19 s. In accordance with these findings, the SolidCAM system is more suitable for production.

Keywords: automated machining; milling strategies; geometrical tolerances

1. Introduction

Industrial production has been accompanied by the development and innovation of manufactured products and manufacturing technologies with the primary aim of overcoming the implemented limits for the whole history. Due to this fact, innovations are an important aspect of competitiveness, which leads to the cycle of acquiring and implementing new knowledge in the practice [1]. Today's trend is to increase productivity and reduce machining costs [2].

Although a level of automation has been achieved in the process planning of 3-axis milling, the present implementation can be improved with the incorporation of some added functionalities. Appropriate tool selection is needed to determine the optimal tool sequences and to adapt the automated feed/speed selection [3]. This technology is increasingly required in practice in automated or robotic form, where computer-assisted technology is increasingly used. These modern technologies are used in various branches of development and production [4].

The importance of the implementation of CAD CAM systems is often described in the literature. The basic sign of CAM is to minimize human intervention in the course of the production process by the exploitation of computer data processed in the main elements of activities [5].

When machining parts, we encounter several factors, the result of which is a lack of precision of parts. These are factors resulting from the CAM system that are caused by the approximation toolpaths. We can also define an error as any deviation in the actual position of the cutting edge of the cutter from the position that was theoretically programmed for the production of the part within the required tolerance [6].

At present, there is an increasing demand for higher dimensional and geometric accuracy and surface treatment of manufactured parts in various areas of the industry. Among the industries that require tighter tolerances to be achieved, typical examples constitute the automotive, aerospace and medical devices industries, including manufacturing trends such as automatic assembly and micro manufacturing. In this regard, realization of the desired part accuracy imposes criteria in the selection of manufacturing systems, e.g., machine tools. Part accuracy is closely related to machining system capability, which in turn is determined by the interaction between machine tool and cutting processes. These are mainly interfaces between the tool and the workpiece, where positional, static, thermal and dynamic accuracy has a significant effect [7]. The manufacturing of high-accuracy parts sets the basic requirement of a robust machining system able to manufacture within the specified tolerances. Although a controlled and repeatable process is important for maintaining quality, machining system accuracy and precision have a determinant role in part accuracy.

Among the challenges that machine and device parts producers face nowadays is the necessity to guarantee a high quality of machined surfaces. Obtaining a high-quality product is an extremely difficult and complex task that entails fulfilling several crucial requirements [8]. The machining parameters are an important part of process plan performance. Equally important in machining is the confidence in the measuring tools, from which part quality characteristics are ascertained [9]. Due to the importance of the finishing stage in the manufacturing processes, the face milling process is the solution that can be used to achieve good surface quality and high accuracy in a short period of time. To achieve the high quality of the desired parts, studying the tool path strategies is inevitable too.

Surface properties constitute a significant measure of the finished component quality, because they influence features such as dimensional accuracy, friction coefficient, wear, post processing requirements, appearance and cost. The quality of the surface produced by milling depends on various technological parameters, such as the cutting conditions as well as the cutting tool and the workpiece specification [10].

Lopez de Lacalle et al. [11] describe the stages of preparation of the milling process and factors impacting the optimization of cutting strategies.

To develop an effective milling process, the programmer must choose the right strategies. Appropriate selection of the tool path is important because it defines the axial depth of cut, which influences the maximum cutting force during machining. Its correct definition controls the overall productivity of the machining process and can usually be influenced by computer production systems. Likewise, the method of milling up or down together with the direction of rotation of the spindle also controls the direction of chip removal [12,13]. When choosing a milling strategy, various factors must be considered, for example, surface roughness or form deviation. The geometry of the tool-related parameters (feed and cutting speed) is equally essential to achieve quality after machining [14].

Among the most used strategies in the production of parts by milling is strategy zigzag, and contour, which we can define using a computer-aided manufacturing system [15]. The zigzag strategy, known as the raster strategy of milling, is a strategy in which the milling cutter moves back and forth along the workpiece in the X-Y plane. In the machining process, we may encounter another problem, such as the generation of toolpaths, which affects not only the performance of the production process but also the quality of machined surfaces [8].

The machining strategy is the methodology used to calculate the operation, where the purpose is to obtain the desired geometric entity in its final form [16]. The efficiency of using a particular toolpath is related to the time required to complete the machining process. The lower limit of the total cutting time for a 2D area is the area divided by the cutting width and the cutting feed rate [17]. Part of the production process is a postprocessor, which provides information about the position of the tool and process parameters [18].

An inseparable part of each manufacturing process is its control. The assessment of manufacturing correctness allows us to define the compatibility between the manufactured product and design requirements. It also allows us to verify the correctness of operation of stages of the manufacturing process [19,20]. Depending on expectations, one can choose the measuring devices which can give metrologically correct results for required time limits. These requirements apply to all control processes connected with the measurement of mass, time or geometrical quantities [21].

In industry, dimensional and geometric accuracy is checked after the end of the production process. A widely used machines for manufacturing process control is the coordinate measuring machine (CMM) [21]. The increasing importance of geometric tolerances points out many questions related to the mathematical method of evaluation, the point sampling strategies in CMM and the manufacturing possibilities and constraints.

Among the most used basic elements/features of a mechanical part are planar elements. The relations between features are defined by dimensions and geometric tolerances, such as flatness, parallelism, perpendicularity and straightness. During the production of the part, dimensional and form deviations from the prescribed value are present [22].

The most used form of deviation for controlling the milling process and checking the machine tools is the flatness [8]. A key issue in flatness measurements is the definition of an appropriate reference plane [16]. Ideally, all points of the measured plane are in one plane [23].

The different types of tolerances show the allowed errors. Considering the errors, the most often used tolerances are the dimensional tolerance, the surface roughness, the shape, position and orientation tolerances (geometric tolerances).

Mikó et al. [24] investigate the dependency between roughness and flatness measured by 12 methods of point sampling strategies. The results show that the evaluated parameters are dependent.

Hazarika et al. [25] studied the influence of surface roughness on parallelism and perpendicularity in milling. The results from the experiment show that surface roughness affects the parallelism and perpendicularity errors, thus confirming the results of Mikó [24].

Mikó and Farkas [26] state the impact of milling machines (milling machine and CNC machining center) on flatness and surface roughness. The results are represented with values of parameters and with the topographical distribution of deviations.

Gapinski et al. [27] evaluate dimensions and geometric tolerances on different types of measuring machines—optical, roundscan, contour measuring machine, CMM and computed tomography. The results clearly show the strengths and weaknesses of measuring devices. The most versatile device is the coordinate measuring machine.

Modern CNC machines or new cutting tool designs provide many possibilities for improving the accuracy of the workpiece surface. In practice, we often encounter the optimization of the machining process, in which, for example, by changing the depth of cut, changing the feed rate, or analyzing the orientation of the workpiece, it is possible to achieve a change in the flatness of the part [28].

Dobrzynski et al. [29] present in their research the impact of machining direction on surface flatness. The result show that length of the face milled section affects the values of flatness deviations; the flatness deviation grows with machining length.

Various contributions examining the flatness of machined surfaces after milling have proven their justification in choosing the right machining strategy for surface quality [30].

Sheth and George [31] investigate the effect of machining parameters on surface roughness and flatness. For the prediction of machining parameters, the results were analyzed by ANOVA. The error of predicted surface roughness was up to 11% and flatness 7%.

Gapinski et al. [32] evaluate the suitability of a coordinate measuring machine to measure the circle radius, position of the center and roundness. For element creation, different fitting methods and numbers of measuring points were used. The authors state that the uncertainty of CMM is the limiting factor.

There are always certain differences between the designed and realized properties that occur during the production process. They are caused by changes in various parameters, such as the machine tool, material, human factors, measuring devices or process parameters. The number of factors affecting the deviation can be reduced by quality control and by the planning of the production process. The acceptable level of deviations is defined as the tolerance of dimensions and geometry. During the measurement with a coordinate measuring machine or any other suitable measuring device, the measuring strategy defines the number of recorded points and the created element [33].

Mikó and Drégelyi-Kiss [34], in three types of milling machines, investigated flatness, perpendicularity and parallelism. The results show that the cutting tool and machine rigidity affect the values of form and shape errors, and the flatness of vertical and horizontal planes is affected by the depth of cut. Additionally, the machine tool, the direction of milling and the position of the surface also affect geometric errors.

Lakota and Görög [35], in their research, analyze the measurement of flatness with seven single- and multi-point measuring methods. The results show that the value of flatness is also affected by the orientation of the profile lines with respect to the machining path and data filtration.

Kawalec and Magdziak [36] investigate the influence of the number of measurement points on the accuracy of the CNC machine tool. The numerical investigation uses from 4 to 200 points, and the results show that the minimum number of points to achieve the minimum deviation is 50.

Mikó [37] presents the factors affecting the measurement of flatness (machining method, strategy, milling parameters, etc.) and the influence of the number of measuring points. The results show that the flatness values are in the range from 0.124–0.0572 mm.

Nowakowski et al. [38] present, in their research, the impact of the depth of cut on surface flatness. Results show the flatness deviation was 6.7 μm between the cutting strategies. The maximum deviation was reported at the beginning and end of the cutting process.

The milling strategy, known as raster, and the use of offset paths, as claimed by Sarma [17], have their advantages and disadvantages. When milling with raster, a shorter toolpath is created, but it is not possible to completely remove traces known as scallop height. By contrast, using a strategy known as offset, it is possible to remove the remaining scallop height to obtain a smooth surface. Toh [39] also described a comprehensive overview of the most common strategies used in the machining process, offset and zig-zag, in his publications.

The aim of the research, conducted by the authors of this work, was to provide the result of a comparison of milling strategies in two different programs and evaluated geometric characteristics of the produced test specimens. The article presents the results of the evaluation of the deviation of flatness, perpendicularity and parallelism. In two different programs, two parts were produced with the same cutting conditions. The main difference was only in the overlap of the tool. The question is the following: What is the relationship among the influence of the program, the control system of the accuracy of part production (dimensional accuracy) and tool overlap in selected geometric deviations? As part of our results, the flatness was determined based on 1945 points by least square (LS). The aim of the article was to verify the effect of the combination of the tools for finishing operations and their overlap.

2. Plan of Experiment

2.1. Materials

In this work, two identical workpieces were produced. The 1st design of the test sample was created in the CAD system SolidWorks and the 2nd in the control system Heidenhain TNC 426. Stock dimensions were programmed in the control system Heidenhain TNC 426 at the beginning of the program in line with stock marked as command BLK FORM.

Test samples were made of aluminum alloy (AlCu4Mg); the dimensions are shown in Figure 1. Table 1 lists the chemical composition of the aluminum alloy. The selected mechanical properties of AlCu4Mg are the following: tensile strength = 105.8 MPa; yield strength = 92.4 MPa; hardness = 73.5 BHN.

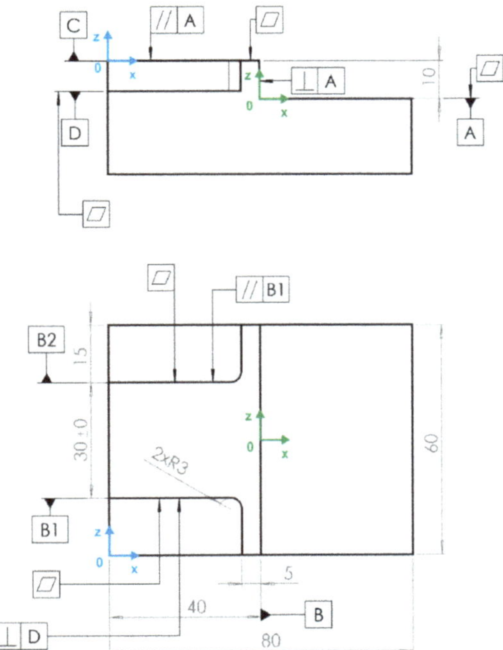

Figure 1. The dimensions and geometrical tolerances of test sample.

Table 1. Chemical composition of AlCu4Mg alloy [17].

Chemical Composition	Content [%]
Cu	4.30
Mg	0.79
Fe	0.26
Si	0.24
Mn	0.3
Ti	0.04
Zn	0.04
Al	balance

2.2. Machining Device, Cutting Tools and Cutting Parameters

The machining of aluminum parts was carried out on the 3-axis milling machine EMCO Concept Mill 155 (EMCO MAIER Ges.m.b.H., Hallein/AUSTRIA) with control system Heidenhain TNC 426 (JOHANNES HEIDENHAIN GmbH, Germany). CAM software today offers several possibilities of strategies for distributing the tool path in the domain of the designed part. Table 2 shows a list of cutting tools and cutting parameters for milling of the AlCu4Mg alloy. The methodology for selecting cutting parameters was chosen based on the recommendations of the tool manufacturer and concerning the parameters of the CNC machine. The maximum RPM of the machine EMCO MILL 155 was 5000. The tool of Ø18 mm in diameter was produced by Korloy (designation AMS 2018S) with two interchangeable cutters plates marked APXT11T3PDR-MA. The other tools were produced by ZPS-FN.

Table 2. Cutting tools and cutting parameters for milling of AlCu4Mg alloy.

Tool Diameter [mm]	Cutting Speed [m·min^{-1}]	Feed per Tooth [mm]	Spindle Frequency [RPM]	Tool Producer	Tool Code
End Mill D 18	270	0.125	4800	Korloy	AMS2018S
End Mill D 14	299	0.021	4800	ZPS-FN	120517
End Mill D 10	232	0.03	4800	ZPS-FN	270618
End Mill D 6	259	0.03	4600	ZPS-FN	273618

The manufacturing process of the part was performed by the face milling, roughing and finishing strategies (Figure 2). For both produced test samples, the same roughing and finishing strategies were used. The coordinate system defined for the milling process is shown in Figure 1 (blue cross). The production of samples is described in the following steps:

- Face milling of the top surface marked Plane C, in Figure 3, where the strategy zig-zag was used.
- Milling (roughing) a flat side surface marked Plane A, A-PER, where strategy contour with offset was used. After this operation, an allowance of 0.3 mm was left, which was removed by a tool with a smaller diameter.
- Milling (roughing) a two-sided surface marked Plane B1, B1-PER_long, Cyl1, B1_short, Plane B2, B2-PER_long, Cyl2, B2_short. After this operation, the allowance of 0.3 mm was left too and was removed by a tool with a smaller diameter.

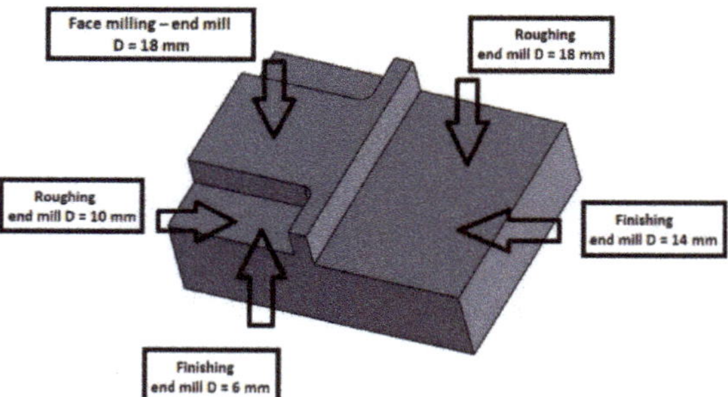

Figure 2. The manufacturing processes.

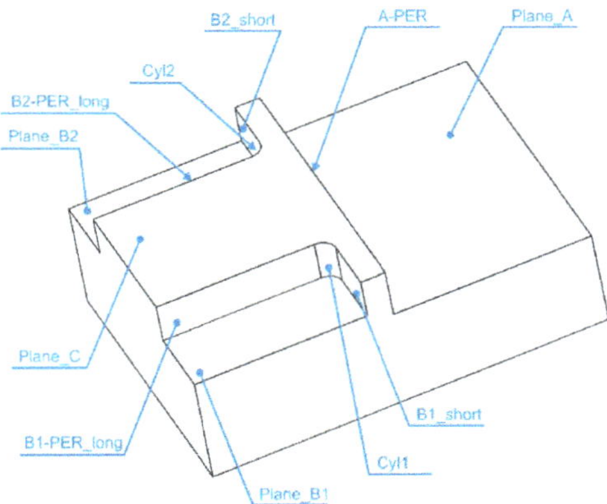

Figure 3. Surfaces used for measuring.

For production two different programs were used. The first was the CAM system Solidcam and the second was the control system Heidenhain TNC 426. The tool path of the tool for the milling of this part was programmed manually in the control system Heidenhain TNC 426. For the produced test samples, both programs used the same strategies and the same cutting parameters. The shape of parts was obtained in two stages: roughing and finishing. After each roughing operation, an allowance of 0.3 mm and a depth of cut for each operation of ap = 1.5 mm were implemented on the surface.

We compared two different programs, where the strategies were programmed, and their effect on the quality of the surface of test samples was evaluated. Surface quality is influenced by several process parameters, such as tool diameter, cutting parameters, work piece material properties, clamping of the part and milling strategy. Among the most important factors is the milling strategy, which influences produced part accuracy.

2.3. Programming NC Program

HEIDENHAIN TNC uses programmable controls that allow the user to define paths for creating contours and pockets for drilling operations, etc. It is possible to program these conventional operations directly on the machine by defining cycles. It is also possible to change the angular position of the spindle, which also corresponds to the definition of certain cycles. In this case, it is a simple creation of programs, where the graphics illustrate all the steps of the machining process of the selected part.

System SolidCAM, on the other hand, is an additional CAD model of the Solidworks system. This computer aided manufacture system offers solutions for all CNC applications. In our case, the solution used was known as 2.5D Milling.

3. Measurement of Geometrical Characteristics

The most popular method of geometrical dimension control is the coordinate measuring technique. This measurement strategy plays an important role in the accuracy of obtained results [23,24]. The number of points collected during the measurement, the calculation of geometric element, filtering and outlier eliminations are the most influencing factors. The setting of these parameters is mostly left to the operator, and significantly affects the value of the flatness and the associated uncertainty [13]. The measurement of surface quality obtained from the CAM system SolidCAM and control system Heidenhain TNC 426 was conducted. After that, geometric characteristics of the produced part were measured with the use of the coordinate measuring machine Carl Zeiss Contura G2 with an RDS articulating probe holder and VAST XXT scanning sensor. The maximum permissible error (MPE_E) of machine was calculated according to ISO 10360-2, where L is the measured length dimension in mm.

$$MPE_E = \left(1.8 + \frac{L}{300}\right)(\mu m) \tag{1}$$

The elements used in the measurement are defined in Figure 3.

Dimension and geometric tolerance evaluation was performed with a coordinate base system consisting of PlaneA (Spatial rotation, Z origin), the PER-A plane (X origin) and the symmetry of the B1-PER_long and B2-PER_long planes (Planar rotation, Y origin). The origin is shown in Figure 1 (green color). In the analysis of production, one length dimension in four ways and 20 geometric tolerances were evaluated. The measured surfaces (Figure 3) describe their identification, surface types and deviations. Two styli were used for the measurement of test samples.

The 1st stylus was used for the measurement of all dimensions and geometrical tolerances except cylindricity. The probe diameter = 1 mm, and the length of stylus = 20 mm. For the measurement, a 50 mm extension was used, because the minimum recommended stylus length is 40 mm. The 2nd stylus with a diameter of 3 mm and length of 50 mm was used for measurement of cylindricity.

4. Evaluations of Geometrical Characteristics

The deviation results obtained by CMM for individual test samples are shown in Table 3. All data are given as deviations from the nominal value. The nominal values for dimension and geometrical tolerances are shown in Figure 1.

Table 3. Coordinate measuring machine (CMM) results.

Name	Deviation SolidCam [mm]	Deviation Heidenhain [mm]
FLT_Plane_A	0.0079	0.0085
FLT_Plane_C	0.0406	0.0139
PAR_C_pd_A	0.0389	0.0379
FLT_Plane_B1	0.0210	0.0093
FLT_Plane_B2	0.0233	0.0106
FLT_A-PER	0.0056	0.0107
PER_A_per	0.0092	0.0159
Distance_LSQ_1X_Y	−0.2299	−0.2056
Distance_LSQ_4X_Y	−0.2179	−0.2486
Distance_LSQ_1Z_Y	−0.2400	−0.2471
Distance_LSQ_10Z_Y	−0.2455	−0.2477
FLT_B1-PER_long	0.0315	0.0156
FLT_B2-PER_long	0.0167	0.0256
PAR_B1_per-long_1	0.0063	0.0045
PAR_B1_per-long_2	0.0081	0.0141
PAR_B1_per-long_3	0.0152	0.0127
PAR_B1_per-long_4	0.0054	0.0126
PER_Plane_B1_PER_long_pd_PER	0.0355	0.0266
PER_B1_short_pd_B1	0.0362	0.0431
PER_B2_short_pd_B2	0.0395	0.0463
PER_Cyl1_Par_to_Plane_A_PER	0.0167	0.0054
PER_Cyl1_Per_to_Plane_A_PER	0.0131	0.0146
PER_Cyl2_Par_to_Plane_A_PER	0.0116	0.0154
PER_Cyl2_Per_to_Plane_A_PER	0.0012	0.0035

According to the same cutting parameters for the parts produced in the CAM system and in the control system Heidenhain TNC 426, the evaluation was divided into six groups:

- Flatness—horizontal planes;
- Flatness—vertical planes;
- Perpendicularity—planes;
- Perpendicularity—cylinders;

- Parallelism;
- Distance.

For the comparison of production, graphs were created in accordance with separate groups, which include the comparison characteristics for both systems. The green color represents SolidCAM, and blue represents the control system Heidenhain TNC 426. The results of flatness measured on the horizontal planes are shown in Figure 4.

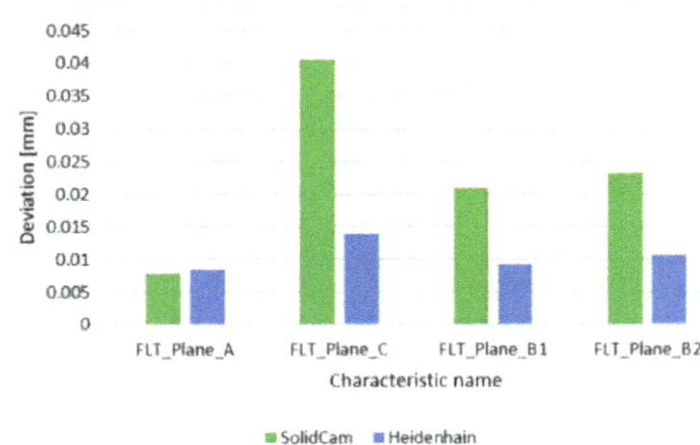

Figure 4. Deviation for flatness—horizontal.

Results of the deviations in three characteristics are higher for SolidCAM than for Heidenhain TNC 426. The smaller deviation is only by 0.6 μm of Plane_A. The surface flatness is marked as Plane_C for both systems, SolidCAM and Heidenhain TNC 426 (Figure 5).

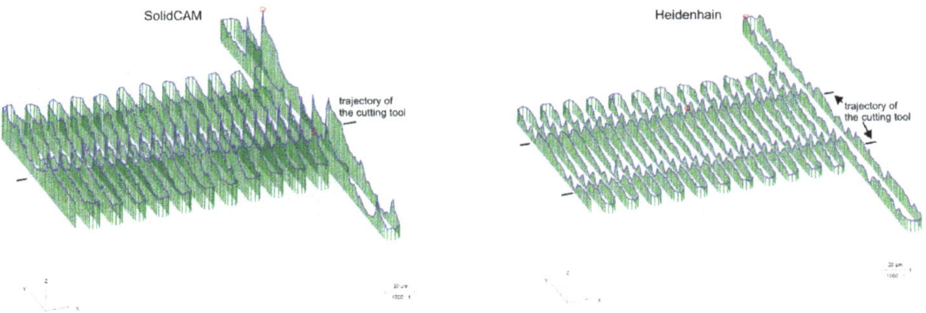

Figure 5. Flatness for Plane_C.

The flatness was measured at a speed of 8 mm/s along a defined path (polyline) without data filtration and outlier elimination. In both cases, the scale was over 1000: 1. The maximum and minimum values of the flatness deviation are marked with red marks. In the case of the Heidenhain system, the process of deviations had no significant differences compared to the SolidCAM system. In both cases, the trajectory of the cutting tool was clearly visible (decrease in the values of deviations in one line).

When comparing the two systems, the difference in the deviations of the flatness can be observed. In both systems, the trajectory of the cutting tool was visible (see Figure 5). In the CAM system, SolidCAM, where the tool overlap was 50%, the trajectory of the cutting tool was clearly visible, marked with lines, but in the Heidenhain control system, where a tool overlap of 70% was used, the height difference was slight. The Heidenhain control system showed a more even distribution of flatness deviations, which can be explained by the overlapping of the tool during the milling process. The denser overlap of the toolpaths during machining increased the number of passes to machine the surface, which increased the milling time, but the surface showed better results in terms of flatness. The comparison between flatness and vertical planes is shown in Figure 6. Flatness measurement tolerances were measured at a speed of 8 mm/s along a defined path (polyline), and a spline filter with wavelength Lc = 2.5 mm was used (roughness Ra = 0.4–3.2 μm).

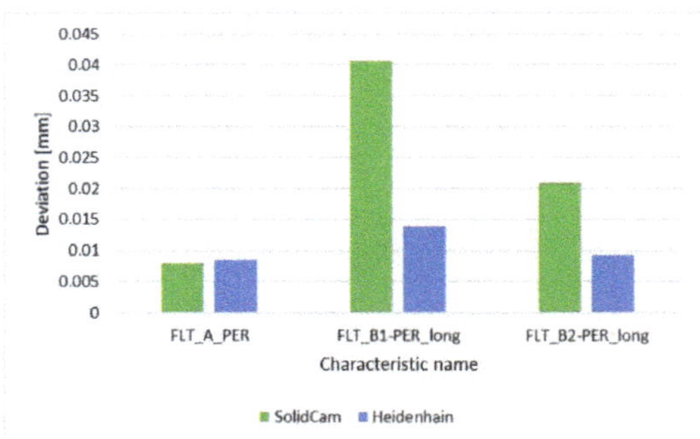

Figure 6. Deviation for flatness—vertical.

In two cases, flatness was better for the CAM system SolidCAM and in one case for Heidenhain TNC 426. After all flatness deviations were compared, globally smaller deviations for Heidenhain TNC 426 were found in four cases and for system SolidCAM in three cases. In the case of the evaluation of perpendicularity, it was necessary to define the reference plane to which the given tolerance would be evaluated.

Table 4 describes the evaluated geometric tolerances and reference planes for the perpendicularity of the planes. All perpendicularity tolerances were measured at a speed of 8 mm/s along a defined path (polyline) and a spline filter with wavelength *Lc* = 2.5 mm.

Table 4. Evaluated geometric tolerances and reference planes for the perpendicularity of the planes.

	Evaluated Plane	**Reference Plane**
PER_A_per	A-PER	Plane_A
PER_Plane_B1_PER_long_pd_PER	B1-PER_long	Plane_A
PER_B1_short_pd_B1	B1_short	Plane_B1
PER_B1_short_pd_B2	B2_short	Plane_B2

The evaluation of the perpendicularity of the planes showed that the plane PER_A_per had the lowest perpendicular deviation, and planes B1_short a B2_short had the highest. Three deviations of four SolidCAM systems had a smaller deviation. The difference between systems SolidCAM and Heidenhain TNC 426 was less than 0.01 mm (Figure 7).

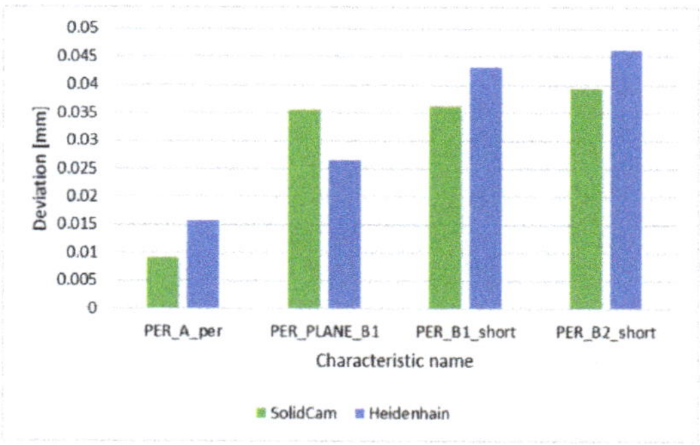

Figure 7. Deviation for perpendicularity—planes. Equal signs: PER_PLANE_B1_PER_long_pd_PER = PER_PLANE_B1, PER_B1_short_pd_B1 = PER_B1_short, PER_B1_short_pd_B2 = PER_B2_short.

The perpendicularity of cylinders was evaluated in two methods: as the perpendicularity of the cylinder to the primary reference plane Plane_A in parallel direction to the secondary reference plane A-PER and as the perpendicularity of the cylinder to the primary reference plane Plane_A in the perpendicular direction to the secondary reference plane A-PER. In three cases, the deviation (Figure 8) was smaller for the SolidCAM system, although only minimally (maximum difference was 0.0038 mm).

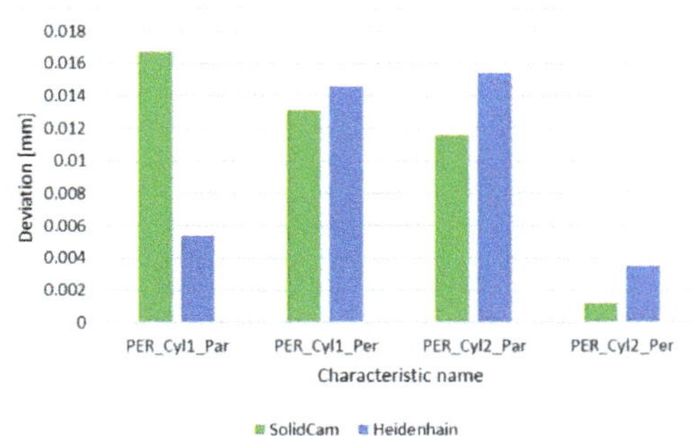

Figure 8. Deviations for perpendicularity—cylinders. Equal signs: PER_Cyl1_Par_to_Plane_A_PER = PER_Cyl1_Par, PER_Cyl1_Per_to_Plane_A_PER = PER_Cyl1_Per, PER_Cyl2_Par_to_Plane_A_PER = PER_Cyl2_Par, PER_Cyl2_Per_to_Plane_A_PER = PER_Cyl2_Per.

The parameters of plane parallelism evaluations are shown in Table 5. The parallelism was measured at a speed of 8 mm/s along a defined path (polyline), and a spline filter with wavelength Lc = 2.5 mm (roughness Ra = 0.4–3.2 μm) was used.

Table 5. Parameters of planes parallelism evaluations.

	Evaluated Plane	**Reference Plane**
PAR_C_pd_A	Plane_C	Plane_A
PAR_B1_per-long_1	B2-PER_long	B1-PER_long
PAR_B1_per-long_2	B2-PER_long	B1-PER_long
PAR_B1_per-long_3	B2-PER_long	B1-PER_long
PAR_B1_per-long_4	B2-PER_long	B1-PER_long

The parallelism deviation of the two horizontal planes (PAR_C_pd_A) showed a significantly higher value than area B2_long to area B1_long. The plane PAR_C_pd_A was measured by the polyline path. The main difference was in the form of a deviation of parallelism. The SolidCAM system showed larger differences than the Heidenhain system.

The parallelism of PAR_B1_per-long was measured using 4 lines. The parallelism of each line from plane B2_long to plane B1_long was evaluated. In two cases, the SolidCAM system had a smaller deviation (Figure 9), and in two cases, it was Heidenhain TNC 426. The maximum difference was less than 0.0072 mm. When comparing the two systems on the PAR_C_pd_A plane, no significant difference in parallel deviations or in their distribution was visible (Figure 10). With the CAM system SOLIDCAM, where the tool overlap was 50%, a more pronounced tool path on the surface could be seen than in the Heidenhain system, where the tool overlap was 70%.

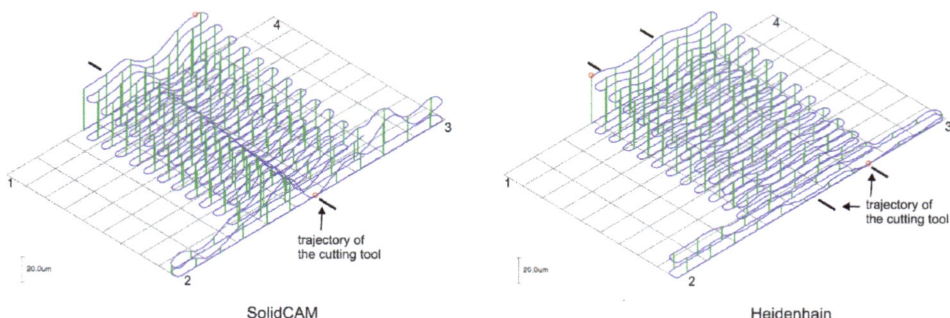

Figure 9. Parallelism of plane PAR_C_pd_A.

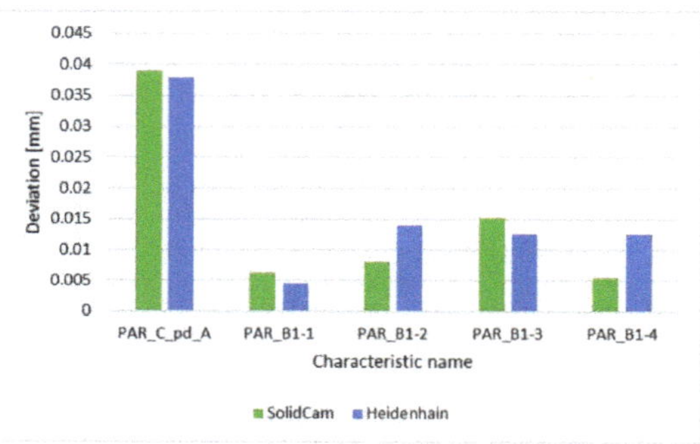

Figure 10. Parallelism of plane PAR_C_pd_A.

When comparing the two systems, the difference in deviations of the parallelism could be seen. In the CAM system SolidCAM, where the tool overlap was 50%, a greater surface waviness was observed than in the case of the Heidenhain system, where the tool overlap was 70%. The Heidenhain control system showed a lower surface waviness of the measured deviations of parallelism compared to the CAM system SolidCAM, where the surface was machined earlier, but the deviations of parallelism were larger in this case. This is demonstrated by the greater surface waviness of the machined surface.

One dimension on the part was evaluated, namely, the distance between planes B1_long and B2_long. The LSQ Featured method was used for measurement. When measuring Distance_LSQ_1X_Y and Distance_LSQ_4X_Y, 4 horizontal lines were created on the given surfaces. The distance between them in the *y*-axis was evaluated. When measuring Distance_LSQ_1Z_Y and Distance_LSQ_10Z_Y, 10 vertical lines were created on the given surfaces. The results are shown in Figure 11.

The difference between results for measuring horizontal lines was minimal, and for perpendicular lines, the difference was close to zero.

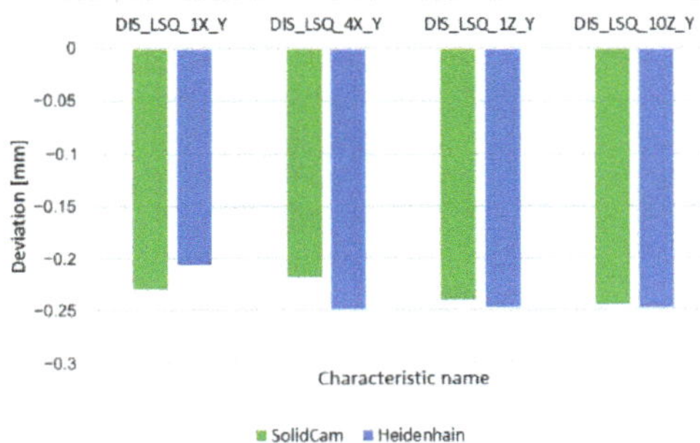

Figure 11. Distance Plane_B1 and Plane_B2.

5. Results

Based on the measurement of the defined elements performed on a coordinate measuring machine, values were obtained for the prescribed dimensions and geometric tolerances of the designed test part. The obtained results led to the following conclusions:

1. The flatness comparison of the surfaces showed that the average flatness value for all surfaces was 0.023 mm for SolidCAM and 0.011 mm for Heidenhain. Average deviations in CAM system were significantly influenced by the flatness of plane Plane_C, which was higher than in system Heidenhain TNC 426. The comparison of the two systems in Plane_C showed the difference in the height deviations of flatness. In the CAM system, SolidCAM, where the tool overlap was 50%, the trajectory of the cutting tool was clearly visible. In the control system Heidenhain, where a tool overlap of 70% was used, the trajectory of tool was still visible, but the deviations from ideal plane were smaller. The Heidenhain control system showed more even height differences, which can be explained by the overlapping of the tool only during the milling process. The denser overlap of the toolpaths during machining increased the number of passes to machine the surface, which increased the milling time, but the surface showed better results in terms of flatness.

2. The comparison of the geometric tolerances of perpendicularity showed that the average value for the SolidCAM system was 0.020 mm and for the Heidenhain TNC 426 system 0.020 mm, so there were no significant differences between them.

3. The parallelism comparison in a global view also did not show a significant difference between the systems. For SolidCAM, the average deviation was 0.015 mm, and for Heidenhain, 0.016 mm. The comparison of the two systems in PAR_C_pd_A showed the difference in the distribution of deviations of parallelism. In the CAM system, SolidCAM, where the tool overlap was 50%, a greater surface waviness was observed than in the case of the Heidenhain system, where the tool overlap was 70%. The Heidenhain control system showed a lower surface waviness of the measured deviations of parallelism compared to the CAM system, SolidCAM, where the surface was machined earlier, but the deviations of parallelism were larger in this case. This was demonstrated by the greater surface waviness of the machined surface.

4. The total average deviation, including all geometric tolerances, was 0.020 mm for SolidCAM and 0.016 mm for Heidenhain TNC 426. The result was significantly affected by flatness, where the SolidCAM system showed significantly higher values in all other comparisons than Heidenhain.

5. For dimensional control, measurements between parallel surfaces B1_per-long and B2_per-long were realized. The average deviation for SolidCam was −0.233 mm, and for Heidenhain, −0.237 mm, so this was an inappreciable difference.

6. There was a significant difference in production time, with SolidCAM 25 min and 30 s, and Heidenhain 48 min and 19 s. In accordance with these findings, the SolidCAM system is more suitable for production.

6. Conclusions

The paper deals with the production accuracy comparison of two different CNC programming systems. Results showed that the highest impact on accuracy is defined by tool overlap in the reference plane.

In the system comparison, the total sample production time was longer in the Solid-CAM system than in Heidenhain TNC 426. The reason for this is that SolidCAM generated the tool path of the tool with the whole model (geometry, shape of the model, accuracy tolerance, etc.) in accordance with Heidenhain TNC 426, where to define the strategy, the system used concrete types of the cycles, which are included in this system.

According to the obtained results, the future plan is to provide further measurements of larger numbers of samples; select another material; and extend the measurements by other types of deviations, such as angularity, position and profile of a surface.

Therefore, it would also be interesting to monitor the impact of milling strategies on the roughness supported by measuring cutting forces.

Author Contributions: Conceptualization, J.V. and T.T.; methodology, J.V. and T.T.; validation, I.Z., J.V. and T.T.; formal analysis, P.F., J.V., T.T. and Ľ.D.; investigation, J.V. and T.T.; resources, J.V.; writing—original draft preparation, J.V. and T.T.; writing—review and editing, J.V., T.T., and Ľ.D.; visualization, I.Z., J.V. and T.T.; supervision, J.Ž. and E.S.; project administration, I.Z. and Ľ.D.; funding acquisition, P.F., J.Ž. and E.S. All authors have read and agreed to the published version of the manuscript.

Funding: This research was funded by The Ministry of Education, Science, Research and Sport of the Slovak Republic, grant numbers VEGA 1/0384/20, VEGA 1/0168/21, VEGA 1/0290/18, KEGA 016TUKE-4/2021, KEGA 041TUKE-4/2019, APVV-16-0283 and ITMS: 26220220182 supported by the Research & Development Operational Programme funded by the ERDF.

Institutional Review Board Statement: Not applicable.

Informed Consent Statement: Not applicable.

Data Availability Statement: Not applicable.

Conflicts of Interest: The authors declare no conflict of interest. The funders had no role in the design of the study; in the collection, analyses, or interpretation of data; in the writing of the manuscript, or in the decision to publish the results.

References

1. Vaško, M.; Sága, M.; Majko, J.; Vaško, A.; Handrik, M. Impact Toughness of FRTP Composites Produced by 3D Printing. *Materials* **2020**, *13*, 5654. [CrossRef]
2. Peterka, J.; Pokorny, P.; Vaclav, S.; Patoprsty, B.; Vozar, M. Modification of Cutting Tools by Drag Finishing. *MM Sci. J.* **2020**, *2020*, 3822–3825. [CrossRef]
3. Turley, S.P.; Diederich, D.M.; Jayanthi, B.K.; Datar, A.; Ligetti, C.B.; Finke, D.A.; Saldana, C.; Joshi, S. Automated Process Planning and CNC-Code Generation. In Proceedings of the IIE Annual Conference and Expo 2014, Montreal, QC, Canada, 31 May–3 June 2014; Institute of Industrial and Systems Engineers (IISE): Norcross, GA, USA, 2014; pp. 2138–2144.
4. Božek, P. Optimiziranje Putanje Robota Kod Točkastog Zavarivanja u Industriji Motornih Vozila—Robot Path Optimization for Spot Welding Applications in Automotive Industry. *Teh. Vjesn.* **2013**, *20*, 913–917.
5. Peterka, J.; Pokorny, P.; Vaclav, S. CAM Strategies and Surfaces Accuracy. *Ann. DAAAM Proc.* **2008**, *19*, 1061–1063.
6. López De Lacalle, L.N.; Lamikiz, A.; Muñoa, J.; Salgado, M.A.; Sánchez, J.A. Improving the High-Speed Finishing of Forming Tools for Advanced High-Strength Steels (AHSS). *Int. J. Adv. Manuf. Technol.* **2006**, *29*, 49–63. [CrossRef]
7. Laspas, T. *Modeling and Measurement of Geometric Error of Machine Tools: Methodology and Implementation*; Royal Institute of Technology: Stockholm, Sweden, 2014.
8. Nadolny, K.; Kapłonek, W. Analysis of Flatness Deviations for Austenitic Stainless Steel Workpieces after Efficient Surface Machining. *Meas. Sci. Rev.* **2014**, *14*, 204–212. [CrossRef]
9. Ali, S.H.R.; Mohamd, O.M. Dimensional and Geometrical Form Accuracy of Circular Pockets Manufactured for Aluminum, Copper and Steel Materials on CNC Milling Machine Using CMM. *Int. J. Eng. Res. Africa* **2015**, *17*, 64–73. [CrossRef]
10. Vakondios, D.; Kyratsis, P.; Yaldiz, S.; Antoniadis, A. Influence of Milling Strategy on the Surface Roughness in Ball End Milling of the Aluminum Alloy Al7075-T6. *Meas. J. Int. Meas. Confed.* **2012**, *45*, 1480–1488. [CrossRef]
11. López de Lacalle, L.N.; Lamikiz, A.; Salgado, M.A.; Herranz, S.; Rivero, A. Process Planning for Reliable High-Speed Machining of Moulds. *Int. J. Prod. Res.* **2002**, *40*, 2789–2809. [CrossRef]

12. Vila, C.; Abellán-Nebot, J.V.; Siller-Carrillo, H.R. Study of Different Cutting Strategies for Sustainable Machining of Hardened Steels. *Procedia Eng.* **2015**, *132*, 1120–1127. [CrossRef]
13. Varga, J.; Stahovec, J.; Beno, J.; Vrabeľ, M. Assessment of Surface Quality for Chosen Milling Strategies When Producing Relief Surfaces. *Adv. Sci. Technol. Res. J.* **2014**, *8*, 37–41. [CrossRef]
14. Chen, Z.C.; Dong, Z.; Vickers, G.W. Automated Surface Subdivision and Tool Path Generation for 3 1/2 1/2-Axis CNC Machining of Sculptured Parts. *Comput. Ind.* **2003**, *50*, 319–331. [CrossRef]
15. Ali, R.; Mia, M.; Khan, A.; Chen, W.; Gupta, M.; Pruncu, C. Multi-Response Optimization of Face Milling Performance Considering Tool Path Strategies in Machining of Al-2024. *Materials* **2019**, *12*, 1013. [CrossRef] [PubMed]
16. Quinsat, Y.; Sabourin, L. Optimal Selection of Machining Direction for Three-Axis Milling of Sculptured Parts. *Int. J. Adv. Manuf. Technol.* **2007**, *33*, 684–692. [CrossRef]
17. Sarma, S.E. Crossing Function and Its Application to Zig-Zag Tool Paths. *CAD Comput. Aided Des.* **1999**, *31*, 881–890. [CrossRef]
18. Bílek, O.; Páč, J.; Lukovics, I.; Čop, J. CNC Machining: An Overview of Available CAM Processors. In *Development in Machining Technology*; Politechnika Krakowska: Krakow, Poland, 2014; pp. 75–89. ISBN 978-83-7242-765-6.
19. Salihu, S.A. Influence of Magnesium Addition on Mechanical Properties and Microstructure of Al-Cu-Mg Alloy. *IOSR J. Pharm. Biol. Sci.* **2012**, *4*, 15–20. [CrossRef]
20. Jalid, A.; Hariri, S.; Laghzale, N.E. Influence of Sample Size on Flatness Estimation and Uncertainty in Three-Dimensional Measurement. *Int. J. Metrol. Qual. Eng.* **2015**, *6*, 102. [CrossRef]
21. Ali, S.H.R.; Mohamed, H.H.; Bedewy, M.K. Identifying Cylinder Liner Wear Using Precise Coordinate Measurements. *Int. J. Precis. Eng. Manuf.* **2009**, *10*, 19–25. [CrossRef]
22. Pathak, V.K.; Kumar, S.; Nayak, C.; Gowripathi Rao, N. Evaluating Geometric Characteristics of Planar Surfaces Using Improved Particle Swarm Optimization. *Meas. Sci. Rev.* **2017**, *17*, 187–196. [CrossRef]
23. Runje, B.; Marković, M.; Lisjak, D.; Medić, S.; Kondić, Ž. Integrated Procedure for Flatness Measurements of Technical Surfaces. *Teh. Vjesn.* **2013**, *20*, 113–116.
24. Mikó, B.; Farkas, G.; Bodonyi, I. Investigation of Points Sampling Strategies in Case of Flatness. *Cut. Tools Technol. Syst.* **2019**, *91*, 143–156. [CrossRef]
25. Hazarika, M.; Dixit, U.S.; Deb, S. Effect of Datum Surface Roughness on Parallelism and Perpendicularity Tolerances in Milling of Prismatic Parts. *Proc. Inst. Mech. Eng. Part B J. Eng. Manuf.* **2010**, *224*, 1377–1388. [CrossRef]
26. Mikó, B.; Farkas, G. Comparison of Flatness and Surface Roughness Parameters When Face Milling and Turning. In *Development in Machining Technology*; Zebala, W., Manková, I., Eds.; Cracow University of Technology: Cracow, Poland, 2017; Volume 7, pp. 18–27.
27. Gapinski, B.; Zachwiej, I.; Kolodziej, A. Comparison of Different Coordinate Measuring Devices for Part Geometry Control. In Proceedings of the Digital Industrial Radiology and Computed Tomography (DIR 2015), Ghent, Belgium, 22–25 June 2015; NDT: Ghent, Belgium, 2015.
28. Rangarajan, A.; Dornfeld, D. Efficient Tool Paths and Part Orientation for Face Milling. *CIRP Ann. Manuf. Technol.* **2004**, *53*, 73–76. [CrossRef]
29. Dobrzynski, M.; Chuchala, D.; Orlowski, K.A. The Effect of Alternative Cutter Paths on Flatness Deviations in the Face Milling of Aluminum Plate Parts. *J. Mach. Eng.* **2018**, *18*, 80–87. [CrossRef]
30. Gusev, V.G.; Morozov, A.V. Deviation from Flatness of Surfaces after Combined Peripheral Grinding. In Proceedings of the MATEC Web of Conferences, Sevastopol, Russia, 10–14 September 2018; EDP Sciences: Les Ulis, France, 2018; Volume 224, p. 01028.
31. Sheth, S.; George, P.M. Experimental Investigation and Prediction of Flatness and Surface Roughness During Face Milling Operation of WCB Material. *Procedia Technol.* **2016**, *23*, 344–351. [CrossRef]
32. Gapinski, M.; Grzelka, P.D.M.; Rucki, P.D.M. The Accuracy Analysis of the Roundness Measurement with Coordinate Measuring Machines. In Proceedings of the XVIII IMEKO World Congress, Metrology for a Sustainable Development, Rio de Janeiro, Brazil, 17–22 September 2006.
33. Mikó, B. Measurement and Evaluation of the Flatness Error of a Milled Plain Surface. In Proceedings of the IOP Conference Series: Materials Science and Engineering, Kecskemét, Hungary, 7–8 June 2018; Institute of Physics Publishing: Bristol, UK, 2018; Volume 448, p. 012007.
34. Mikó, B.; Drégelyi-Kiss, Á. Study on Tolerance of Shape and Orientation in Case of Shoulder Milling. In *Development in Machining Technology*; Zębala, W., Mankova, I., Eds.; Cracow University of Technology: Cracow, Poland, 2015; pp. 136–150.
35. Lakota, S.; Görög, A. Flatness Measurement by Multi-Point Methods and by Scanning Methods. *Ad Alta J. Interdiscip. Res.* **2011**, *1*, 124–127.
36. Kawalec, A.; Magdziak, M. An Influence of the Number of Measurement Points on the Accuracy of Measurements of Free-Form Surfaces on CNC Machine Tool. *Adv. Manuf. Sci. Technol.* **2011**, *35*, 17–27.
37. Mikó, B. Assessment of Flatness Error by Regression Analysis. *Meas. J. Int. Meas. Confed.* **2021**, *171*, 108720. [CrossRef]
38. Nowakowski, L.; Skrzyniarz, M.; Blasiak, S.; Bartoszuk, M. Influence of the Cutting Strategy on the Temperature and Surface Flatness of the Workpiece in Face Milling. *Materials* **2020**, *13*, 4542. [CrossRef]
39. Toh, C.K. A Study of the Effects of Cutter Path Strategies and Orientations in Milling. *J. Mater. Process. Technol.* **2004**, *152*, 346–356. [CrossRef]

MDPI

St. Alban-Anlage 66

4052 Basel

Switzerland

Tel. +41 61 683 77 34

Fax +41 61 302 89 18

www.mdpi.com

Applied Sciences Editorial Office

E-mail: applsci@mdpi.com

www.mdpi.com/journal/applsci